Basic Hotel and Motel Premises Protection. Special Security Situations. Fire Protection and Evacuation Procedures. Internal Theft Control. Fraud Committed Against the Hotel. Food and Beverage Controls and Procedures. Help. Summary.

7. **Campus Security and Crime Prevention** 126

Instituting the Program. Whistle Program. Rape Awareness Program. Neighborhood Watch on the College Campus. Thefts in the Library. Bicycle Theft Prevention. Administration Office Security Programs. Operation Identification. Intrusion Alarms. *Safety in Numbers* Program. *Lock Your Door* Program. Follow-up to Thefts. Flyers and Posters. Conclusion. Appendix 7a: Directed Patrols. Appendix 7b: Campus Security Manual.

8. **Hospital Security** 149

Rationale for Hospital Security. Unique Aspects of Hospitals. Health Care Security Expanding. Hospital Security Vulnerabilities. Parking. Summary

9. **Museum, Library, and Archive Security** 165

The Problem. Foundations of a Security Program. Personnel Security. Security Responsibility. Physical Security. Fire Protection. Guards. Transporting Collectibles.

10. **High-Rise Office Building Security** 173

Intrusion Detection System. Access Control System. Television Systems. Security Guard Patrol System. Fire Control System. Environmental Control System. Emergency Power. Summary.

PART THREE / LOSS PREVENTION PLANNING AND CONTROLS 179

11. **Computer Security** 181

Personnel Screening of Key Employees. Vulnerabilities of Computers. Summary of Security Approaches. Organizing for Security. Security Management. Physical and Environmental Security. Security of Data Communications. Computer Systems Security. Data Security. Security Mechanisms. Appendix 11a: Computer Security Checklist.

12. **Planning for Emergencies and Disasters** 204

Emergency Team. Emergency Evacuation.

Protecting Vital Records Controlling the Computer Facility. Public Information. Emergency Equipment Location. Training. Appendix 12a: Checklist—Preparedness for Data Center Disaster.

13. **Bomb Threats and Search Techniques** 221

Purpose of Calls. Preparation. Evacuation. How to Prepare. Written Threats. Bomb Search Techniques. Room Search. Suspicious Object Located. Letter and Package Bombs. Buildings—Their Problems.

14. **Applicant Screening** 234

The Polygraph and the Psychological Stress Evaluator (PSE). Laws Affecting the Employment Relationship and the Screening and Hiring Process. The Applicant Screening Process. Information Sources. Appendix 14a: Sample Forms and Letters.

Security Applications
in Industry and Institutions

Security Applications in Industry and Institutions

Lawrence J. Fennelly
Editor

Butterworth–Heinemann
Boston   London   Oxford   Singapore   Sydney   Toronto   Wellington

Copyright © 1992 by Butterworth Publishers, a division of Reed Publishing (USA) Inc.
All rights reserved.

No part of this publication may be reproduced, stored in a retrieval system, or transmitted, in any form or by any means, electronic, mechanical, photocopying, recording, or otherwise, without the prior written permission of the publisher.

Library of Congress Cataloging-in-Publication Data
Security applications in industry and institutions / Lawrence J.
  Fennelly, editor.
    p. cm.
  Nineteen chapters excerpted from: Handbook of loss prevention and crime prevention.
    2nd ed. c1982.
  Includes bibliographical references and index.
  ISBN 0-7506-9389-4 (hard : alk. paper)
  1. Private security systems.   2. Security systems.
3. Industry—Security measures.   4. Crime prevention and architecture.   5. Burglary protection.
I. Fennelly, Lawrence J., 1940-
II. Title.
HV8290.H34525 1992   91-46108
658.47—dc20   CIP

British Library Cataloguing in Publication Data
Security applications in industry and institutions.
  I. Fennelly, Lawrence J. *1940–*
  658.47
  ISBN 0750693894

Butterworth Publishers
80 Montvale Avenue
Stoneham, MA 02180

10 9 8 7 6 5 4 3 2 1

Printed in the United States of America

Contributors

Security Applications in Industry and Institutions

R. E. Anderson, President, REA Associates Inc., McAllen, TX

Robert L. Barnard, Technical Security Consultant, Annandale, Va

Norman Bates, J.D., CPP, Liability Consultants, Framingham, MA

John M. Carroll, Department of Computer Science, University of Western Ontario, Canada

James Cleary, Jr., SCI Inc., Clearwater, FL

Russell Colling, CPP, Director of Security, Hospital Shared Services of Colorado, Denver, CO

Rex D. Davis, Director, Bureau of Alcohol and Tobacco and Firearms, Washington, D.C.

Charles J. Diecidue, CPP, Vice-President, Special Services, American Stock Exchange, New York, NY

Kevin P. Fennelly, Educational and Training Coordinator, *The Record,* Hackensack, NJ

Lawrence J. Fennelly, Sgt., Crime Prevention Specialist, Harvard University Police Department, Cambridge, MA

Eugene Finneran, CPP, Titan Security Ltd., Chicago, IL

Robert J. Fischer, Chairman, Law Enforcement Administration, Western Illinois University, Macomb, IL

B. M. Gray II, National Crime Prevention Council, Washington, D.C.

Philip P. Purpura, CPP, College Educator, Consultant, Florence, SC

Albert Seedman, B.B.A., M.P.A., Corporate Vice-President, Loss Prevention, Alexanders, Inc., Brooklyn, NY

Gerard Shirar, CPP, Manager, Security/Safety, American International Group, Manchester, NH

Louis A. Tyska, CPP, Security Director, Revlon, Inc., Edison, NJ

Joseph Wyllie, Joe Wyllie & Associates, Hicksville, NY

Contents

Preface   xi

PART ONE / INTRODUCTION   1

1. **History and Principles of Crime Prevention and Security**   3

   Crime Prevention Defined. HIstorical Overview. English Heritage. Instructions. Policing in America. Current Public Service Efforts. Community Relations. Organization and Administration. A National Educational Campaign. A Status Report. Private Security Development. Tools and Knowledge. Crime Risk Management. Security Tools. Other Knowledge Areas. The Future. App. 1a, App.1b

PART TWO / INSTITUTIONAL AND INDUSTRIAL SECURITY   29

2. **Retail Security**   31

   Dealing with Employee Theft. Theft of Cash by Employees. Undercover Investigators. Detaining Suspected Employees. Dealing with Shoplifting. Security Hardware. Loss Prevention Procedures. Methods of Shoplifting. Apprehending the Shoplifter. Checks and Credit Cards. Guidelines for Cashiers. Refund Operators. Appendix 2a: Loss Prevention Techniques. Appendix 2b: Control of Credit Card Fraud Checklist.

3. **The Shoplifting Problem**   47

   Who Shoplifts. The Extent of Shoplifting Losses. Preventing Shoplifting by Sales Employee Vigilance. Prosecution of Shoplifting Cases Deters Future Shoplifting Attempts. Reducing the Civil Liability Risk Factors When Handling Shoplifting Vases. The Bottom Line.

4. **Transportation-Distribution Theft and Loss Prevention Plan**   63

   Prevention Plan. Pilferage. Theft. Organized Crime. Cargo Package and Movement Controls. Appendix 4a: Cargo Security Checklist. Appendix 4b: Personnel Security Checklist. Appendix 4c: Physical Security Check list. Appendix 4d: Inspection Report Forms. Appendix 4e: Documentation.

5. **Bank Security**   93

   The Bank Protection Act of 1968: Public Law 90-389. Current Risks Affecting Financial Institutions. Role of the Bank Security Officer. Security Devices. Security Training. Critical Problems in Bank Security. Appendix 5a: The Bank Protection Act of 1968. Appendix 5b: Minimum Security Devices and Procedures [12 CFR 21]. Appendix 5c: Tear Gas/Dye Pac Research. Appendix 5d: Crime Prevention Techniques.

6. **Hotel Security**   105

   The Role of Security in a Hotel. Innkeeper's Liability.

Basic Hotel and Motel Premises Protection. Special Security Situations. Fire Protection and Evacuation Procedures. Internal Theft Control. Fraud Committed Against the Hotel. Food and Beverage Controls and Procedures. Help. Summary.

7. **Campus Security and Crime Prevention** 126

    Instituting the Program. Whistle Program. Rape Awareness Program. Neighborhood Watch on the College Campus. Thefts in the Library. Bicycle Theft Prevention. Administration Office Security Programs. Operation Identification. Intrusion Alarms. *Safety in Numbers* Program. *Lock Your Door* Program. Follow-up to Thefts. Flyers and Posters. Conclusion. Appendix 7a: Directed Patrols. Appendix 7b: Campus Security Manual.

8. **Hospital Security** 149

    Rationale for Hospital Security. Unique Aspects of Hospitals. Health Care Security Expanding. Hospital Security Vulnerabilities. Parking. Summary

9. **Museum, Library, and Archive Security** 165

    The Problem. Foundations of a Security Program. Personnel Security. Security Responsibility. Physical Security. Fire Protection. Guards. Transporting Collectibles.

10. **High-Rise Office Building Security** 173

    Intrusion Detection System. Access Control System. Television Systems. Security Guard Patrol System. Fire Control System. Environmental Control System. Emergency Power. Summary.

PART THREE / LOSS PREVENTION PLANNING AND CONTROLS 179

11. **Computer Security** 181

    Personnel Screening of Key Employees. Vulnerabilities of Computers. Summary of Security Approaches. Organizing for Security. Security Management. Physical and Environmental Security. Security of Data Communications. Computer Systems Security. Data Security. Security Mechanisms. Appendix 11a: Computer Security Checklist.

12. **Planning for Emergencies and Disasters** 204

    Emergency Team. Emergency Evacuation.

    Protecting Vital Records Controlling the Computer Facility. Public Information. Emergency Equipment Location. Training. Appendix 12a: Checklist—Preparedness for Data Center Disaster.

13. **Bomb Threats and Search Techniques** 221

    Purpose of Calls. Preparation. Evacuation. How to Prepare. Written Threats. Bomb Search Techniques. Room Search. Suspicious Object Located. Letter and Package Bombs. Buildings—Their Problems.

14. **Applicant Screening** 234

    The Polygraph and the Psychological Stress Evaluator (PSE). Laws Affecting the Employment Relationship and the Screening and Hiring Process. The Applicant Screening Process. Information Sources. Appendix 14a: Sample Forms and Letters.

15.	**Internal Theft Controls**	245

What is Honesty? The Dishonest Employee. Procedural Controls. When Controls Fail. Appendix 15a: Loss Prevention Checklists.

16.	**Executive Protection Planning**	266

Rise of Terrorism in U.S. Formulating a Protection Plan. Home and Office Survey. Risk Management Analysis. Protection in Transit. Everyday Security Awareness. If an Abduction Occurs.

17.	**Operations of a Guard Force**	273

Security Guards of the Future. Guards at a Nuclear Power Plant. Executive Protection. Security Services for Air Travel. Liabilities Connected with Guard Force. Power and Authority of the Security Guards. Missouri. Training. Report Writing. Weapons Safety. Safety. Bomb Threats. Fire Protection. Fire Prevention. Emergency Medical Assistance. Controlling Bleeding. Call for Assistance. Reporting a Medical Case. Guard Supervision.

18.	**Civil Law: The Controller for Private Security**	292

Tort Law: Source of Power and Limits. Arrest. Detention. Interrogation. Search and Seizure. Exclusionary Rule. Use of Force. Civil Liability. Contract Security and Liability.

19.	**Public Relations and the Media**	303

Media Relations. Exclusives. News Conferences. Handouts and Press Kits. What Is News? Writing Your News Release. Public Speaking. Lecturing and Making Community Presentations. Proper Use of Flyers and Handouts. Selling Security within the Organization.

Index   313

Preface and Acknowledgments

Crime prevention and loss prevention are closely allied fields differing primarily in their forms of management and sources of authority. A crime prevention officer is a public servant who possesses police powers. A loss prevention manager or security director works in the private sector and receives whatever authority he possesses from his employer, but he is not granted the powers of a public law enforcement officer. The job descriptions of these two positions, however, are quite similar and the skills required are identical. Prevention is the focus in both cases. The crime prevention officer and the loss prevention manager are expected to assess crime vulnerability—no matter whether it is of a residential area, small business, college campus, hospital, or a corporation—and recommend cost-effective security measures. Security problems are common to both the public sector and the private sector and so, too, are solutions.

Because prevention is the key concept in controlling losses and crime, we have assembled a collection of security applications for professionals in industry and institutions. The 19 chapters in this book, *Security Applications in Industry and Institutions,* were culled from an earlier Butterworths publication, *Handbook of Loss Prevention and Crime Prevention, Second Edition.* This pared-down volume will better serve crime prevention and loss prevention practitioners as well as students of police science.

This collection of material includes separate chapters on history and principles of crime prevention and security, retail, bank, and computer security, and public relations and the media.

A special thanks goes to those individuals and groups who aided and supported publication: Stephen Allen, Thomas Chuda, John Hunter, Albert Janjigian, American Society for Industrial Security, Door and Hardware Institute, National Crime Prevention Institute, Texas Crime Prevention Institute, and Assets Protection.

Thanks, also, the Joseph Barry, CPP, Fred Buck, Dennis Devlin, and M. Yee, who reviewed various chapters.

PART ONE
INTRODUCTION

Chapter 1
History and Principles of Crime Prevention and Security

B.M. GRAY II

A common thread runs through security issues ranging from protecting a nuclear plant to the reduction of frauds against the elderly; from protection of a high ranking business executive to the reduction of the incidence of home burglaries. This common bond is the need for crime prevention action. In each case, the objective is to reduce the chance that a crime will occur and to minimize the loss suffered in the event one does take place.

The prevention of crime must become a concern —indeed a philosophy—ingrained in every citizen. Why? Because crime affects everyone's lifestyle, from their ability to walk on the street safely to the amount of insurance premiums paid on cars and homes. More pointedly and importantly, law enforcement officers and executives must attune their thinking to a preventive rather than reactive mode if they are to successfully carry out their goal of protecting life and property. The publicly-paid police officer or sheriff is concerned with the safety and welfare of the general public while the private security officer is concerned with the welfare of a specific client. Yet the objectives of stopping a crime before it occurs and minimizing the consequences after a crime happens remain the same in all cases. Both are concerned with crime prevention, avoidance and control.

The practice of crime prevention is much more than simply applying techniques to a problem once it is discovered. Planning is required since it is an approach that recommends identifying and stopping a problem before it occurs, of reducing hazards in given situations, and of forecasting of rather than reacting to circumstances.

While it is simple to state the philosophy of crime prevention, its implementation becomes more difficult because of people's normal tendency to procrastinate about safety issues of all types and the psychological dissonance that occurs whenever a negative concern is brought to their attention. The crime prevention manager of corporation or community must surmount the psychological barrier of "It won't happen to me," particularly in the business environment. Even so, the reality of crime in our society provides constant motivation for both community crime prevention efforts and business loss prevention programs.

Crime Prevention Defined

Ask any ten persons—lay citizens, law enforcement professionals, or security personnel—to define the term *crime prevention*, and you will get at least five different responses. Those responses will probably not differ a great deal in substance from those found in the writings of criminological scholars, ranging from the global to the finite. A virtual quagmire of definitions exists, varying in nature from juvenile delinquency prevention to increasing prison sentences, and from surgically altering the capacity of criminals to offend to opportunity reduction efforts which focus on hardening the potential targets.

Definitions used by informed practitioners also vary but are usually practical and self-explanatory. The same type of activity may be called direct, primary, or mechanical. The police talk in terms of crime prevention while security professionals discuss the field as *loss prevention* or the *protection of assets*.

All of these terms refer to preventive and mitigating actions which can be initiated, managed, and controlled, and are therefore usually limited and specific in nature. These terms are contrasted to the indirect, tertiary, and corrective activities that deal more with the management of socioeconomic factors and other far-ranging variables that affect crime.

The most common set of definitions now used by police crime prevention officers was developed in 1976. It uses terminology analogous to the medical field where practitioners are concerned with just as serious and disparate a problem as crime: the prevention and management of disease. This definition allows actions to be sorted into three major categories—primary, secondary, and tertiary.

> Primary prevention techniques are directed at modification of criminogenic conditions in the physical and social environment at large, such as citizen education programs, household and business security precautions, and modification of the physical environment to reduce criminal opportunity. Secondary crime prevention is directed at early identification and intervention in the lives of individuals or groups in criminogenic circumstances. Secondary crime prevention techniques include patrol peace-keeping actions, diversion, pre-delinquent, screening, educational intervention programs, employee screening, and crime location analysis for neighborhood education and modification programs. Tertiary prevention is directed at the prevention of recidivism through post adjudication diversion, reform, rehabilitation, incapacitation, hiring ex-offenders, aftercare services and other techniques.[1]

This discussion focuses on primary and secondary crime prevention since they can more likely be influenced by the management decisions of community and business leaders, law enforcement and security managers. The specific definition used hereafter is that "crime prevention is the anticipation, recognition and appraisal of a crime risk and initiation of action to remove or reduce it."[2]

Analysis of this definition provides the crime prevention manager with an outline of functional concerns. First, crime risks must be identified. Then, they have to be categorized into priorities, since theoretically, all persons can be harmed and all property stolen.

This categorizing begins the anticipation stage of crime prevention because it requires an assessment of probability regarding the risk of the potential crime target. What are the chances that potential articles in a store will be shoplifted, a home broken into or an executive kidnapped? The process requires the recognition not only of the target but also of the possible method(s) of attack. For example, banks lose more money each year to employee theft than to robberies committed by strangers. This process then leads to the appraisal stage of assessing specific options to reduce the risk prior to taking the final action step. These last stages, appraisal and action, are critical. Here, decisions must be made regarding the worth of the potential target(s) and the effort that will be expended to protect it. An assessment of value is required in order for crime prevention actions to be cost effective (and therefore implemented by the authorities in command), whether they are elected public officials or members of the board of directors of a corporation.

Using this definition of crime prevention requires the practice of reducing opportunity rather than attempting to deal with the potential criminal's desire or ability to commit a crime. While various psychological and sociological theories exist concerning ways to reduce the need or desire of certain individuals to commit crime (better educational systems, full employment, reducing racism and discrimination, for example), none provides the immediacy needed to protect assets or reduce crime now. The ability of a criminal to commit a crime is controlled by the criminal's own ingenuity and access to the tools required. The only element that a potential victim controls, whether an individual, neighborhood, or business, is the opportunity for the crime to occur.

Historical Overview

Webster's Dictionary defines crime as "an act committed in violation of a law prohibiting it, an act omitted in violation of a law ordering it, or an offense against morality." People have been concerned with crime since its beginning. The earliest cave dwellers fought among each other to protect themselves and their property; they eventually banded together in tribes for mutual protection. Families and clans have often banded together for mutual protection purposes, with the premise that there is strength in numbers.

The administration of justice practiced centuries ago was more brutal and physical than that in America today. Force—not mores, folkways or law —was the determining factor in righteousness.

The cultures within various societies dictated their practices to a large degree. Dixon and Rogers trace the development of crime prevention[3]:

> Punishment of the offender, either to "correct" him to serve some other social purpose has been but one response. In preliterate cultures, in contrast, we find

"certain motives and attitudes which apparently preceded the punitive reactions to law breaking but were not, in themselves, punishment; desire to annihilate an enemy of the group, sacrifice to appease or fend off the wrath of the gods, social hygiene measures to rid the community of pollution, self redress in cases of private injury and surprise and disgust at the person who injured his own family. Deliberate and "just" infliction of pain by the group in its corporate capacity was not invented until later.

It was not until central authority figures such as kings came into power that some consistency was established regarding crime and punishment. The first codification of law occurred with Hammurabi's Code around 1800 B.C. It contained a statement of required and prohibited activities and actually was the originator of the concept of restitution through inclusions such as requiring a builder to replace a house if the one he built collapsed.[4]

The history of crime prevention even incorporates the Bible. For example, we now urge witnesses to come forward to testify in court in order to assure that fair trials, with all evidence revealed, take place. Leviticus 5:1 states: "If a person hears a solemn adjuration to give evidence as a witness to something he has seen or heard and does not declare what he knows, he commits a sin and must accept responsibility."

English Heritage

Much U.S. heritage regarding crime prevention and criminal justice derives from England. Our common law and statutory laws are based upon the traditions and precedents of that country, with modifications emanating from the Constitution and the Bill of Rights.

British police traditions influence American policing and philosophy toward crime prevention in the public sector. The British rejected formal, sworn, and paid police for centuries and did not begin their current system until the mid-1700s when Sir Henry Fielding became a London Magistrate. Not only did Fielding want to stamp out crime but he also wanted to prevent future outbreaks of crime. He believed that the community could no longer police itself without paid agents. Fielding declared that to achieve his objectives he needed the help and cooperation of the public, the removal of conditions in which crime flourished, and a strong police force. Police executives in the twentieth century still subscribe to these three necessities.

In 1829, the leadership of Sir Robert Peel forcibly marked the beginning of the police role in crime prevention from a policy perspective. "The first order for the Metropolitan Police was a triumph of clarity, simplicity, and vision:

Instructions

The following General Instructions for the different ranks of the Police Force do not apply to every variety of circumstances and roles that may occur in the performance of police duty. Something must depend upon the intelligence and discretion of individuals; and the degree to which they show themselves possessed of these qualities and to their zeal, activity and judgment, on all occasions, will be their claim of future promotion and reward.

It should be understood, at the outset, that the principal object to be attained is the prevention of crime.

> To this great end every effort of the police is to be directed. The security of person and property, the preservation of public tranquillity, and all the other objects of a Police Establishment, will thus be better effected, than by the detection and punishment of the offender, after he has succeeded in committing the crime. This should constantly be kept in mind by every member of the police force, as the guide for his own conduct. Officers and Police Constables should endeavor to distinguish themselves by such vigilance and activity, as may render it extremely difficult for anyone to commit a crime within that portion of the town under their charge.
>
> When in any Division offenses are frequently committed there must be a reason to suspect that the Police is not in that Division properly conducted. The absence of crime will be considered the best proof of the complete efficiency of the Police. In Divisions, where this security and good order have been effected, the officers and men belonging to it may feel assured that such conduct will be noticed by rewards and promotion.[5]

The British lost their perspective regarding the need to emphasize prevention during the ensuing years. In time, priority was given to the investigation of crimes and the apprehension of offenders. This was the model adopted largely by U.S. policing. At first, community volunteers took turns on patrol; then paid personnel took over the duties in the eastern cities. The rough-and-ready sheriffs of the West and the stories that extolled their capacity to deal with criminals gave rise to the tough image the law enforcement officer still maintains today, despite the fact that over eighty percent of police assignments in the United States entail providing services and ordering maintenance rather than being directly crime-related in nature.

The British police resumed their interest in primary crime prevention practices in 1963 with the establishment of the Home Office Crime Prevention Training Center (HOCPTC) in Stafford, England. This action followed a great deal of local programming and public education campaigns on methods to reduce burglaries and other types of opportunity crimes. There also had been a national public education campaign conducted during the 1950s that led to a growing liaison between the police, the private security sector, and insurance companies.

The purpose of the HOCPTC was to provide training to members of all police forces in the United Kingdom. This training continues today and includes both introductory and advanced sessions.

Policing in America

American policing had been developing concurrent to the British but with greater emphasis on criminal investigation and apprehension. Prevention activities were carried out primarily through the use of highly visible patrol techniques.

The concept that the police were simply agents of the community selected to do the community's work seemed to be largely forgotten. Society seemed to be totally delegating its obligations to the police instead of maintaining a partnership of responsibility. Although the police actively sought cooperation from the public in reporting crimes and serving as witnesses, the message from far too many elected officials and police chiefs was that increasing the staff of the police force was all that was needed to reduce crime. Emphasis on the citizens' responsibility to become involved in a variety of community affairs concerning crime, and especially taking action (for example, to reduce residential burglary), did not become national in scope until the mid-1970s.

Emphasis on the role of police as crime prevention counselors to the public, as program developers and managers, and as teachers of public policy, began after the National Crime Prevention Institute (NCPI) was created at the University of Louisville in Kentucky. This institute, modeled after the British crime prevention center, was begun in 1971 with major funding assistance from the Law Enforcement Assistance Administration (LEAA) of the U.S. Department of Justice. The LEAA provided training to several thousand police crime prevention officers, police managers, and city/county/state officials from across the country. It was the catalyst needed to promote the concept of police/community interaction to reduce crime through opportunity reduction.

Following the lead of the NCPI, the Texas Crime Prevention Institute was established at the Southwest Texas State University, and Macomb Community College in Michigan began offering courses for police and security personnel. There are now training programs in more than 20 states to prepare law enforcement personnel to assume their responsibilities as crime prevention specialists.

Current Public Service Efforts

The role of the police in crime prevention remains one of great controversy. While some groups believe the police should be aggressive leaders, others advocate a lesser role, causing each community's program to differ in style and strategy. Too many law enforcement chief executives across the nation speak of the necessity for public law enforcement agencies to be the leaders of crime prevention efforts but fail to offer career path incentives to their officers, cut funding to their crime prevention units, or even disband them in times of shrinking budgets. Those executives who are maintaining an emphasis on citizen action in collaboration with the police will reap significant benefits in the future.

> It is a cardinal principal of democratic societies that ultimate responsibilities for peace, good order and law observance rests with the community of citizens of that society, not with an organized police force.

Although the complexity of modern societies usually dictates that policing efforts be coordinated and directed by a force of paid professionals, their responsibility is derivative of the community's. Their role is to supplement and aid community efforts, not supplant them. The powers permitted to these police must be carefully defined and limited.

A community which abandons its basic duty to police itself, leaving all responsibility to a professional police service will soon find that the police can hope to provide no more than a modicum of public order and security, and this only through such repressive measures that the basic liberties of a free people are eroded and the very democracy that the police seek to preserve is endangered.

Only if the proper balance is maintained between the basic responsibility of the community and the derivative responsibility of the police can a safe and orderly society be preserved with the least burden on individual rights and freedom.

It is unfortunate, therefore, that the history

of twentieth century urban policing in the United States is a consistent record of errors by the police service to assume a disproportionate share of the responsibility for maintaining social control and the concurrent abandonment by U.S. communities of their portion of this duty. The result has been an increasing lawlessness that even increasingly repressive measures have been unable to curb.

The delicate balance between the traditional roles of the community and the police needs to be restored. Peacekeeping must again become a joint police/community effort to stand any reasonable chance for lasting success. In this respect, the Lincoln (Nebraska) Police benefit from serving a community which is vitally interested in assuring a high level of safety in a secure public order and assumes the responsibility for policing itself. The fundamental mission of the Lincoln Police, therefore, is to provide the leadership and professional support required to sustain and improve the community's efforts, and to develop a balanced and cooperative police/community campaign against lawless and disorderly behavior.

Community Relations

If the control of crime and other disorders is to be a joint effort between the police and the citizens of Lincoln, then the keystone of this effort must be cordial, respectful and trusting relations between the police and the community. Every effort must be made, and no opportunity overlooked, to strengthen this police/community bond. Every action by the police department and its members must be promptly judged. The police must endeavor to ensure that all citizens of the City of Lincoln view the police department as an integral part of the community and not as an organization outside or estranged for it.[6]

Helping to sustain the interest in primary prevention programs of law enforcement agencies and citizen's groups are state level crime prevention programs and associations, along with the enthusiastic support of experienced crime prevention officers in hundreds of departments across the United States. Many of these practitioners belong to the fledgling International Society of Crime Prevention Practitioners, an organization of growing influence in the field.

There are no national surveys to determine the total number of law enforcement agencies with active crime prevention programs; but there are numerous indicators. We do know that over 30 state-level crime prevention programs and slightly more than 30 state-wide crime prevention associations operate under the auspices of the governor or attorney general.[8] Both work directly with community leaders, police executives, and police crime prevention officers. These organizations promote and aid the coordination of local crime prevention programs conducted by law enforcement agencies and often serve as a clearinghouse for information. Training programs are held periodically, along with award programs to recognize outstanding citizen, corporate, media, and law enforcement efforts. Materials are standardized for each state. Policy issues about such topics as building security codes, programs for youth, and methods of reducing false alarms may be resolved. This constant process of development is critical to an evolving field.

On the national level, the International Association of Chiefs of Police adopted a resolution at their 1980 annual conference which essentially states that all police agencies should consider crime prevention as a standard police function and should, where appropriate, form specialized crime prevention units. While a resolution of this kind does not spur action, it does provide a reference point for chiefs and sheriffs to convince their elected officials of the need to keep crime prevention programs funded (or even start them). This emphasis has been enhanced by the Commission on Accreditation for Law Enforcement Agencies, which has developed standards for crime prevention as a normal operation of law enforcement agencies.

Beyond stating their commitment to crime prevention, law enforcement agencies should establish specific policies, goals, and objectives by which their commitment can be realized. Some agencies may need to create a crime prevention component; others may merely designate an individual as responsible for crime prevention activities. In all agencies, it should be understood that all officers are responsible for achieving agency crime prevention goals and should assist others in this regard. In addition, the use of nonsworn officers and the elderly should be encouraged, and programs geared toward youth should be developed.

Data-based plans should be developed to indicate the types of crimes that pose the greatest threat to the community and show where (geographically) criminal activity of this type is most prevalent. The plans behind specific crime prevention programs should be designed with evaluation in mind. Programs must be evaluated annually to determine whether they have achieved stated goals.

The development of neighborhood watch groups should be fostered, as should provision of security surveys and other related services. Once established, neighborhood watch groups should not be allowed to "wither on the vine;" a maintenance plan should be established and adhered to. The harassment of citizens who participate in such programs should be a chief concern of those responsible for maintenance activities.

Organization and Administration

- A written directive states that the agency is committed to the development of perpetuation of community crime prevention programs.
- A written directive establishes the agency's crime prevention component and defines the relationships between all organizational elements of the agency in pursuing crime prevention activities.
- The agency has at least one employee whose responsibilities include planning and coordinating crime prevention activities.
- A written directive requires the crime prevention component to have access to foreign language specialists.
- A written directive establishes the agency's crime prevention priority programs and provides for the following:

    the targeting of programs by crime type and geographic area based on an empirical analysis of local crime data
    the evaluation of all crime prevention programs at least annually

- A written device requires the agency to assist in organizing crime prevention groups in residential areas targeted for such activity, as well as on request.
- The agency promotes crime prevention programs for all citizens and proprietors in areas targeted for such activity.
- A written directive requires the agency's crime prevention practitioner to maintain liaison with interested community groups.
- A written directive requires the agency to provide crime prevention input into development and/or revision of zoning policies, building codes, fire codes, and residential and commercial construction building permits.

These national efforts are reinforced by the continuation of the National Neighborhood Watch program of the National Sheriffs Association, even after federal grant dollars were no longer available.

Also, the Federal Bureau of Investigation, Police Executive Research Forum, National Organization of Black Law Enforcement Executives, International Union of Police Associations, and the Military Police of the U.S. Air Force and Army are members of the National Crime Prevention Coalition.

A National Educational Campaign

McGruff, the animated crime prevention dog, has developed as the national symbol for crime prevention in less than a decade. This friendly but informed spokesdog captured the attention of the U.S. public because of the support of the electronic and print media. The time and space donated to McGruff's public service educational messages allowed citizens of all ages to learn how to reduce the probability that they will become victims of crime.

The National Citizens' Crime Prevention Campaign, as it is formally called, is a true public/private partnership for the prevention of crime. The campaign consists of four key elements:

1. *Public service advertising* developed by DFS Dorland Worldwide and the Advertising Council, Inc. (both of whom volunteer their services and are paid only out-of-pocket costs) receives over $30 million annually in donated media space and time.
2. *Materials* reinforce the advertising. These range from attractive booklets, to carefully created kits of camera-ready materials for police and corporations to print, to original research published in the form of monographs.
3. *Technical assistance* is provided to businesses, police departments, community organizations, state government offices of crime prevention, and state crime prevention associations. Assistance takes the form of conferences, one-on-one consultations, and access to data collected and managed by the Computerized Information Center of the National Crime Prevention Council (NCPC).
4. *The Crime Prevention Coalition*, begun in 1980 by 19 national organizations to support the campaign, grew to 105 by 1986. Guided by a steering committee, the Coalition is composed of national constituency groups ranging from the American Society for Industrial Security and the National Retail Merchants Association to the General Federation of Women's Clubs and the Boys' Clubs. Federal agencies include the Federal

Bureau of Investigation, the Postal Inspection, and all four branches of the military.

The NCPC, a national nonprofit organization headquartered in Washington, D.C., manages and staffs the entire campaign. Funded primarily by the U.S. Department of Justice, Bureau of Justice Assistance, the NCPC is also active in raising private funds to maintain and expand the Council's activities to prevent people from becoming victims of crime.

What has this campaign meant to the United States? There is now an effective symbol, a psychological trigger to cause people to think of crime prevention and to recognize quality educational materials. Children, corporate leaders, community policy-makers, and older Americans are all learning about the realities of crime and law enforcement and that they do not have to live in fear of being a crime victim or to tolerate aberrant behavior—if they join together to make their communities and workplaces safe. The campaign's personal protective (to watch out) and community participative (to help out) messages were formed by consensus of the Coalition and are slowly becoming ingrained in the public's psyche: a critical first step to individual and community action to reduce crime.

Whatever the efforts at state and national levels, the primary point of contact and action to reduce crime occurs at the local level. This is where crime prevention practices occur, through the stimulation of individuals, neighborhoods, and businesses to get involved and do something positive to reduce crime. The leadership and commitment of the local law enforcement chief executive is critical to achieving success.

Given the volume and disparity of issues facing police managers, myriad philosophies and programs still exist around the nation. We find that some effective, well-designed, and clearly focused crime prevention programs organized by law enforcement officers are expanding while budget cuts force others to reduce programs or close down. Those that remain show the prudence of conducting careful needs analysis studies prior to any type of action and of conducting ongoing evaluation efforts to determine what is working and what is extraneous in terms of both community perception and actual crime/loss reduction. Those that are sensitive to community needs and are producing visible results will survive.

Crime prevention takes place at the neighborhood level. The neighborhood and more specifically the block is the natural organizing part for such activities as block watch, helping hands, and civilian patrols. Crime prevention is best organized through existing community groups such as churches, civic organizations, and youth organizations. These organizations are multi-issue groups and are therefore likely to endure. It is also easier for them to enlist volunteer support and community resources and crime fighting activities because they have already established communication networks and credibility in the community. Block watch is the major building of a successful crime prevention program. It is almost always combined with home security surveys and operation identifications, but it is the watch activities organized through block clubs that result in neighbors getting to know one another. Whether initiated by community groups or police, crime prevention is most successful where there is a strong feeling of police community cooperation. Many police departments are not organized to even accommodate, let alone reward crime prevention activities. To change this, incentives for doing crime prevention work must be built into the merit and promotional system. To keep volunteers involved and interested, they must have meaningful tasks that make their continuing contributions to law enforcement apparent. Volunteers can help with many of the administrative, analytical and organizational and information sharing functions that are necessary in police department.[9]

The developments in the Detroit Police Department during 1980 exemplify this idea, as does the development of programs in California.

During 1980, the economy of Detroit suffered greatly, leading to the layoff of hundreds of police officers. At the same time the layoffs occurred, the crime prevention section's strength was doubled to approximately 125 persons operating out of over 50 storefronts. This increased staffing was initiated by the Chief of Police, William Hart, because of the documented success of past crime prevention programs under the leadership of Inspector James Humphrey, an enlightened manager. After tactical and strategic planning was completed, neighborhood street crimes and burglaries decreased by 30 to 40 percent in districts aggressively organized by the crime prevention officers. Cooperation between residents, business owners, and the police reduced crime and fear of crime.

California benefits from crime prevention leadership from offices of both the governor and the attorney general in media programs, materials, special events, and ongoing research. A corporate group, the Stop Crime Coalition, has joined with the government to design and implement programs for this diverse state, which has resulted in participation of over 75 percent of California's law enforcement agencies.

A Status Report

The NCPC conducted a survey of crime prevention practitioners participating in its Computerized Information Center in 1986. The findings, contained in the first *Crime Prevention: Status and Trends Report*, are of significance.[10]

Survey of Practitioners

- Based on practitioners' estimates of the public demand for their services, net demand for crime prevention appears to be on the increase.
- Crime prevention has reduced crime and the fear of crime in the eyes of practitioners, having an impact on all types of crime—and it has the potential to make even greater reductions in crime and fear.
- Although practitioners saw actual incidence of crime decreasing, in the main, for their communities, they estimated citizens' fear of crime as stabilized rather than decreasing with crime's down-trend. This may reflect perception lags.
- Funding of crime prevention programs, evaluating them, and documenting results form the greatest problems facing practitioners.
- New programs being added to practitioners' repertories are predominately addressed to children and youth, business, neighborhood watch, and closely allied topics, although a smattering of other issues surfaced.
- Training needs for the field include updating knowledge in general, learning effective ways to sustain citizen interest, and developing topical skills.
- Citizens still see crime prevention as primarily a law enforcement responsibility, but there is a shift toward citizens' accepting of a greater portion of the obligation.

Ten Experts Look at Crime Prevention

Additionally, the NCPC interviewed ten knowledgeable crime practitioners and identified the following themes. They noted that because crime prevention draws from diverse disciplines, and because it has not yet developed a clear corps of adherents, identifying the "experts" who can speak from a national perspective is a difficult task.

- The problems which confront the crime prevention field today have less to do with whether crime prevention works (there was clear consensus that it does) and more to do with how to document its effectiveness and develop public and institutional support at all levels.
- This theme was reflected in three general problem areas highlighted by the interviewees:
  1. Law enforcement must see crime prevention more clearly as a priority.
  2. Citizens must understand that they can, by individual and collective action, reduce the risk and incidence of crime.
  3. Crime prevention is community-building and needs to be inverted from a negative into a positive opportunity in the eyes of citizens, law enforcement, and policy makers.
- Operational problems which impede crime prevention practitioners include ineffective or inefficient planning, and difficulties securing and transferring such support as exists into needed resources.
- Crime prevention is nonetheless on the brink of some extraordinary opportunities to become a major social force for the improvement of communities across the country.
- The audiences crime prevention has not yet reached—but which it must reach to increase its impact—are teens and residents of high-crime areas, two groups whose chief need is to develop a positive sense of stake and investment in their communities.
- Crime prevention by the end of this century should be a positive adjunct of community-building and civic betterment. As a standard agenda item for civic groups and an integral part of the community's governance structure, it can help build community cohesion and spirit as it helps communities prevent crime.[10]

These efforts at the local, state, and national level are enhanced dramatically by the resources and knowledge of leaders in the private security field. A brief review of developments in the private security field is appropriate since "in effect, crime prevention appears to be no more than a consolidation of security techniques and procedures within the public law enforcement field."[11]

Private Security Development

> Private security makes an invaluable contribution to the public weal. Its absence would be disastrous to law and order, and the operation of the criminal justice system. It continues [to be] the largest available untapped, un-

structured, dormant resource for the prevention and control of crime.[12]

The roots of the private security field can be traced back to the colonists' desire to protect their settlements. They banded together in mutual protection groups, and each able-bodied male took his turn as watchman. This was very much like the system used in England and stayed in effect until the early 1800s. The watchmen who were drafted into service were responsible for sounding the alarm, taking action against thieves, marauding Indians, fires, and other intrusions on the safety and security of the residents. Naturally, many able-bodied men drafted chose to hire someone else to take the tour of duty. Thus, contract security was born.

The first full-time, publicly funded police agency was established in New York City in 1844. Other cities followed its lead, but despite the development of paid public police, a need for private assistance remained. The first large-scale corporate effort was begun by Allan Pinkerton during the 1850s. He developed the first of many firms that would provide protective and investigative services to a client for a fee.

A variety of existing national security firms began during the mid- and late-1800s. Edwin Holmes started the first central office burglar alarm operation in 1858. American District Telegraph formed in 1874. In 1891 Washington P. Brink delivered his first payroll, originating the concept of armored car service.

During the same period, railroad companies became large proprietary security users with the establishment of their own security forces to deal with robbers and saboteurs. Armed men were used to protect silver, gold, and other precious metals coming from Western states such as Nevada and Colorado. Labor disputes often involved the use of force, both by unions and businesses, giving rise to security companies like Baker Industries in the early 1900s. The William J. Burns International Detective Agency was formed in 1909.

World War I caused growth in the private security field. This period of industrialization, urban growth, and wartime sabotage and espionage spurred the use of preventive measures by corporations and government. The Depression years showed the least growth for private security.

Again, the tidings of war spurred the industry. World War II created a need for security in an unprecedented manner.

Prior to the awarding of national defense contracts, the Federal Government required that munitions contractors implement stringent and comprehensive security measures to protect classified materials and defense secrets from sabotage and espionage. The FBI assisted in establishing these security programs. Additionally, the Government granted the status of auxiliary military police to more than 200,000 plant watchmen. Their primary duties included protection of war goods and products, supplies, equipment and personnel. Local law enforcement agencies were responsible for their training. As a result of the heightened emphasis on security within the government/military sphere, industry became increasingly aware of the need for plant security, and its value in protection of their assets.[13]

The private security field has developed rapidly since World War II. The increase in crime, the cost of crime, the ineffectiveness of our criminal justice system, and the expansion of population and industry in the United States have all contributed to the growth of private sector enforcement. Increasingly efficient technological developments regarding security devices and systems, the requirements of insurance companies prior to underwriting the insurance desired by business, and the civil liabilities that might be suffered if inadequate security is provided to the public are also catalytic agents to the growth.

The professionalism of both proprietary and contract security personnel also is increasing. While there is no average to be cited regarding training or competency, the more than thirty security trade associations and security committees of major national associations, such as the American Bankers Association, constantly upgrade their programming. Finally, the American Society for Industrial Security (ASIS), with over 15,000 members, exists "to further the objective of crime prevention and the protection of assets."[14] It does so through professional publications, workshops, and seminars. ASIS has developed a Certified Protection Professional testing program to further increase the professionalism of the field; there is also a certification program for security trainers. The Canadian Society for Industrial Security (CSIS) is also establishing a certification program for security professionals.

The private security field can be a source of assistance to the public law enforcement sector. Both are concerned with (and exist because of) the need to protect life and property. Yet there are areas in need of cooperation, as noted in the *Hallcrest Report*:

Research findings noted that in 1980 over 1 million persons were employed for private security purposes and that expenditures on private security and public law enforcement exceeded those one-third. Further:

• Private security resources, both expenditures and

employment, now exceed those of law enforcement and will continue to increase as resources for public law enforcement.
- Business, industry, and institutions together spend more than $20 billion annually for security in their organizations.
- Both police and security managers are receptive to the ideas that private security respond to some minor criminal incidents occurring on the property it protects and that some non-crime-related police tasks be contracted out to the private sector.
- There is limited interaction and cooperation thus far between the public police and the private security industry in crime prevention and public safety.
- Law enforcement executives tend to view private security programs as largely ineffective in reducing crime; they rate its performance generally low.
- The quality of security personnel is a major concern to the police, who favor (as does most of the security industry) state legislation to license and upgrade the quality of security personnel.
- Two major problems hamper police security relationships—off-duty police moonlighting in private security jobs and the excessive number of false burglary alarms to which police must respond.
- Crime reporting is a low priority for security managers—sometimes as a policy the police are rarely called upon to investigate such crimes as internal theft and fraud.
- Businesses and institutions divert many criminal acts from the public justice system by resolving the incidents internally. Little is known about these "private justice systems."

Despite the expanded role of the police in crime prevention in recent years, it appears that the private sector will bear an increased prevention role while law enforcement concentrates more heavily on violent crimes and crime response. Economic realities are forcing law enforcement to seek ways to reduce workloads.

Major Recommendations

- Upgrade private security. Upgrading the quality of security personnel was the most frequent recommendation made by both police and security managers who were surveyed. Both groups overwhelmingly agree on the need for a state-wide regulatory statute for contract security, plus mandatory criminal background checks and minimum levels of training for both proprietary and contract security officers. In addition, adoption of standards, codes of ethics, and model licensing, certification, and contract performance specifications are recommended.
- Increase police knowledge of private security. Seminars, training materials, designation of security of liaison officers, inventories of security firms, and other mechanisms are recommended to develop a greater awareness by police of the role and resources of private security in their communities.
- Expand interaction. Recommended strategies include identification of specialized investigative resources and establishment of private security that are available to complement police investigations, use of joint task forces for investigation of major or recurring losses, and development of official policies for sharing investigative information.
- Experiment with transfer of police functions. Research and demonstrations are recommended to isolate police activities that do not require police authority, identify areas where contracting may be effective, and explore legal mechanisms and special officer status for security personnel. Special attention should be given to contract burglar alarm response in the private sector, including measuring whether the deterrent value of response rises from police authority or merely from attention that is quick, uniformed, and armed.[15]

Tools and Knowledge

The operating assumptions of crime prevention, developed at the National Crime Prevention Institute in 1978, are important to the practitioner. These ten assumptions provide a basis for a wide variety of programming:

1. Potential crime victims or those responsible for them must be helped to take action which reduces their vulnerability to crimes and which reduces their likelihood of injury or loss, should a crime occur.
2. At the same time, it must be recognized that potential victims (and those responsible for them) are limited in the action they can take by the limits of their control over their environment.
3. The environment to be controlled is that of the potential victim, not of the potential criminal.
4. Direct control over the victim's environment can nevertheless affect criminal motivation in

that reduced criminal opportunity means less temptation to commit offenses and learn criminal behavior and, consequently, fewer offenders. In this sense, crime prevention is a practical rather than a moralistic approach to reducing criminal motivation. The intent is to discourage the offender.

5. The traditional approaches used by the criminal justice system (such as punishment and rehabilitation capabilities of courts and prisons, and the investigative and apprehension functions for police), can increase the risk perceived by the criminal, and thus have a significant (although secondary) role in criminal opportunity reduction.
6. Law enforcement agencies have a primary role in crime prevention to the extent that they are effective in providing opportunity reduction education, information, and guidance to the public and to various organizations, institutions, and agencies in the community.
7. Many skill and interest groups need to operate in an active and coordinated fashion if crime prevention is to be effective in a community-wide sense.
8. Crime prevention can be both a cause and an effect of efforts to revitalize urban and rural communities.
9. The knowledge of crime prevention is interdisciplinary and is in a continual process of discovery, as well as discarding misinformation. There must be a continual sifting and integration of discoveries as well as a constant sharing of new knowledge among practitioners.
10. Crime prevention strategies and techniques must remain flexible and specific. What will work for one crime in one place may not work for the same crime in another place. Crime prevention is a "thinking person's" practice, and countermeasures must be taken after a thorough analysis of the problem, not before.[16]

These assumptions were developed as an expansion of the theory of opportunity reduction which had been taught at NCPI since 1971. That theory had six components which hold true today:

1. Criminal behavior is learned behavior.
2. Reducing criminal opportunity reduces the opportunity to learn criminal behavior.
3. Criminal opportunity can be lessened by improved security measures and by increasing the level of surveillance on the part of the general public.
4. Long-range crime prevention will not be achieved unless criminal opportunities are reduced on a national basis.
5. The police are in a pivotal position and as such they should be trained in crime prevention and become involved in the preplanning of any community activity where their services will later be called for.
6. Insurance, security hardware, and other areas of business and industry involved in crime prevention programs must exchange information with the police.[17]

The broadest description of tools of the trade focuses on research education and cooperation. Without these, technical knowledge would cease to be of benefit. Crucial to the understanding of crime and loss prevention theory and practice is risk management.

Crime Risk Management

The crime prevention practitioner must possess a wide array of skills to be successful. New technologies and equipment in this developing field must be tested and implemented, evaluated and modified, as appropriate. The practitioner must operate as a manager, not just of people, but of circumstance. This is accomplished by maintaining an up-to-date knowledge of the field and the tools necessary to reduce criminal opportunity. Foremost among the skills needed is the concept of crime risk management.

The ability to assess potential risks accurately is an absolute necessity for success in crime prevention. Concurrently, the capability to develop cost-effective crime risk management systems is crucial. If the decision-maker (in the public or private sector) doesn't perceive that benefits will accrue from acting to prevent a potential problem, no action will occur.

The concept of risk management is derived from the business world. Crime is categorized as a pure risk as contrasted to the dynamic risk assumed in business. A pure risk is one in which there is no potential of benefits being derived or profit obtained. Other types of pure risks are floods, fires, or natural disasters. Comparatively, a dynamic risk is one that can produce gain or profit. This is exemplified by investing in the stock market, or in conducting a retail business in hopes that customers will buy the goods offered for sale and a profit will be earned.

Crime prevention programming can be generalized and comprehensive in nature and applicable to an individual, a community, or an entire city. It may be specifically directed to one target. Crime risk management can be applied only when a central authority can make trade-off decisions as a business manager who is motivated to protect a potential target, such as a computer facility or valuable gems.

Crime and incident analysis is one of the first tools to be applied in the assessment of risks. The basic investigative questions of what crimes are occurring, their location (where) and methods of commission (how) should be asked. This allows intelligent conjecture of frequency regarding the potential crime and allows the manager or client to begin to determine the level of vulnerability. This is the beginning phase of crime and target-specific planning. It allows for awareness of the result of actions and indicates if crime displacement (by time, type, or location) is occurring.

When assessing vulnerability and the response to risk, the PML factors must be considered. PML stands for *possible maximum loss* and the *probable maximum loss*, which differ greatly. Possible maximum loss is the maximum loss that would be sustained if a given target or combination of targets were totally removed, destroyed, or both. In a retail store, for example, the possible maximum loss would be the store's entire stock. Probable maximum loss on the other hand, refers to the amount of loss a target would be likely to sustain.[18] This is an important distinction when one is assigning priorities and determining cost benefit ratios in order to make crime and loss prevention decisions.

Once this process is completed, one can apply the five principal crime risk management methods: (1) risk elimination or avoidance, (2) risk reduction, (3) risk spreading, (4) risk transfer, and (5) risk acceptance. A mixture of these methods is quite normal in reducing criminal opportunity and reducing the PML, since no single method meets all needs.

- **Risk Avoidance**. This involves the removal of the target, such as dealing through direct deposit systems instead of handling cash and negotiable documents.
- **Risk Reduction**. This technique calls for minimizing the potential loss as much as possible—for example, not allowing over a set dollar amount to accumulate in a retail clerk's cash drawer before it is removed by the manager.
- **Risk Spreading**. The potential target(s) is spread over as large an area as possible in order to reduce the loss if a crime occurs—for example, precious gems being kept in several small vaults in different locations in a jewelry store instead of one large vault.
- **Risk Transfer**. Perhaps the most common is the concept of transferring the risk to other parties, particularly insurance companies. Another example which combines risk reduction, spreading and transfer would be that of depositing valuables in an insured safety deposit box at a bank.
- **Risk Acceptance**. There may be times when a risk simply must be accepted, as exemplified by the coin or art collector who refuses to be separated from the collection. Measures can be taken to reduce, spread, or even transfer part of the risk but some must be accepted by the act of displaying valuable articles.

These methods of crime risk management are critical to the crime prevention manager. Without them there is no systematic way of assessing the alternatives available when the vulnerability and importance of a potential crime have been assessed.

Security Tools

While the purpose of reducing criminal opportunity is to prevent a crime from occurring, administrative policies and electronic and mechanical devices are also intended to detect a crime in progress in order to allow apprehension of the offender if the crime occurs. The security and crime prevention manager is concerned with what are referred to as the four *D*s of crime prevention: *deterring* the criminal attack, *detecting* the attack if it occurs so that authorities can respond, *delaying* the attack to allow time for an apprehension to be made or to frustrate the offender into leaving before crime success is achieved, and *denying* access to selected targets.

The objectives of crime prevention are achieved through a variety of strategies and techniques. The techniques combine the application of electronic, physical, and procedural security.

"Physical security places barriers on the path of the potential attacker to deter him from attacking, delay him if he decides to attack and deny him access to high value targets even if he succeeds in penetrating the security system."[19] Examples of physical security include walls, doors, locks, and safes. Electronic security is used to detect and deter an offender. Examples of electronic security include intrusion detection systems (alarms), closed circuit televisions, and automatic cameras. Some systems combine the two, such as electronically activated access control systems that open doors to admit entry of authorized persons.

According to the NCPI Student Notebook[20]:

> Procedural security measures can deter, delay, detect or deny by restricting authorized access to targets by requiring that several persons cooperate in any effort to reach a target; by providing for formal or informal observation of targets; by reducing, dividing and spreading, or eliminating the targets themselves; and by many other similar approaches.

An example of the application of procedural controls is the separation of shipping and receiving in a warehouse setting. Procedural controls are one of the most valuable tools available to reduce internal theft crimes in business.

The greatest protection of assets is provided when a comprehensively designed security system integrally regulates an appropriate mix of electronic, physical, and procedural security measures.

Other Knowledge Areas

In addition to security knowledge, a crime prevention manager must possess planning and communications skills. Actions will not be successful, under normal conditions, unless they are well planned and effectively communicated or *sold* to those who must implement them. In order to obtain positive and productive responses to security recommendations, the client (public or private) must have the knowledge, justification, and opportunity to act. This is true whether the client is an individual homeowner considering joining a neighborhood watch program or a business manager deciding whether or not a loss prevention program is worth installing or maintaining based on a cost benefit analysis.

The public service aspects of crime prevention require additional skills, knowledge and ability if activities are to be successful. The practitioner in the public arena must also possess the skills of a community organizer if anything more than a *flash in the pan* is to be achieved. Neighbors, business associates, and whole communities must agree on mutually beneficial programming and reinforce one another if long-term success is to be enjoyed. Implementation of a *buddy alarm system* for businesses and *neighborhood watch* in specific geographical areas are but two examples. The public service practitioner must sell the concept and then reinforce it with result-oriented reports to the public if interest is to be maintained. Therefore, the skills of, or access to, a public relations specialist can help.

Finally, knowledge of the art of management beyond supervision is a requisite for success. The practitioner must develop program goals, priorities, objectives, and strategies. The basic crime prevention strategies of: (1) public awareness, (2) crime risk management counseling, (3) education and counseling services for specific groups, (4) target and crime-specific programs, (5) environmental design, and (6) surveillance and reporting, must be applied to clients with vastly different needs and attitudes.

The Future

There is no question that the American public is concerned about crime and willing to take action to control crime's incidence and harm. In 1986, the U.S. prison population exceeded 500,000 for the first time in history. Concurrently, over 20 percent of the population lives in a community with a neighborhood watch program. The image of police doing a good or excellent job in the community increased in the public's eye despite the fact that the overall clearance rate for major crimes has not increased substantially. Why? Perhaps because the average citizen has had enough and has communicated to the elected leadership that crime must be dealt with more effectively than in the past.

However, while the public is taking a hard line toward crime, we see a reported increase in drug abuse. The problem is growing so severe that an increasing number of community and corporate leaders are calling it an epidemic. Not just marijuana use, we see increase in use of PCP, heroin, and especially the various forms of cocaine, including the insidious *crack*. The problems of drug use will not only affect communities in terms of street crime but will increase both internal and external thefts from businesses, thus creating an increased need for psychological honesty testing, background investigations, use of the polygraph, and other asset protection techniques.

Two types of crime are going to have a particularly dramatic impact on security plans and activities: terrorist attacks and computer-related crimes. The first is extremely frightening, regardless of the actual probability of victimization, because of the immense threat to life. (Witness the dramatic reduction in American tourism in Europe during 1986.) The second is critical because of our society's ever-increasing reliance on data processing and electronic fund transferring systems. The proliferation of technology far exceeds our use of systematic safeguards.

Further, the investigative and prosecutorial functions of our criminal justice system are not yet geared to respond effectively to these high technology offenses.

Another societal influence on the future of crime prevention is the demographic patterning of our population. Concurrent with the increased role of women in the workforce, there are significant numbers of children under 12 years of age at home alone for several hours each day. These so-called latch key children, or children in self care, must be taught how to care for themselves not only regarding crime prevention but also in the areas of fire, poison, and accident prevention. This will result in local police presenting even more educational programs in the schools and school boards adopting required crime prevention curricula.

Two final major social phenomena that will affect our society are the continuing development of an underclass of minority, economically dependent persons, and the expected baby boom. If we have a class of citizens whose value structure is not such that they control their behavior toward others, and if we have a growth of males in the crime prone years of 14 to 24, street crime is going to be affected.

> Just as the public schools are anticipating increased enrollment for the 1990s, so should the Criminal Justice System and the security field anticipate and prepare for a rise in crime in the 1990s.[21]

So, in the future we will be seeing:

- an increase in public concern about effective crime prevention techniques
- more community programs directed from a social action perspective to encourage community cohesion
- school education curricula at the elementary, middle, and senior high levels with an emphasis on safety for the younger years and law-related education for the older students
- an emphasis on reducing crimes against business as companies adopt increasingly tighter budgets with *lean and mean* staff sizes, including the use of high-technology mechanical, electronic, and procedural safeguards
- a focus on residential burglary reduction as alarm systems become less expensive, resulting in more stringent standards for police response and a growth in private security armed response units in middle-and upper-income areas
- an offer from the insurance industry of economic incentives to home owners and automobile owners to encourage the use of effective locking devices and alarms
- design of the environment for security becoming increasingly important in regard to corporate facilities, housing, recreation areas, and urban centers

Ten knowledge areas required for law enforcement officials and private security crime and loss prevention specialists will remain basic to proficiency. Content knowledge under each will change with technology and research. They are:

1. the history and principles of crime and loss prevention
2. crime risk management
3. environmental design
4. fixed and movable barriers
5. surveillance detection systems
6. citizen participation
7. information dissemination
8. crime-specific programming
9. target-specific programming
10. program development and assessment

Persons demonstrating competency in these areas have a bright future in prevention programming for either the public or private sector.

References

1. Paul J. Brantingham and Frederick L. Fraust, "A Conceptual Model of Crime Prevention," *Crime and Delinquency* 22 (1976): 284–296.
2. Arthur Kingsbury, "Functions of the Crime Prevention Officer," (Home Office Crime Prevention Center, Ph. D. diss., Stafford, England, 1976).
3. Roy W. Dixon and D.W. Rogers, *Understanding Crime Prevention*, (Lexington, Kentucky, NCPI Press, 1978), p. 2-1.
4. Ibid., p. 2-2.
5. Ibid., pp. 2-4–2-5.
6. Chief B. Dean Leitner, Statement of Mission of the Lincoln Police Department in a letter to the National Crime Prevention Council, April 1986.
7. Commission on Accreditation for Law Enforcement Agencies, Inc., *The Standards Manual of the Law Enforcement Agency Accreditation Program*, August 1983, pp. 45-1–45-2.
8. Mailing list of the National Crime Prevention Council, 1986.
9. Leonard A. Sipes, Jr., "Community Involvement and Community Crime Prevention; Reflections on the History, Present and Future Implications of Community Crime Prevention" (Unpublished monograph, 1986).
10. "Crime Prevention: Status and Trends," A Report by the National Crime Prevention Council, Sponsored by the Florence V. Foundation, 1986.

11. Kingsbury, "Functions of the Crime Prevention Officer," p. 33.
12. Milton/Lipson, *On Guard: The Business of Private Security* (NY: Quadrangle/The New York Times Co., 1975), p. vii.
13. National Advisory Committee on Criminal Justice Standards and Goals, Report of the Task Force on Private Security, Washington, D.C.: (U.S. Government Printing Office, 1976), p. 31.
14. Ibid., p. 38.
15. Hallcrest Report
16. Dixon and Rogers, *Understanding Crime Prevention*, p. 2-12–2-13.
17. NCPI Student Notebook; Louisville, Kentucky; undated.
18. Ibid., p. 4-7.
19. Ibid., p. 5-1.
20. Ibid., pp. 5-2–5-3.
21. "Are We Ready For the Future?", *Security and Distributing Marketing*, (Boston: Cahners Publishing Company, 1986), p. 55.

Appendix 1a
Crime Prevention Glossary*

The following list of terms are included to assist the reader and to facilitate common terminology usage. It should be noted that the definitions of the terms listed reflect their utilization in the area of community crime prevention.

Behavior Modification. A change in behavior patterns brought about by education or training in crime prevention principles and/or techniques.

Burglary. The unlawful entry or trespass within a structure with the intent to commit a crime therein. (Related terms: Breaking and Entering, Housebreaking, Unlawful Entry.)

Citizen Awareness. A state in which the members of the community are cognizant of a situation (in this case, a crime risk), and possible solutions or methods of dealing with it. (Related term: Citizen Education)

Citizen Crime Reporting. Includes encouraging citizens, either witnesses or victims, to report crimes and to assist police in apprehension of the offenders. Should provide procedures for making such reports accurate and useful. (Related term: Community Involvement)

Citizen Community Patrols. A concept in which residents of a community (residential, business, etc.) organize into a group and provide security patrols of the environs of that community.

Community Action Programs. Programs undertaken by the residents of a community with a specific goal in mind. (Related term: Community Involvement)

Community Crime Prevention. Direct involvement of a single sector or a combination of sectors in a community in the planning, funding, implementation, and operation of a crime prevention program; usually stressing cooperation with the local criminal justice system. (Related term: Community Involvement)

Community Anticrime Efforts. Actions taken by the members of a community to prevent crime or to increase the amount of cooperation with the police in reporting crimes.

Crime Control Program. Various programs dealing with the control of prevention of specific types of crime, based on patterns of occurrences and known offenders and victims.

Crime Displacement. Theory which states that criminals denied the opportunity to commit crimes in a certain area will move to other areas or to other crimes. (Related term: Mercury Effect)

Crime Prediction. A system of predicting future crime patterns and trends using past patterns and trends as indicators.

Crime Prevention. The anticipation, recognition, and analysis of a crime risk, and the initiation of some action to remove or reduce it.

Crime Prevention Training. The training of crime prevention practitioners in the theory and principles of crime prevention.

Crime Resistance. Term used by the Federal Bureau of Investigation (FBI) and other law enforcement and criminal justice agencies which is synonymous with crime prevention.

Crime Specific Countermeasures. Specific strategies which are intended to prevent certain crimes (e.g., an anti-burglary campaign).

Defensible Space. A term for the range of mechanisms—real and symbolic barriers, strongly defined areas of influence, improved opportunities for surveillance—that combine to bring an environment under the control of its residents. (Related term: Environmental Design)

Deterrence. A concept which holds that the threat of punishment or the denial of opportunity will forestall the criminal or delinquent act. (Related term: Punitive Crime Prevention)

Dynamic Risk. A risk situation which carries the potential for both benefit and cost or loss. Normally considered to be the type of risk that is inherent in doing business. (Related terms: Risk Management, Risk Assessment, Pure Risk)

Environmental Design. Selectively controlling variables in the planning, design, and effective use of physical space to create physical and social conditions which will promote citizen surveillance, reduce criminal opportunity, and increase the risk of apprehension and arrest.

Information Transfer. A means by which professionals and practitioners exchange ideas, concepts, and programatical information to facilitate the development and the practice of crime prevention.

Mechanical Crime Prevention. Concept developed by Lejins based upon opportunity reduction (emphasis on target hardening).

Media Campaigns. The use of mass media (radio, television, newspapers, etc.) as part of a public awareness program.

---

*From the Director of Community Crime Prevention Programs, U.S. Department of Justice, LEAA, December 1978.

Neighborhood Watch. A community action program administered by the National Sheriff's Association which encourages neighborhood residents to organize with a purpose of neighborhood security, and cooperation with the criminal justice system. (Related terms: Block Watch, Radio Watch, Community Action Programs)

Operation Identification. A program in which citizens mark property for identification purposes. Intent of program is to facilitate the recovery and return of stolen property, and to provide a deterrent to potential offenders. (Related term: Property Identification)

Opportunity Reduction. The removal of opportunity, a necessary ingredient for the commission of crime, by making a potential target of attack inaccessible or unattractive and by making the attack itself dangerous or unprofitable to the criminal.

Physical Crime Prevention. Prevention of anticipated crimes or delinquencies by placing obstacles in the way of potential offenders so that it becomes difficult or impossible for them to perpetrate the offense. (Related term: Opportunity Reduction)

Private Security. Self-employed individuals and privately funded business entities and organizations who provide security-related services to a restricted clientele for a fee, for the individual or entity that retains or employs them, or for themselves in order to protect their persons, private property, or interests from varied hazards. (Related terms: Security Systems, Deterrence)

Property Identification. See Operation Identification.

Pure Risk. A risk situation in which there is no possibility for benefit, only cost or loss (e.g., fire, flood, etc.). (Related terms: Risk Management, Risk Assessment, Dynamic Risk)

Radio Watch. A community action program in which citizens organize to facilitate the reporting of criminal activity through the use of radio transmissions. This usually involves the use of citizen band radios. (Related terms: Neighborhood Watch, Block Watch, Community Action Programs)

Residential Security. Security concept utilizing mechanical, electronic, procedural, and other methods to protect the home from crime and other hazards. (Related terms: Physical Crime Prevention, Operation Identification, Security Systems)

Risk Management. The anticipation, recognition, and appraisal of a risk and the initiation of some action to remove the risk or reduce the potential loss from it to an acceptable level. (NCPI) (Related terms: Dynamic Risk, Pure Risk, Risk Assessment)

Robbery. Unlawful and forcible taking of property in the possession of another, from his person or immediate presence, and against his will by use of violence or threat of violence, and with the intent to steal.

Security Codes and Ordinances. Laws which require that buildings, recreational facilities, streets, and other public areas meet certain security requirements. (Related terms: Environmental Design, Defensible Space, Alarm and Deadbolt Laws)

Security Surveys. Surveys of residences, businesses, public buildings and other facilities for the purpose of evaluating the degree of security present in order to make recommendations for physical and procedural improvement. (Related term: Vulnerability Analysis/Assessment)

Security Systems. Methods and materials employed to ensure the protection of persons and premises against encroachment. (Related terms: Alarm Systems, Security Hardware, Lighting, Target Hardening)

Street Lighting. A concern to the crime prevention practitioner. Street lighting is an important component of environmental security. (Related terms: Defensible Space, Deterrents, Environmental Design)

Target Hardening. Concept of opportunity reduction which seeks to deter the criminal act by making a potential target of attack inaccessible or unattractive and by making the attack itself dangerous or unprofitable to the criminal. (Related terms: Opportunity Reduction, Security Hardware, Security Systems)

Target Specific Measures. Specific strategies intended to protect a certain premise from crime risk or safety hazards.

Vigilante. Those who extralegally assume authority for summary action, professedly to keep order and to punish crime because of the alleged inadequacy or failure of the usual law enforcement agencies.

Whistle Stop. A program designed to provide women with a means of signaling for help in the event of an attack. A secondary purpose is to increase public awareness and public participation in crime prevention programs.

Youth Services Programs. (As applied to crime prevention.) Programs which focus on youths and attempt to divert them from criminal activities. Some of these deal with children in schools, others deal with students who have dropped out, and others are based in the community. (Related term: Juvenile Delinquency Prevention)

Appendix 1b

Salesmanship: A Critical Element in Crime Prevention*

CHARLES M. GIRARD, Ph.D.

The Purposes and Benefits of Sales Training in Crime Prevention

As every law enforcement official involved in crime prevention soon learns, effective selling is an essential ingredient in a successful program. Thus, a crime prevention officer must possess the knowledge and skills of selling in order to be an effective participant in a unit. Moreover, it will be the purpose of this section to assist the crime prevention supervisor and, in turn, subordinate personnel, in understanding sales methods and techniques.

The first step in this process will be to define salesmanship. For purposes of this discussion, it will be defined as "the art and science of selling."[1] Like the police officer, the techniques and skills used by the salesperson must be so well learned that they are automatic because if either must stop to ponder each move, the situations faced will quickly pass by.

As with sales in general, selling crime prevention is quite complex since an officer must deal with many facets of human nature. In addition, a significant proportion of one's efforts will prove unsuccessful.[2] Moreover, it takes a particular kind of individual to be able to carry off the sales aspect of crime prevention yet be able to respond to common but often short-term failures with a constantly renewed sense of challenge. The remainder of this section is presented with these qualities in mind.

---

*Reprinted with permission from "Salesmanship: A Critical Element in Crime Prevention," Dr. Charles M. Giraud, from *Administration Crime Prevention Course: Student Handbook* (Koepsell & Giraud and Associates, Inc., 1975).

Consultative Selling: The Role of the Crime Prevention Officer

The Nature of Consultative Selling

Consultative selling or the selling of the crime prevention concept requires a crime prevention officer to become a professional marketeer rather than just a law enforcement officer and professional crime prevention specialist. There are distinct differences, however, between consultative selling (which generally involves a service) and sales in general (which almost always involves a product). As one text notes, consultative selling involves the following:[3]

- The consultative salesman views himself as the "manager of a personal service business." In this role he has two objectives: to help his clients maximize the profitability of their operation (or, in crime prevention terms, to maximize the odds that his clients will not be victimized); and, thereby, to maximize the profitability of his own activities (or, to maximize the opportunity of implementing a crime prevention program).
- The consultative sales approach requires planning. That is, those who are involved in selling crime prevention services must identify the type of clients with which they will work and must detail how the problems of his clients will be approached.
- Finally, the consultative selling approach requires "innovation;" that is, the salesman must identify a combination of approaches that will satisfy the needs and wants of a particular client.

Each of these elements can easily be related to the activities of a crime prevention unit. To be effective in consultative sales, however, a number of requisites must be met.

Prerequisites to Consultative Sales

With regard to performance in consultative sales (as well as in other forms of sales), there are several prerequisites to success. The first is knowledge regarding the service to be provided. When discussing crime prevention with various clients (from the chief, to the head of patrol, to individuals and citizen groups), one must project a firm knowledge and understanding of the concept. It is also important to project the feeling that the subject is important, that it can provide the answers to certain crime problems; and, that if the service is provided (and accepted), the chance of criminal victimization will be lessened.

Other prerequisites to the effective sales of crime prevention include the following:

- *Attitude.* Without the proper attitude, anyone attempting to sell crime prevention or oneself will not succeed.
- *Habit.* To be effective, it is necessary that a crime prevention officer work hard and wisely in relation to a predetermined plan.
- *Skill.* To successfully sell crime prevention, practice will be necessary to develop sales skills.[4]

A few additional comments regarding these points follow:

Attitude

A crime prevention officer with a positive attitude breeds confidence and enthusiasm into everything he does. The officer is proud of law enforcement, proud of the unit, proud of the crime prevention concept itself, and proud of the job as a crime prevention officer. Without a positive attitude about oneself as well as the concept of crime prevention, it will be difficult for an officer to pay more than lip service to the idea within the community. If negativism is the prevailing attitude among the officers in a unit, it may be incumbent upon the supervisor to carefully consider if they are right for the crime prevention job. Since crime prevention places an officer in the public eye perhaps more than any other police function, a negative attitude can do a grave disservice both to the crime prevention concept and to the department as a whole.

Work Habits and Self Management

Many studies have shown that the lack of good planning and organization is the most frequent cause of failure among sales personnel.[5] The underlying reasons for this fact have an interesting parallel with crime prevention. That is, salespeople do not have a natural knack of organizing their time. Their job takes them in many directions and demands a flexible schedule. This is all the more reason why they must be capable of planning their activities and spending their time where it will pay the greatest dividends. In fact, those involved in training salespeople recommend that they:

1. plan each discussion or presentation in advance
2. plan how to handle objections
3. plan an efficient itinerary
4. schedule appointments so as to avoid excessive waiting time
5. schedule time to do paper work
6. systematically look for new prospects or groups to contact

These habits have direct applicability to a crime prevention operation. For example, there is a clear need to develop and plan for program implementation in relation to specific objectives and work tasks. And, since time and resources will always be in limited supply, it is essential that these resources be maximized to increase the chances of program success. Moreover, each of the above points can directly relate to maximizing resource utilization.

Skill Areas

For purposes of this discussion, *skill* is defined as the ability to accomplish a specific sales objective—such as convincing a businessowner to invest in better security hardware.[6] To expand this ability, practice will be necessary. The question remains, however: What should one practice to improve his skills in the area of selling? Answers to this question are as follows:

The skill of understanding people. It is important that a crime prevention officer develop a knack for explaining a proposition—say, the installation of improved security hardware—so that the person (or people) with whom one is dealing can quickly grasp the direct benefit of such a measure. Understanding people is, of course, an intrinsic part of the law enforcement official's function. In the role of a crime *prevention* officer, however, it is necessary to look at the individual from a different perspective—no longer as a potential criminal, but rather as someone who is being provided a service.

The skill of qualifying prospects. This element in selling requires one to carefully identify who is to be approached and on what grounds. That is, it will be necessary to first determine who needs, or could best utilize, the services of a crime prevention unit (for example, portions of a community most severely affected by crime). Second, it must be determined if they should be contacted to sell them improved security hardware, a neighborhood watch program, a property marking program, etc.

The skill of determining client needs. This skill cannot be overemphasized. In order to sell crime prevention services, an officer must be able to relate to the citizen. To relate to the citizen, one must understand their problems and needs. That means that the officer must understand the crime problems in an area; why victimization rates are high (i.e., poor security, citizen awareness, etc.); why action has not yet been taken (i.e., residents in low-income areas might not be able to afford improved security hardware); and, what can be done to help in the variety of situations one is likely to face. Thus, it is imperative that the officer examine area crime data, socioeconomic information, etc.; and come armed with an arsenal of crime prevention programs and techniques (as well as arguments for their use). It must be stressed that security and crime prevention will be a *low order need* of most citizens, so an officer's arguments must be strong and convincing in order to succeed.

The skill of making a presentation. Demonstration and the ability to converse freely and easily with people is an important asset to the crime prevention officer. Unfortunately, this skill can be perfected in only one way—practice.

The skill of listening. Good sellers listen more than they talk. Good crime prevention officers should do the same. That is, they should listen not only to the words individuals are using, but to what they are really saying, what they really mean and why they are saying it. Only in this way will the officer be able to gain a clue as to how to appeal to the person.

The skill of demonstrating crime prevention. Successful sellers and successful crime prevention officers must be able to handle their service or concept and demonstrate what it is, how it works, and why its use will be beneficial. For example, shimming a lock with a credit card or demonstrating how easy it is to gain entry through a window is a good technique to demonstrate the need for crime prevention. On the positive side, the demonstration of cylinder locks or window pins could help illustrate the use and benefit of prevention.

The skill of handling objections. A wide range of objections will likely arise during any discussion of crime prevention. Unfortunately, no tailor-made answers are available. However, by comparing notes with other crime prevention officers to help prepare for such objections, and by responding to them positively (i.e., as an indicator of interest), they can be much more effectively treated.

The skill of closing the sale. In sales jargon, closing a sale is getting the order or convincing an individual to make a final decision to go along with what is being sold. A closing should be sought whenever, and as soon as, possible. Waiting until the next time may create an even more difficult task.

Elaboration of several of these points, as well as others, is presented in the remainder of this appendix.

The Three Steps of Sales

An Overview

A crime prevention sales approach can, in some ways, be related to the process a doctor might use in analyzing the problems or needs of a patient. For example, a doctor first inquires about the background and medical history of a patient, then reviews current symptoms. Finally, a treatment is prescribed. Of course, it then is up to the patient to accept or discard the prescribed treatment.[7]

A crime prevention officer must take many of the same steps. For example, when arriving at a home or business, one must be familiar with the socioeconomic and crime-related background of the neighborhood, the business, or the individuals. Next, one must learn about the contact's current problems or needs (or symptoms), such as recent victimization or fear of victimization. Finally, the officer must prescribe a remedy—and, in particular, one which the contact is both capable and desirous of using.

In the following discussion, this process is described more fully within the context of three major steps:

1. The preapproach
2. The approach
3. The presentation: Creating interest

Step One: The Preapproach in Sales

Developing Information on the Individual, Business, or Organization

A crime prevention officer's potential will clearly suffer if not thoroughly familiar with the individuals and organizations with which one must deal. A variety of sources and a number of facts are available to help in this process. One primary source is the crime database. Before reviewing the particular types of information needed, however, it may be helpful to discuss why preapproach knowledge is important to selling crime prevention.

There are generally four ways in which information about people and their problems will be helpful in selling crime prevention. First, it will permit the crime prevention officer to have something in common with contacts and, thus, an initial rapport will be more easily established (i.e., the nature of a business, the crime problem in a community, etc.). Second, such prior knowledge will permit the officer to be prepared to answer many of the questions that are likely to arise (i.e., what are the predominant types of crimes committed in an area? If it is burglary, how has entry usually been gained?). Third, by coming prepared with knowledge, facts, and personal presence, an officer will project an air of professionalism and competence that will tend to make an audience far more attentive. Finally, pulling together each of these points, the officer will be far more likely to gain the respect of people contacted. In the case of crime prevention, respect—drawn from an officer's knowledge, understanding and rapport with contacts—will be a key ingredient in the acceptance and use of final recommendations.

In addition to the data included in a crime database, a variety of other subjects and types of information will be helpful to the crime prevention officer. For example, documents available through the U.S. Census Bureau will provide information (by census tract or neighborhood) on such factors as family size and income, educational attainment, average age, and housing type (i.e., single or multiple family units, owner or renter occupied). Information provided by local chambers of commerce, as well as commercial directories, developed by private organizations in a community will provide information on businesses and industries (i.e., type of product manufactured, number of employees, etc.). Still more information can be gathered on a personal basis while conducting security surveys or speaking to neighborhood groups that will add to an officer's preapproach knowledge (i.e., problems concerning victimization and so on).

Making Appointments

Time is a valuable resource for any crime prevention officer. The systematic arrangement of appointments is one method of maximizing all available time. In particular, once it is determined upon which portion of a community or which type of business a program is to concentrate, several methods of arranging appointments can be used, including: telephone contacts, personal calls, cards or letters, and using third parties. A few guidelines regarding each of these approaches follows[8]:

Even though the telephone may be impersonal, it does have the advantage of being both convenient and inexpensive for both the crime prevention officer and the person or organization contacted. In using this tool as an introductory device, however, calls should be restricted to about one minute. Do not try to sell crime prevention over the phone—just make the appointment. The officer should know exactly what she or he is going to say and should be definite regarding time.

If postcards and/or letters are used to inform the public about a program or to promote personal appointments, it should be recognized that such mailing will be competing with many other similar communications. Thus, they should incorporate the following points if they are to be at all successful:

- They should be brief.
- They should be personal and signed by the crime prevention officer, supervisor, or chief.
- They should be interesting.
- They should be positive in approach.

The personal call is used by some units as a method of selling crime prevention. In this approach, officers select a target area and simply go door-to-door to deliver free brochures and to discuss the need for and merits of crime prevention. As requested, security surveys are also provided as part of this approach. While this can be a useful technique, it does have certain shortcomings; i.e., even if people are at home, it may frequently be inconvenient for them to spend time with the officer. Thus, nonproductive time may be substantial.

The final technique involves the use of a third party. For example, satisfied customers such as homeowners, neighborhood groups, and business groups may be used to help an officer arrange interviews or appointments with friends or business associates. This method has merit in that it builds your prestige in crime prevention and demonstrates to others that such services are valuable.

Step Two: The Approach

The Nature of the Approach Phase

The first few minutes of a discussion with a person are of the utmost importance because they set the overall tone for the future. The approach, which may last up to five minutes, is thus aimed at securing and holding an individual's attention. During this period, the officer should:

- Introduce oneself so that the individual feels it is worthwhile to listen.
- Sell oneself, the concept of crime prevention, and the police department.
- Secure and hold the individual's attention.

Although these steps may be difficult to an inexperienced crime prevention seller, one may be more successful if aware of what people will look for during the first few minutes of an interview. Generally, during the approach phase, people will size up the officer. Therefore, the image that is projected is important (i.e., appearance and manner, as well as the points discussed in the preapproach phase). As a means of improving their officers' appearance, various crime prevention agencies have experimented with the use of traditional uniforms, business clothing, official blazers, etc. No one method, however, has proven to be the best. In some cases, it has been found that by approaching people in uniform, they are more receptive and, in a neighborhood environment, are more willing to answer the door (i.e., some plainclothes officers have been mistaken as other kinds of solicitors and have found an inordinate number of people "not at home"). On the other hand, the use of blazers and sports jackets in the business environment has also proven beneficial. Regardless of the approach you select, however, it is important to look sharp.

Other factors to consider in the early moments of a discussion include presence (i.e., an officer should speak factually, confidently and in a relaxed conversational manner); the opening speech (i.e., it must have the ability to capture attention and interest); and style (i.e., do not sit down until invited to do so and wait until the individual offers his hand before offering yours). The presentation of a business card may also be helpful; however, an officer should learn quickly whether offering a card in the beginning of a discussion or at the end will be most suitable to the personal method of presentation.

Useful Approach Techniques

Over the years, sales specialists have developed a variety of approach techniques that can be instructive to the crime prevention officer. Most of these techniques are based on the personal needs and desires of the people a crime prevention officer will contact. Of particular importance, however, will be personal needs which include love, food, and security. Of special relevance is, of course, security.

To relate effectively to people's basic security needs, however, will require the use of varied approaches in varied situations. As a basis for determining the most appropriate approach, the following checklist is presented.

Benefit approach. Simply stress the benefits of crime prevention to meeting one's security needs.

Curiosity approach. This technique requires that an officer raise a question that sparks the curiosity of the contact. For example, "Do you know the most common way that a burglar gains entry into homes in this area?"

Dramatic approach. This approach requires an officer to use dramatics in support of opening remarks. For example, slip the lock on the front door.

Factual approach. The use of statistics can be an eye opening technique. For example, "Did you know that one in every eight families was the victim of a burglary in this neighborhood last year?"

The "I'm here to help you" approach. This is a direct offer to help. For example, an officer may indicate that he will spend as much time at a business or residence as necessary to be of immediate assistance.

Introductory approach. This usually involves a letter of introduction, an introductory card or note, or a testimonial from someone in the community who has benefited through the use of crime prevention techniques.

Opinion approach. This approach asks an individual's opinion with a question such as "What is your reaction to the rampant crime that is occurring in this neighborhood?" Obviously, this is a simple approach, but it may be disarming (and even open the police department up to criticism), yet it may be effective in quickly gaining the interest of the contact.

Problemsolving approach. Through preapproach research, an officer will be familiar with the crime problems of an area. In the problemsolving approach, one may open by stating how to help a resident, for example, to increase the security of doors and windows.

Praise or compliment approach. A compliment or honest praise is usually well-received. For example, an officer might open by complimenting a resident on the value of a family dog as a good alarm system, or a businessperson on the sound thinking displayed by a well-lighted and visible safe. This approach often succeeds in bringing a contact over to the officer's side because it demonstrates that the contact has already taken some positive steps in the direction of security.

Question approach. This approach involves the use of an interesting fact about crime prevention, framed as a question. For example, "Did you know that crime prevention can help keep your family safe?"

Shock approach. A startling, factual statement can make an individual sit up and take notice. For example, "Three burglaries have taken place on this block in the last two weeks!"

Building a Comfortable and Positive Atmosphere

The creation of a comfortable and positive environment is another key component in a successful approach. This means that the crime prevention officer must study each contact and establish a feeling of warmth or good will so that the contact will begin to discuss problems.

This approach may be unusual to a police officer who is used to commanding respect and expecting responses to questions. However, if the officer approaches crime prevention autocratically, and uses a hardline manner (techniques frequently necessary in traditional police operations), he may quickly end the conversation. Thus, the officer must be acutely aware of the feelings, moods, and temperaments of the people contacted. This will require constant study and close observation, since variations in people and temperament are endless. Further, whether individuals are pleasant or rude, the officer must approach each encounter as a challenge and win over the various personality types. Some helpful hints in achieving this end are as follows:

- Appear pleasant, in good spirits, and smile.
- Call the individual by name.
- Take a sincere interest in the individual and their surroundings; they will react more favorably if they believe an officer is concerned about their problems.
- Be a good listener, the more the individual talks the more the officer will learn about specific problem and needs.

Near the end of the approach the officer should strive to accomplish two things:

1. Establish rapport and get the interview on a relaxed basis as soon as possible.
2. Get the individual to begin discussing needs so the officer can complete the balance of the discussion.

Achievement of these objectives may require some small talk to break the tension. It is important, however, that the officer begin asking questions without being abrupt. One should get the individual into a yes frame of mind by asking questions that require a yes answer. This is known in selling as yes building. For example, when talking with a businessperson, ask whether or not it is best to take every reasonable step to prevent someone gaining an unfair advantage over a person in a business. Similarly, you might query a resident as to whether or not he is concerned with protecting property and family.

After experimenting with a number of approaches and techniques, a crime prevention officer soon determines those which feel most comfortable. One should then stick with that approach, for if a seller attempts to use techniques with which he feels uncomfortable, he will be seen through quickly and will be hardpressed to even begin selling the concept of crime prevention.

Step Three: The Presentation— Creating Interest

Planning What You Will Do and Say

Planning what one will say and do is as important as the actual presentation. The more completely one plans steps, the better the chances will be of persuading an audience to accept and implement recommendations. Regardless of the techniques employed, however, it should always be remembered that the end result of a presentation is to motivate people into action. To do this, the officer must fulfill the needs of the contact, inform and persuade, and finally, assure the person that crime prevention is the best choice to make.

One technique to help achieve this end is to determine how others might respond or react to various aproaches and techniques. This process can be aided if an officer asks the following questions:

- How can I fit my presentation or comments around the particular needs of an individual?
- How can I adapt my presentation to these interests?
- How can I shape the listener's questions or objections so that he agrees that crime prevention is the best choice?
- How can I finally convince this person?

In answering these questions, the officer takes the position of the individual addressed. In so doing, one should be aware of the following points. First, an individual will normally be responsive if dissatisfied with the present security situation. Thus, the officer must demonstrate the current deficiencies.

Second, an individual will question whether or not crime prevention will have a positive impact on these deficiencies. Therefore, the officer must show clearly how the concept can help.

Third, the individual will question whether or not the officer's crime prevention claims can be backed up. In this instance, the officer should bring to bear both logical and statistical support. For example, one might state that most residential burglaries are crimes of opportunity committed by amateurs out for a quick hit and that studies have proven that if such persons can be denied entry—through the use of improved security hardware—for as little as four minutes, they will almost always abort the attempt; and, finally, that the security measures and equipment that will be discussed by the officer will delay forced entry to such a level.

Another approach might be to quote statistics regarding the reduced rates of victimization among residents or businesses that have installed better security hardware, participated in a property marking program, etc.

Fourth, an individual will question what it will cost to implement the recommended measures. At this stage, the officer must quickly assess, for example, the value of the property to be secured, the financial capacity of the contract, and make recommendations that take these factors into account (i.e., the value of security measures should not exceed the value of the property to be secured; and, a person's ability to afford the measures will have a clear and direct bearing on this acceptance of an officer's recommendations).

Finally, an individual will question whether or not to accept and act on the officer's recommendations. Moreover, an officer must carefully assess what the market will bear. Regardless of the soundness of recommendations, recommendations that are not implemented are of absolutely no value. So, if the officer feels that a contact will actually use no more than a broomstick to secure a sliding door, or nails or pins to secure a double hung window, the officer should accept that fact and go for the sale. Even though marginal, such steps will improve the security of a family who might otherwise have not taken any action at all.

Essentials of a Good Presentation

The ideal presentation is rarely achieved, but a crime prevention officer should make each presentation as good as possible. Because of the variety of situations one will face, it is impossible to provide a blueprint of what a prototype presentation should contain. However, certain guidelines can be offered.

First, each presentation should be as complete and understandable as possible, so the individual knows exactly why and what is being proposed. A presentation need not be reduced to minute detail, but should include:

- An initial outline of the topics to be discussed and the conclusions to be reached.
- Simplified step-by-step explanations of complex points.
- The presentation of new and important information in small doses.
- Explanations related by examples to the problems of the person or group being addressed.
- A summary of points and conclusions reached.

A second guideline for a sound persentation calls for the anticipation of objections to the extent that the objections—and their rebuttals—can be built into the presentation. Experience has shown that if an officer can list all of the possible objections to the points he will raise, write them down and develop appropriate rebuttals and responses, that officer will be in an advantageous position during a presentation.

References

1. Robert F. Vizza, ed., *The New Handbook of Sales Training* (Englewood Cliffs, N.J.: Prentice-Hall, Inc., 1968), p. 134.
2. *Ibid.*, p. 134.
3. Mack Hanan, James Cribbin and Herman Heiser, *Consultative Selling*, (Washington, D.C.: American Management Association, Inc., 1970), pp. 12–14.
4. Vizza, *Handbook of Sales Training*, p. 129.

5. *Ibid.*, p. 129.
6. *Ibid.*, p. 130
7. The Southwestern Company, *Home Learning Division Sales Manual* (Nashville, Tennessee: The Southwestern Company, n.d.), pp. 18–19.
8. Lowell Anderson and Sam Dermengian, *Salesmanship: A Programmed Guide* (Englewood Cliffs, N.J.: Prentice-Hall Inc., 1971), pp. 92–93. Unless otherwise noted, the discussion in Appendix 36b is based on information provided in this source.

PART TWO
INSTITUTIONAL AND INDUSTRIAL SECURITY

Chapter 2
Retail Security

ALBERT A. SEEDMAN, B.B.A., M.P.A.

Losses in the retail industry have a staggering effect on the economy, through higher prices, business failures, and loss of jobs. These losses are mainly due to a variety of crimes—burglary, robbery, and larceny (shoplifting and employee theft).

Retail stores take their physical inventory at intervals of a year or less. The merchandise that remains unaccounted for when the inventory is taken is commonly called shrinkage. Shrinkage can be attributed to shoplifting, employee theft, and poor recordkeeping within the company.

The National Coalition to Prevent Shoplifting has recently stated that "shoplifting is the nation's most expensive crime." It is estimated that shoplifting alone cost in excess of $16 billion in 1979. By contrast, during the same year, bank robbery losses in the United States amounted to $47.5 million.

Shoplifting is widely publicized and most people are aware of its impact. However, security experts generally agree that employee theft causes equal or greater losses. In recent years, excessive shrinkage has been responsible for an ever-increasing number of business failures.

Retailers know that they must devote a portion of their energy and resources to fight the causes of shrinkage. This chapter is devoted to an in-depth examination of shrinkage and what can be done to reduce it.

Dealing with Employee Theft

The retail industry provides employment for all kinds of people. To some it is their lifelong career, and to others it is a stop on the way up or on the way down. There is a constant turnover of personnel because of holiday buildups, changes in availability of students and homemakers, or any number of other reasons.

This flow of people into and out of retailing is the cause for some of the loss-related problems, because untrained people make errors frequently.

Part of the internal theft problem is that retailing is a low-paying industry. People who are new and earn minimum wages are neither loyal nor sympathetic to their companies. If they think they can get away with it, they will steal. Today, more employees are stealing than ever before.

Preemployment Screening

The first line of defense against theft is to make every effort to employ only honest people. All too frequently, when many jobs need to be filled for the holiday rush, just about anyone will do. Former employers and references are not checked, nor is any real screening done. Even though the intention is to hire a person for only a few weeks, it is poor business to employ anyone without doing *some* kind of investigation.

Every job-seeker should be required to submit a completed application form. Because the Civil Rights Act of 1964, and similar legislation which guarantees equal employment opportunities, have limited what one may be entitled to know about an applicant, it has become increasingly difficult to screen out undesirable people. Questions about age, health, marital status, arrests, religion, and race are prohibited. While some of these subjects certainly have nothing to do with honesty, they may be partially relevant as predictors of potential for job performance. Although it is permissible to ask if the applicant has been convicted of a crime, it may be discriminatory to deny a job if the crime in question is totally unrelated to job performance.

Perhaps the most valuable information on the application form is the documentation of previous employment, dates worked, and reasons for leaving. A personal interview often reveals inconsistencies or untruths about these matters. Most companies follow up by mailing a form letter to previous employers. Frequently, these requests are not answered; even when they are, many companies will not go on record about acts of dishonesty or other serious matters. The personnel officer can be more successful by checking employers and references by telephone. Even if the facts are not totally or directly revealed, innuendos, hesitation, and tone of voice will convey much information of value.

Sometimes, particularly for an important or high-risk job, it may be advisable to have a private agency do a more thorough background check. Bank, credit card companies, hospitals, and other places have extensive files for people who know how to go about such investigations.

Ordinarily, financial information about individuals is not disseminated except in conjunction with applications for credit. Since a person's credit rating is an excellent indicator of some of the desirable traits of an employee, such information enlightens the screening process. In some metropolitan areas, the major retailers participate in mutual protective associations where they maintain centralized records of shoplifters and dishonest employees. Member stores use these records when screening job applicants and also when seeking prior activity of persons apprehended while shoplifting. Nonmembers do not have access to these files. Incidentally, the Fair Credit Reporting Act requires that such files must be constantly purged of any information that is over seven years old.

Criminal records usually are not available to inquiries from private industry. Most police departments are prohibited by law from disseminating such information outside of official channels. Federal law prohibits any criminal justice agency that receives federal funds from supplying such information. However, in some localities, particularly small cities, such data can often be obtained.

Polygraph

The retail industry makes extensive use of the polygraph as a screening tool and for investigating specific acts of suspected theft. If a company uses this device, the application for employment should state that this is so and that the applicant agrees to submit to such test when requested.

The polygraph has been proven to be a very accurate instrument, but nonetheless, it remains a controversial topic for use in private industry. Today, in a number of states, the use of the polygraph in connection with employment is illegal or otherwise restricted. Conversely, nineteen states have officially recognized the polygraph for commercial purposes by having enacted licensing legislation which sets standards for examiners and testing.

Psychological Stress Evaluator

There is another so-called lie detection instrument available today, which functions on different bodily reactions than the polygraph. The polygraph traces the reactions of the heart, lungs, and bodily electrical systems to a series of questions. The Psychological Stress Evaluator (PSE) detects physiological changes in voice patterns for the same purposes. The PSE has run into more criticism than the polygraph, and is more widely banned and less well known. As would be expected, the proponents of the polygraph have attacked the accuracy and validity of the PSE with a fair degree of success.

Many companies make use of commercially available questionnaires designed to detect such undesirable traits as dishonesty, alcoholism, and a propensity for violence. One of the principal benefits a company can derive from the use of any of these devices or tests is the establishing of a reputation that thieves are being kept out.

Dishonesty Profile

The problem of employee theft was undertaken in a research project at the University of Minnesota in the late 1970s. The preliminary findings of this study were released late in 1979.[1] A profile of retail employee thieves was constructed from responses to a questionnaire. The average thief is described as a white male between the ages of 16 and 25, with the youngest people being the most involved in stealing. This profile goes into further detail about marital status, total household income, and an employee's personal concerns. It was found that employees who steal were most concerned with finances and career training.

The findings go on to say that those who steal the most are paid the least and are dissatisfied with one or more aspects of their work. Dissatisfaction concerns their particular job, workload, supervisors, the companies for which they work, or opportunities for promotion. Additionally, it was found that theft was highest in work places where employees thought they would not be caught. Conversely, the greatest deterrent to theft was a high probability of being caught.

Successful screening of prospective employees cannot rely on emphasizing any individual technique. The proper combination of all procedures and accumulated data to fit each case separately will assist in making intelligent hiring decisions. However, in the real world, there will always be some level of dishonest employees. Success in keeping employee theft at a minimum will depend, in addition to screening, upon company policy and systems of inventory and cash control.

Preventing Theft of Merchandise by Employees

There is no way to be certain of the extent of employee theft, since inventory shrinkage can be caused by shoplifters and other reasons. Because employees have constant access to merchandise, the methods that they use to steal it are so varied that they stretch the imagination.

The best way to prevent such theft is to discourage it from the outset. This means there should be positive programs to promote employee morale and working conditions. This also means having rigid systems for control and accountability of merchandise. People are basically honest and will remain that way if they know they are working for a concerned and well-run organization.

These programs and systems do not operate by themselves. They must be constantly enforced by supervisors and other representatives of management. The average employee probably appreciates the opportunity to work in a well managed atmosphere.

Supervisory personnel must be involved and concerned if theft of merchandise is to be prevented. In companies which can afford to hire specially assigned security people, the task is delegated, but the responsibility remains with management nevertheless.

Controlling Employee Traffic

Ordinarily, employees come to work and leave by the same entrance. In large stores, this entrance may be patrolled by a security guard. When this is the case, it can be routine practice to examine all parcels being carried out by employees. In small stores, this can be a problem. It should be a routine matter to require employees to permit such inspections, by designated persons, when leaving work. Unless this right of inspection is reserved by management, it will be difficult to put it into effect at any arbitrary time. It is best as a company rule and explained as such to new employees at time of hiring. If this practice is engaged in by a retailer, it must apply to everyone and not be enforced in a discriminatory manner. Repairmen, contractors, and other non-employees at work within the store should also be subject to the same regulation.

Burglar Alarms

The subject of burglar alarms on doors is covered elsewhere in this volume, but special mention must be made in connection with the comings and goings of employees. If there is a burglar alarm installation, the alarm should be put back in operating condition after the employees enter. The alarm should not be shut off until the store is open for business. This will help prevent anyone from carrying any merchandise out during such hours when porters are cleaning or clerks readying the store for business. Conversely, the same procedure should be followed at night when the store closes and until the last employee leaves. It is good practice to require at least two people to unlock the door in the morning and to lock it again at night.

Many small stores, particularly those that are located within shopping centers which are protected by security guards, have no burglar alarms. At the very least, the doors should be equipped with locks which provide a printed record of opening and closing hours.

These records of opening and closing, whether maintained by a central alarm company or within the time lock, must be examined and reviewed on a daily basis by management. Any deviations from prescribed times for opening and closing must be thoroughly investigated.

Personal Property

In all stores, large or small, employees should be provided with lockers or other safe places to store their coats, handbags and packages. These articles should never be allowed to be kept on the selling floor or in stock areas.

Stockrooms

Stockrooms are favorite places for people to find remote corners to waste time or steal merchandise.

It is common practice for employees to discard worn articles of their own clothing and substitute new merchandise in its place. Once having removed price tickets or other identifying data, it is difficult to establish proof of theft. Stockrooms should therefore be kept locked as much as possible. This will result in a reduction of opportunity to steal and will improve service, since the employees will spend more time on the selling floor. Very often, customers intent upon stealing will wander into the stock areas. Locked doors will of course prevent this, but all employees should be instructed to be alert for any strange faces in stockrooms.

Repair and Cleaning Personnel

Contractors, telephone repairmen, porters, window cleaners and other outside workers should be under observation while they are at work within a store. If security personnel are available, this is a routine assignment; otherwise an awareness must be prevalent among the store staff.

Trash Disposal

A favorite method of theft is for some employees to hide merchandise in empty boxes or bags to be discarded. Then when the refuse is put out at the end of the day, they return to pick up the loot. To prevent this, all refuse must be inspected before it is carried out of the store.

Locks and Keys

It is not the intent of this chapter to go into detail about locks and keys. Safeguarding keys is of major importance, and those employees who are given keys must be thoroughly advised of the implications. When keys are lost, locks must be replaced. All keys should be collected at the end of the day and safely stored. The exception to this rule applies to keys to outside entrances. When an employee who has been entrusted with important keys is discharged, some locks may have to be replaced.

Unauthorized Markdowns

A common method of theft that many people do not regard as dishonesty is the purchase of merchandise at a self-determined discount. A common practice in the retail industry is to mark down the selling price of merchandise for a variety of reasons; out-of-season, old style, or special sales. Markdowns create a lot of paperwork, because records must be adjusted to reflect the newly-created values. Failure to do this will create a shortage since the merchandise, when sold, will be turned into fewer dollars than reflected in the records.

Some employees, however, want the new style today rather than six months from now when it might be marked down. For some, it may be a simple matter to change tickets or just prices. However they manage to do it, they create shortages unless the company records reflect these price changes. One way to minimize such activity is to frequently inspect items that employees set aside for themselves. Usually they must wait for payday to make their purchases and they prepare their selections in advance.

Receiving and Shipping Departments

Receiving and shipping operations are particularly vulnerable to theft. Receiving clerks and truckmen can easily arrange for a shipment to be short. All incoming and outgoing shipments should be counted and verified by at least two people besides the trucker. All counts must be matched against bills of lading for accuracy. Shortages or overages must be entered on receiving records and also noted on any documents that serve as receipts for truckers.

There are certain types of merchandise that may be particularly vulnerable to theft, i.e., cameras, furs, and drugs. Such items should be safeguarded in special cages or containers, and moved off receiving platforms without any delay.

Trucks

Company-owned trucks that transport merchandise between warehouses and stores should be sealed and locked. The keys for these locks should never be entrusted to truck drivers. Keys should remain at the various facilities. When trucks are unlocked, the seal numbers should be recorded and verified. Discrepancies must be immediately investigated, as must all shortages of merchandise in transit. To facilitate verification of seal numbers, they should be documented and enclosed within the locked trucks. As an additional safeguard, the person in charge at the point of origin should make certain that the proper person at the point of destination is advised of seal numbers, size of cargo, and other necessary data. As an additional precaution, truck bodies should be physically inspected from time to time to ascertain that there are no loose boards or removable panels.

Cargo theft has, for many years, been one of the mainstays of organized crime. Trucks are often hijacked when they carry the type of merchandise that can easily be disposed of, i.e., liquor, television sets, and furs. The police have often found the truck drivers themselves to be involved, or at least to be reluctant witnesses. There are many types of communications and alarm systems available which can help deter hijackings.

When truck drivers are involved, criminal investigations have not been too successful. Strong labor unions frown upon any questioning of their members, and usually have the clout to impose their will. Investigation of hijackings is most successful if left to the police specialists who have the proper sources of information, and the legal authority to obtain search warrants.

Warehouses

Big retail chains usually have one or more warehouses where merchandise is received to be stored and later shipped to their retail outlets. It is common practice for merchandise to be price-ticketed at such facilities. Because employees are usually the only people allowed into warehouses, the control of merchandise should be fairly simple as compared to a store that is open to the public. Truck drivers should not be allowed into the warehouse. Receiving docks should not be used as shipping docks. There should be separate storage facilities to safeguard merchandise that is most often easily stolen. These facilities should be adequately locked and alarmed. Very often a closed circuit television system is used to good advantage.

Warehouse environments are conducive to such activities as narcotics trafficking and gambling. These, in turn, promote debt and the necessity to steal. Efforts must be made to identify and get rid of anyone involved in these activities. Management can never feel secure, because these activities are so lucrative they will continue to crop up in spite of vigilance.

Handouts

One of the most common methods used by employees to steal merchandise is to hand it over to so-called customers. Cashiers and salespeople whose job requires the wrapping of purchases and making the sale are the employees who can do this with minimum risk of being caught. Close supervision and surveillance of suspected violators are the best methods of catching these people in the act. The usual practice of surprise audits will not be successful in detecting anyone who engages in this type of theft.

Theft of Cash by Employees

Cashiers

The basic function of a retail organization is to sell merchandise and receive cash in exchange. This cash may be in the form of currency, checks, or credit card charges. Experience has proven that some previously honest people succumb to temptation when there is an endless flow of cash going through their hands. The methods of stealing at the cash register are so numerous and often so innovative that any listing of these acts could not be considered all inclusive. Nevertheless, some well known and often practiced ways to steal money are:

1. Cashier rings the sale for a smaller amount than price ticket indicates and pockets the difference.
2. Cashier does not ring the sale and pockets the entire transaction.
3. Cashier fails to give a receipt to the customer, later voiding the sale and pocketing the cash.
4. Cashier merely voids the sale.
5. Cashier does not close register drawer after sale, thus avoiding the necessity of ringing the next sale. Cashier is thus able to make correct change to customer. Register will now contain some cash that has not been recorded on register tape.
6. Cashier takes some cash out of register, or allows accomplice to do this. The cashier then notifies supervisor that a "till-tapper" has "hit" the register.
7. Where more than one employee is assigned to the same cash drawer, the cashier merely removes some cash and creates an easily detected shortage. The cashier relies upon any subsequent investigation failing to single out the culprit.
8. Cashier accepts a check from an accomplice and fails to follow company procedure to detect if check is bad. The check is subsequently found to be useless.
9. Cashier accepts a good check from customer and later raises the amount. Cashier pockets the difference.

10. A credit card sale is made to an accomplice. Company procedures are not followed and the cashier fraudulently enters authorization for the sale.
11. Cashier makes a credit card sale to a legitimate customer with a good card. However, cashier imprints additional charge slips with the card if customer is not alert. Later, cashier substitutes these slips for cash.
12. Cashier making a credit card sale learns that the card has been reported lost or stolen. The card is not returned to the customer who is happy to get away without being detained. The cashier uses this card to create false sales and removes the like amount of cash.
13. Cashier makes sales to friends at discount prices.

Prevention of Cashier Theft

Close supervision of cashiers at work will either deter or reveal acts of theft. Supervision can be extended to include continuous observation by closed circuit television or frequent observation by store detectives from hidden observation posts. Cash receipts, copies of sales slips, and register tapes must be audited daily. This auditing operation should be done as soon as possible after each day's receipts are closed and tallied. Many times the audits fall behind by a week or more which makes it difficult to conduct a proper investigation. By the time a substantial cash shortage is discovered, or a particular suspicious pattern detected, it is learned that the cashier has either quit or been fired.

There should be a continuous policy of frequent test counts or audits of cash register contents during the workday. Frequent voids, no sales, and unexplained shortages or overages are cause for further action. Suspected cashiers should be switched to different registers each day and a meticulous record maintained of their cash activities. Surveillance should be maintained of such cashiers by experienced store detectives who are familiar with cashiering operations. A successful investigative technique is to assign integrity shoppers to test all cashiers and particularly those who are suspected of theft.

Integrity Shopping

Honesty shopping tests are conducted by specially trained people who go into stores posing as customers. These shoppers are usually contracted from outside agencies, but in large retail chains, they may be in-house personnel working for the security department. Shoppers vary their techniques according to the type of cash registers in use and the particular method of cashiering operations.

Often the shoppers are supplied with advance information concerning the specific cashiers who should be shopped. Often cashiers are shopped at random with the expectation that dishonesty will be detected.

The basic test is the *even money buy*. All others are variations of this test. The shopper approaches the cashier with the selected article and pays for it with the exact amount of money including any sales tax. The shopper appears to be in a hurry and leaves without waiting for a receipt or the usual wrapping. This affords a dishonest cashier the opportunity to neglect to ring the sale and keep the cash.

More complicated and sophisticated buys require the shopper to be able to identify the transaction on the register tapes later on. The sale just previous to the buy is observed by a second shopper and the amount of the sale is noted for identification. Then when the buy is made and not properly rung up, the register tape will supply documentation of the theft. An *uneven buy* requires the cashier to make change. The shopper making an uneven buy usually follows an even buy for purposes of observing what is being done with the cash. There are variations of buys such as *double buys* and *exchange buys*. Shoppers should mark *buy* money bills or record the serial numbers in advance. This facilitates proper identification of the evidence when necessary.

Cash Offices

Large stores, particularly department stores, have cash offices where receipts are brought each day, bank deposits prepared, cashier's daily working funds prepared and issued, coin received from banks, and money safeguarded overnight. The constant receipt and disbursing of money makes it difficult to conduct an error-proof audit in such offices. Small shortages are usually no cause for serious alarm and cash room employees have been known to steal small sums of money on a frequent basis over a long period of time. In an effort to reduce such thefts, at least two people should be assigned to work in such offices at all times.

The construction and layout of money rooms should be planned to offer adequate resistance to robbers and burglars. Money rooms must have strong safes and both burglar and holdup alarms.

People who are hired to work in these rooms should be carefully screened. Whenever a cash office employee is terminated, the safe combinations should be changed.

Cash shortages in these offices require an immediate investigation which can include one or more of these suggested steps:

1. Thoroughly audit all cash in the entire room.
2. Assign an undercover operator to the room.
3. Deliberately create an overage in receipts turned in by cashiers.
4. Track individuals' working schedules to match against occurrence of shortages.
5. In states where permissible, schedule polygraph examinations.
6. Try to determine lifestyle of employees.
7. If investigation is unsuccessful, and losses continue, rotate personnel out of the cash office.

Refunds

Refunding is an area of retail operations that is subject to theft by both employees and customers. Comments about refunding theft are included in the section that deals with shoplifting.

Undercover Investigators

Undercover investigation is a concept of law enforcement which is used by official agencies at all levels of government. It is no less effective, when properly employed in private industry. Often an undercover operator may be the only source of intelligence that can have any effect in combatting theft.

Undercover operatives can be supplied by contract security agencies or they can be employees of the company's security department. The operatives are placed in stores through the usual hiring procedures. Very often they apply for jobs on their own, but sometimes they come as referrals from an invented friend of the store manager. The fewer number of people in the company who are in the know, the greater their chances of success. Having been placed, the undercover operatives (UCs) perform their jobs (stock, sales, cashiering, etc.) while being alert to what the people around them are doing. They report suspected acts of dishonesty, as well as violations of regulations or any practices that are detrimental to the good order and efficiency of the store.

Many such investigations are conducted over a considerable period of time. Relationships and confidence must be developed; this requires a considerable amount of patience by the operatives. The best information may be a long time in coming and it cannot be pushed by the UCs. Any conversation or actions by UCs which could be construed as entrapment must be scrupulously avoided.

While information about dishonesty is most important, anything that may have an effect on shrinkage is also valuable. Information pertaining to poor working conditions, violation of policies, and supervisory inadequacies are brought to the attention of management.

Usually, the undercover information about theft is used merely as a basis to conduct an independent investigation. If the existence of the operative can remain secret, the assignment can be continued and the agent's safety not jeopardized. However, if the case is important and independent evidence cannot be developed, it may be necessary to blow the UCs cover.

Detaining Suspected Employees

An excellent way to deter theft is to have a rigid policy of prosecuting all employee thieves. Detention of suspected employees must be conducted so as to meet all legal requirements. When the suspect is summoned to the office for questioning, such legalities may become important issues at a subsequent time. The employee may claim to have been unwillfully detained. The employer has the right to sit anyone down and talk during working hours. If the employee has been detained and accused of theft, the situation should be construed as an arrest. If there is no intention to make an arrest, the employee should be told that he or she can walk out at any time. Even if the employee doesn't walk out, the interview should be concluded as quickly as possible.

Most investigations of employee theft are conducted after the fact and the evidence is not available. Many companies proceed with such investigations in the hope that they will obtain a confession and have the employee agree to make restitution. Such procedure should only be undertaken after consulting an attorney. The suspected employee might later claim to have been coerced into making the confession, and claim that payments are being extorted from him.

Even when it is possible to catch the employee in the act, reasonable care must be taken to make a proper case. Although the hiding of merchandise or money within the store is reasonable grounds to detain someone, the suspect may be able to convince a judge or jury otherwise. It is always safest to wait until the property is being removed from the premises, and then the eyewitness to the theft becomes the complainant or the arresting officer. The subject may be frisked and searched.

Sometimes it may be necessary to search the suspect's locker or desk. It is wise to have a standing company rule concerning the right to such searches. If such a policy does not exist, depending upon which of the states in our country we are operating in, the legality may be confusing. Personal lockers ordinarily cannot be searched without consent or a search warrant.

The suspect can be terminated or prosecuted. In either case, a signed confession should be obtained as well as a resignation and a waiver releasing the company from liability. In the case of minors, the binding legality of such documents is often challenged by their parents. When the employee is to be prosecuted, the police should be notified without delay.

Dealing with Shoplifting

The dollar losses to American business from shoplifting and related crimes can only be estimated. Retailers do not call upon the police and the FBI at the end of the year to have the shrinkage figures included in the crime reports. Therefore, the police do not have a real role in fighting this problem. At best, the police act as a conduit in helping the retailer feed the shoplifter into the criminal justice system. But the prosecutors and the courts regard shoplifting too lightly and thereby encourage and perpetuate this crime which reaches into the pockets of everyone in this country. Ever-rising rates of shoplifting contribute to price increases that the retailer needs to survive. Even then, unless retailers take measures to fight shoplifting, they may eventually suffer too many losses and be forced out of business.

Security Force

Although it is unrealistic to expect to completely eliminate crime, the proper use of people and equipment should have some effect in reducing crime. The presence of uniformed guards in a store is visible evidence of concern and effort by management to do something about shoplifting. Unless these guards are properly trained and supervised, their effectiveness may not reach the level that is expected. Productivity of a guard force cannot be precisely measured, but the effects of deterrence should ultimately be reflected in a reduction of attempts at shoplifting.

Ordinarily, budgetary considerations do not allow an unlimited guard force. Their deployment must be carefully considered and they should be posted at places where they can do the most good. In retail operations, these locations are most likely at receiving platforms, employee entrances, and customer doors.

The actual apprehension of shoplifters involves legal problems which should dictate that these actions be left to employees who are specially trained in laws relating to arrest, use of force and evidence. In large stores, these functions are delegated to plainclothes store detectives. In smaller stores, owners, managers and sales people sometimes become involved with shoplifters, but it is recommended that they seek police assistance without any delay.

Store detectives, as well as all other retail employees, should know shoplifting techniques. Shoplifters cannot be picked out of a crowd at random. Most of them blend very well with legitimate customers. Any representative of the store who stops a suspect should have witnessed the actual taking of the article being stolen, and be absolutely certain that it is in the possession of the person being detained. In some states, a suspected shoplifter may not be stopped until practically outside of the store. In other states, the secreting of the merchandise can be construed as intent to steal. If the apprehension is contrary to the law of the particular state, that may be the basis for a false arrest suit.

Most shoplifting is accomplished by the thief placing the merchandise in a bag or box brought into the store for such purpose. Another common practice is to hide the loot in pockets or within one's clothing. Many professional shoplifters have specially prepared boxes or articles of clothing to facilitate the swift taking of the items they have targeted. Successful observation of professionals requires excellent powers of observation and well developed surveillance techniques.

In recent years, a heretofore unknown element has become prevalent in shoplifting apprehensions. Many shoplifters are carrying weapons today, in contrast to years past. Employees who come into cantact with shoplifters must know how to protect themselves and how to keep a situation from escalating into danger for other employees and customers.

In addition to usual customer areas, a lot of shoplifting takes place in the fitting rooms where customers try on articles of clothing that they are considering for purchase. There should be attendants assigned to these rooms to observe everything being carried in or out, and also to deter any theft while persons are in the fitting booths.

Nonsecurity Employees

Many retail stores, particularly the smaller ones, do not employ people specifically for security purposes. However, even if there are guards and detectives, any security effort will produce results if the entire staff is aware of the problems of theft and knows how to combat them. With proper training, almost anyone can be expected to detect the presence of a shoplifter. Usually the best action that a sales or stock person can take is to make her or his presence obvious enough to prevent the commission of the theft. Unless such employees have specific knowledge of the legal requirements to apprehend, their role should emphasize prevention.

Security Hardware

The war on theft in retail stores has fostered the growth of a vast industry that supplies numerous scientific devices for such purposes. Cameras and closed circuit television are widely used. Two-way mirrors are often strategically installed to enable constant surveillance of selling floors. Miniature radios (walkie-talkies) and pocket beepers are also commonly used in stores.

In recent years, the article surveillance tag has come upon the scene. These tags are affixed to merchandise so that they are virtually impossible to remove without a special device. If such merchandise is being carried out of the store while the tag is still attached, an alarm is sounded. The existing systems utilize one of three types of technology: magnetics, low radio frequency waves, and microwaves. Each system has some advantages and disadvantages, but on an overall basis, these surveillance tags have been accepted as an effective means of reducing shoplifting.

In passing, it should be mentioned that some retailers have made effective use of chains, steel cables, and locks to protect their merchandise.

Loss Prevention Procedures

1. Keep small articles of high value behind counters and locked.
2. Return merchandise to stock after showing to customers.
3. Keep aisles clear. Do not create high displays, thus blocking the view.
4. Inspect contents of any boxes when making a sale.
5. Ascertain that price tickets are firmly affixed.
6. Employees should be available to serve all customers.
7. Use article surveillance tags, chains, mirrors and cameras.
8. Keep floor clear of discarded sales receipts.
9. Inspect boxes and take frequent counts of merchandise in high shortage departments.
10. Use store detectives, if possible.
11. Maintain constant training program for all employees.
12. Keep nonemployees out of stock areas or offices.
13. Do not allow teenagers to loiter.

Methods of Shoplifting

1. Stolen merchandise is concealed in bags, boxes, or purses. Booster boxes may be used by professionals.
2. Many people conceal articles in pockets or under clothing they are wearing. Oversized garments are sometimes worn to make concealment easier. Professionals outfit their coats with inside pockets and hooks.
3. Price tickets are cut off and stolen articles are carried or worn out of the store in plain sight. Small articles may merely be secreted in the palm of the hand.
4. A recent professional technique is to wrap stolen garments around the legs, tuck into stretch socks, and hide under trouser legs.
5. Another technique that has become common is for two or more shoplifters to simultaneously grab a number of hanging articles off a rack and run out the nearest door.

Apprehending the Shoplifter

Shoplifting is a form of larceny. Many states have adopted specific statutes to deal with shoplifting as a separate crime. In some jurisdictions, the stolen merchandise must be taken out of the store to justify an arrest. In other states, concealment of merchandise by the suspect while still in the store may justify an arrest. It is essential that any security employee or others who are authorized to act when a theft is attempted must be well schooled in local and state criminal statutes.

In many states, the retailer has been afforded some measure of protection from lawsuits for false arrest, providing any action taken against suspected shoplifters was based upon reasonable cause.

A retail employee who attempts to take a suspect into custody should be a witness to the actual taking and concealment of the article being stolen. An excellent motto for such personnel to follow is, "If you didn't see it, it didn't happen."

The suspect should be approached quietly so as not to draw the attention of other shoppers. It is wise to frisk for weapons as soon as possible. It is important to stress self-defense techniques in training sessions, because shoplifters have often been found carrying deadly weapons. People who have been previously arrested and released on parole or probation may resist being taken into custody when caught shoplifting. Once the suspect is disarmed, or it is determined that no weapons are being carried, the search for the evidence of shoplifting can be undertaken.

If uniformed guards are stationed at the customer doors, they should assist the store detectives when shoplifters are apprehended. Such assistance will generally reduce the possibilities of physical resistance because of the presence of authority represented by the uniform.

When the stolen items have been recovered, the retailer should apprehend the shoplifter without further delay. One can take either of two routes with the suspect: release or prosecute. If shoplifting is to be discouraged, the culprit should be prosecuted and punished. Many retailers are reluctant to prosecute because it is costly in terms of time needed to appear in court. Additionally, shoplifting is not viewed as a serious crime in many courts, and punishment may be nonexistent or minimal, which is disheartening to the shopkeeper.

In larger stores where many shoplifters are picked up, it is impractical to prefer charges against everyone. It is self-defeating to detain the entire store security force. Depending upon the amount of police involvement necessary to prosecute a shoplifer, it would be advisable to avoid calling on the police for *nickel and dime* cases. The last thing that a retail establishment wants is to wear thin their ability to call upon the police.

Guidelines can be established for prosecution or release. Professionals, recidivists, those who assault security employees, and those who steal above a fixed dollar value, should be arrested. At this time all necessary information regarding the incident should be recorded (see Figure 2–1). These guidelines should be flexible as dictated by the circumstances of any incident.

If the ratailer exercises the option of releasing the

CASE HISTORY – PROTECTION DEPARTMENT

File No.

Date

Name	Identification:
Address:	Maiden Name:
Apt. No.: Tel. No.:	Alias:
City:	Occupation:
Resides with:	Employed By:
Former Address:	Address:
	How long: Earnings:

Age	Date of Birth			Birthplace				Citizen	Lineage or Extraction
Sex	Color	Height	Weight	Build	Hair	Eyes	Complexion	Marital Status	No. of Children

Personal History - Names, Addresses and Business Connections of Family and Friends

Figure 2–1. Sample case history form.

Purchases made by subject		Reason for Act given by Subject		
Place and time first observed		Cash on Person	Searched by	Checked by
Observed by (Witness to operation)	Special Officer Assisting	Came to Office (about)	Police Notified	Left Office
Subject approached (place and time)		Signed out	Prosecuted	Juv. Aid Bur.
PREVIOUS RECORD		Names of Police Officers		
		Interviewed by		
PREVIOUS ARREST RECORD		Authorizing Signature		

Dept.	Quan.	Description of Merchandise	Unit Price	Value Dollars	Cents

Manner of Operation

Signature of Individual Submitting Report

Figure 2–1. *Continued*

shoplifter, it is advisable to obtain some documentation to protect the store from a subsequent lawsuit. Such documentation is usually in the form of a signed confession and a signed general release concerning the incident (see Figures 2–2 and 2–3).

Sometimes a suspect is detained and the subsequent search fails to reveal any stolen merchandise. This may be due to either poor security work or smart shoplifting. It is suggested that such suspects be released as soon as possible. Any documents such shoplifters may be forced to sign before release will be of no value as a defense in a lawsuit. The best

```
I _____ residing at
_____
and knowing that I need make no statement, do of my own free will and not being
under the influence of fear produced by threats of anyone, and without any threat
or promise of any kind, admit and declare that on the _____ day of _____,
19___, I took from the possession of _____ Department
Store, located in _____ county, at _____
_____
        (street address)                    (city, state, zip code)
without making payment therefore and with intent to take for my own use or
disposition, certain articles of personal property of which _____
Department Store is the owner, described as follows:
_____
_____
_____
_____
_____
_____
_____
_____
_____
_____
said articles being of the value of _____ dollars and _____ cents.
                    Dated, _____ day of _____ 19 ____ .
WITNESS: _____ SIGNATURE _____
WITNESS: _____ ADDRESS: _____
WITNESS: _____           _____
```

Figure 2.2. Sample statement form.

defense which the retailer has to offer will prove to be based upon the reasonableness of any action that was taken.

Checks and Credit Cards

Checks

Stealing from retail stores is not confined to shoplifting. To many thieves, it seems a lot easier to pass a bad check or to use a stolen credit card than to carry stolen merchandise. Concerning checks, a retail store should establish rigid procedures for employees to follow:

1. Accept only personal checks payable to the store.
2. Inspect checks to ascertain:
   (a) No postdated or undated checks
   (b) Written in ink
   (c) No erasures or corrections
   (d) Written amount and numerical amount are identical

```
                    GENERAL RELEASE
I, _____ residing at:
_____
do hereby totally release and discharge _____
its officers, directors, agents, employees, successors and assigns, of all claims, debts,
demands, obligations and causes of action whatsoever which I have against them for
any cause whatsoever up to the present time.

DATE: _____    _____ 19___
STORE: _____
                                _____
                                    SIGNATURE
WITNESSES:
_____
_____
```

Figure 2-3. Sample general release form.

(e) Compare signature with identification
(f) Be careful if it is a starter check
(g) Check should be for exact amount of the sale
3. Require valid identification, e.g., driver's license, other licenses, Medicare card.
4. Use extra care on out-of-town checks.
5. Call bank if amount is very large.
6. Be careful if customer is unconcerned about price.

If a check does not clear, the following procedures should be implemented:

1. For insufficient funds:
   (a) Redeposit
   (b) Notify maker
   (c) Turn over to collection agency
   (d) Check with local police regarding prosecution.
2. For closed accounts and forgeries:
   (a) Try to identify the check passer
   (b) Maintain a list of names and aliases on bad checks
   (c) Exchange names with other retailers
   (d) Be alert for refunds
   (e) Prosecute if identification is made.

Credit Cards

Suggested procedures to reduce losses:

1. Examine card closely to ascertain that it has not expired.
2. Imprinted name and signature must match.
3. Consult listing of stolen or otherwise fraudulent cards.
4. Obtain authorization for card acceptance by terminal or telephone.
5. If card is issued by an out-of-state bank, be aware of time lag before such card appears on hot list.

Guidelines for Cashiers

1. Keep register closed between sales.
2. When leaving a station, lock the register and take the key.
3. Don't count contents in view of customers.
4. If large amount of cash has accumulated, request a cash pick up.
5. Don't argue with a customer who claims to be shortchanged. Request assistance.
6. Know counterfeit money.
7. Know the prices of merchandise in the store.
8. If any argument or confusion arises, lock the cash drawer.

Refund Operators

If a store has a liberal refund policy, shoplifters will find it more profitable to return stolen goods than to sell them to a fence or others. Often the stolen merchandise is refunded without it ever having been taken out of the store. Some common methods used by refunders are:

1. Purchases merchandise and obtains sales slip. Steals identical item and uses sales slip for refund.
2. Steals merchandise, removes price tickets and claims item was a gift.
3. Looks for sales slips on floor and then appropriates properly priced item for refund.
4. Creates counterfeit sales slips.
5. Steals merchandise, returns same for even exchange claiming it is damaged or wrong size. Gets proper documents with exchange. Goes to refunds.

Some suggested policies which stores should follow to help reduce such losses are:

1. No refunds permitted without presentation of price tickets and sales receipts.
2. Purchases made by check should not be refunded until after a reasonable time for check to clear.
3. All slips and tickets should be carefully examined for authenticity.
4. Verify that the merchandise is handled by the store and that the price is correct.
5. Credit card purchases will be refunded in a similar manner.
6. Refunds should be paid by check through the mail.

References

1. *Theft by Employees in Work Organizations*, Department of Sociology, University of Minnesota, 1979.

Appendix 2a
Loss Prevention Techniques*

The Shoplifter Seeks Privacy

- Display merchandise so that employees can easily watch the movement of customers. Avoid narrow, cluttered aisles and maintain an open, neat appearance throughout the store. Avoid long, unbroken aisles and leave space between display cases. Keep displays at a medium height so that employees can see across the top of them. Arrange items neatly so that anything missing can be quickly noticed.
- Make sure the entire store is well lit.
- Optical, mechanical, and electronic devices can be used against shoplifers. Convex mirrors should be placed so that several areas of the store can be seen in one glance. Peepholes and one-way mirrors are also useful for observation of customers. Consider the use of a closed circuit television surveillance system.
- Place cash registers away from doors, but in clear view from the outside. The cashier should have an unobstructed view of the store.
- Divide the store into sections and assign each employee the responsibility of covering specific sections. Never leave a section unattended. Schedule employees' working hours with adequate floor coverage in mind.
- Encourage employees to circulate throughout their sections and to serve all customers as promptly as possible. If a customer enters the store while another is being helped, the newcomer's presence should be acknowledged with the words, "I'll be right with you."
- Develop a warning system for alerting employees that a shoplifter is suspected.
- Uniformed personnel, store detectives and guards are useful in deterring shoplifters.

*Reprinted with permission of Texas Crime Prevention Institute, San Marcos, Texas.

Don't Make It Easy on Shoplifters

- Control entry and exit at the store. Consider one-way turnstiles or separate doors for entry and exit.
- Fasten down or otherwise secure small appliances used for display purposes. If merchandise is sold in pairs, display only one of the pair.
- Remove empty hangers from clothing racks after articles of clothing have been purchased. Require customers to check articles of clothing when entering and leaving fitting rooms. Post signs indicating the maximum number of clothing articles allowed in a dressing room at one time. Many stores have found three items to be the maximum controllable limit.
- Do not place more than one valuable item on a counter at the same time. Keep the more expensive items in locked display cases or counters, and never leave a display case unlocked.
- Establish clear cash register procedures. The register should be open only while it is actually being used. The cash drawer should be closed before merchandise is packaged. Cashiers should be alerted to avoid distractions from other customers while helping someone at the cash register. Keep cash registers locked and remove the key when they are not in use.
- At check out counters, look inside items such as trash cans, ice chests, tool boxes, shoe boxes, and purses for concealed merchandise.
- Cashiers should double check the price of any item which seems to be inaccurately marked.
- To aid in observation and to restrict exit points, check stands and check out lanes should be closed or blocked off when not in use.
- Give customers receipts for all purchases. Do not make refunds without requiring a sales receipt. Keep the store clear of discarded sales receipts.
- Post signs in the store which warn that shoplifters will be prosecuted.

What to Look For

- Watch the customers' hands and eyes. Jittery eyes and nervous hands often indicate a shoplifter.
- Keep a watchful eye on customers who just wander about the store.
- Watch out for customers who avoid the attention of employees.
- Keep an eye on people who wear baggy clothes or heavy outer garments out of season.
- Shoplifters sometimes enter the store carrying bundles, bags, boxes, brief-cases, top coats, umbrellas, oversized packages, musical instruments or books to conceal merchandise. Post signs that require every customer to check packages at the door or with a checker before they shop. Seal packages which are sold so that they cannot be used to conceal other items.
- Look out for customers who try to divert the clerk's attention. Shoplifters sometimes ask for more articles than the clerk can control, disrupt a display, or simply engage a clerk in conversation while an accomplice does the thieving.
- Shoplifting techniques are as varied as the imagination. Train employees to be alert and to watch for signs that could indicate a shoplifter.
- Remember, a store with a reputation for good security and a tough prosecution policy is a less likely target for shoplifters.

Appendix 2b

Control of Credit Card Fraud Checklist

VICTOR HAROLD

To aid merchants in the prevention of loss through credit card chargebacks, unauthorized returns, and fraud, this section uncovers the problem areas:

1. Are you monitoring your credit card system consistently to be sure it is free of chargebacks?
2. Does your bank notify you in advance of a chargeback, describing it, which will enable you to timely refute it?
3. When a credit is necessary, do you obtain a letter from the cardholder stating the reason for the chargeback?
4. Do you work at chargebacks to attempt to get them reversed?
5. Have you a policy for resubmitting unjustified chargebacks?
6. Have you a training program to detect and stop fraud for all personnel handling credit transactions?
7. Is the training program updated frequently? Via memos? Meetings?
8. Do you prevent the return of items purchased elsewhere?
9. Have you a system to detect the return and seizure of items shoplifted from your store?
10. Is senior management kept aware of fraud losses by amounts and type?
11. Is there at least one management person who receives frequent training in loss prevention and fraud control?
12. Have you determined the feasibility of obtaining authorization numbers on all credit card transactions, regardless of amount?
13. Are the employees issuing credit aware of the fact that credit company computers are updated daily with new credit information, often weeks ahead of the published *hot card lists*?
14. If your facility issues credit through coupons, order forms, or telephone, do you obtain the name of the bank on the customer's card?
15. The first three to six numbers on Visa and Master-card identify the issuing bank. Have

you several copies of the directory which identifies the bank with the number? (bin number directory)
16. On suspicious or marginal transactions, do you check and match the bank issuing number (see #15) with the bin number directory?
17. In a questionable transaction, do you call the bank to verify and match the name and address of the cardholder with the customer?
18. Do you know whom to contact directly at the banks if you need immediate credit assistance?
19. Are two pieces of identification requested for most credit transactions?
20. Are your employees instructed absolutely to match up the signature on the credit cards with the transaction slips?
21. If you use mailing lists, are credit risks purged?
22. Do you subscribe to bad name lists?
23. Are the bad names added to your lists?
24. Does the mailing list company from whom you purchase the list certify that credit risks have been removed?
25. Are your employees tearing up charge slip carbons in front of the customer?
26. Are you obtaining home and office telephone numbers and verifying with information that the telephone number and purchaser match?
27. Do you call the purchaser of a larger order to confirm that they placed the order?
28. Forgery, which is prosecutable, occurs during credit card fraud. Do you have the customer sign the purchase after the I.D. has been checked and after the authorization code has been written on the charge slip?
29. Are you aware that box numbers and similar nonstreet address orders may be placed by inmates? Have you a system for checking?
30. Does every transaction have an order number?
31. Has your accountant worked out a procedure to avoid being debited twice for the same transaction?
32. Does the sales receipt indicate that the transaction is a credit purchase?
33. Do you cross-record the sales receipt number and the credit transaction?
34. Are the customer I.D. numbers written on the sales slip or credit slip?
35. Is the security of the blank credit slips and other credit forms strictly supervised and not handed out carelessly?
36. If credit fraud via mail is suspected, do you know how to report it?
37. Do you routinely request proof of delivery from common carriers such as UPS?
38. If a money claim has to be made against a carrier, have you a procedure for an immediate filing?
39. Are proofs of delivery kept in the customer's account?
40. Are frequent customers assigned a permanent number?
41. If your credit business is growing, have you originated a processing system enabling unified control of all transactions?
42. Have you a procedure which determines that returned merchandise is properly credited and sent back to inventory or, if unsaleable, properly secured for return to the manufacturer?
43. Have you a system to monitor inventory and sales of high-cost, smaller, and easily sold items?
44. Have you a complete up-to-date list of agencies and contacts to whom credit card fraud is reported?
45. Are employees who handle credit bonded?
46. Do you subscribe to newsletters and bulletins which provide information on credit card fraud?
47. Do the merchants and law departments in the community share credit card information?
48. Are credit card data kept very secure?
49. Have you a file, preferably computerized, of negative names to which orders are routinely matched?
50. Are any of your employees on the negative list?
51. Do you shop banks for the lowest possible discount rate?
52. Do you use more than one bank for credit card transactions?

Chapter 3
The Shoplifting Problem*

JAMES CLEARY, JR.

Who Shoplifts

There is no typical shoplifter. Based on apprehensions there are three categories of shoplifters. The most numerous is the *average citizen shoplifter*. They account for about 75 percent of all persons picked up for shoplifting. The next category is the *full-time amateur shoplifter*, who makes up about 20 percent of the total. The final group is the *professional shoplifter* or *booster* as they are called on the street. This group makes up no more than 5 percent of the shoplifters, and that estimate may be high.

The Average Citizen Shoplifter

An article in the January 1982 issue of *Ladies Home Journal* is a good place to begin discussing the average citizen shoplifter. A photograph at the beginning of the article, entitled "When 'Honest' Women Steal—A Middle-Class Crime Wave," shows an attractive and well-dressed woman standing at a counter in a fashionable department store. She is dropping a $40 gold compact into her purse. The article says that this college-educated woman has two elementary school children. She had her checkbook and eight credit cards, as well as $80 in cash, on her at the time she shoplifted the compact. The article goes on to point out that from 25 to 33 percent of all persons apprehended shoplifting are women between the ages of twenty and fifty. According to the article, the only larger identifiable group of shoplifters is teenagers.

Moving from the world of popular journalism to documented studies of persons arrested for shoplifting is helpful in answering the question—*Who shoplifts?* In Travis County (Austin) Texas, 1,000 persons were interviewed after being caught shoplifting. None of them had a prior criminal record so they were placed in a pretrial diversion program by the prosecutor and the court. Consider some of the findings based on anonymous interviews with these average citizen shoplifters.[1]

Sex: Female, 52 percent; male, 48 percent
Decided to shoplift: before entered store, 32 percent; after entered store, 68 percent
Reason given for shoplifting:
    Convenient, easy mark, won't miss it, 36 percent
    Angry, ripoff system, get even, 25 percent
    For the thrill, 14 percent
    No money but wanted the item, 12 percent
    Line too long, did not want to wait, 9 percent
    Other reasons (depression, personal problems, mental illness, and innocence), 9 percent
Items shoplifted:
    Cosmetics, 23.6 percent
    Clothing, 16.6 percent
    Miscellaneous, 15.9 percent
    Food, 14.6 percent
    Medicine, 12.7 percent
    Tools and hardware, 11.4 percent
    Household products, 5.2 percent
Did you have the money to pay? Yes, 88 percent

A similar survey was conducted in Pennsylvania with 1,700 persons picked up for shoplifting over a five-year period.[2] None had a prior conviction for shoplifting. Consider some of the more important findings of this survey:

The average value of a single theft was $26.35.
The average person apprehended had shoplifted 98.8 times before being caught.
58.3 percent were female and 41.7 percent were male.
57 percent (or 969 out of 1,700) said they had shop-

---
*From *Prosecuting the Shoplifter*, by James Cleary, Jr., J.D. (Stoneham, MA: Butterworths, 1986).

lifted at least once per month before being apprehended.

The age breakdown was as follows: juveniles, 17.7 percent; persons between ages sixteen and twenty, 28.7 percent; persons between ages twenty-one and fifty-four, 44.3 percent; and persons over age fifty-five, 9.3 percent.

Answer to question "Why did you shoplift?"
"Stores write off the losses on their taxes."
"I've spent plenty of money in that store and I'm entitled to get a little of it back."

This well-documented study concluded there are three categories of shoplifters. The *soft-core shoplifter* is a person who steals for personal use regularly, even habitually, and has never been prosecuted. This person has no criminal record and is not considered by society as a thief. They make up probably 75 percent of the people who shoplift. The *hard-core shoplifter* is a person who steals for personal use but has a criminal record, either for shoplifting or for other offenses.

The *professional shoplifter* is a person who steals regularly to resell the merchandise for money to support a drug habit. The findings thus uphold the traditional view that from two-thirds to three-fourths of the people shoplifting on any given day are these soft core or average citizen shoplifters. The fact that should concern retailers is this—probably 80 percent of these average citizen shoplifters steal with regularity. Of the 1,700 first offenders interviewed, 57 percent said they shoplifted at least once per month before being caught. The study concluded that shoplifting among a large percentage of the average citizen types is habitual.

What can merchants learn from this information about who shoplifts? The most crucial fact is that the average citizen types shoplift regularly. Like the woman in the *Ladies Home Journal* article shoplifting the $40 compact, these people conceal merchandise in pockets, purses, shopping bags, and under clothing. Then they walk out of stores without paying. They *do not use gimmicks* like booster boxes (ones with a false bottom) or booster coats (ones with pockets sewn inside the linings) and they do not put merchandise between the legs and walk out of stores. I am not saying people do not shoplift by using these gimmicks because some do. The point is simply this—the largest number of people shoplifting on any given day are these average citizen types who do not use gimmicks to steal.

The average citizen type shoplifter steals relatively little each time (average value only $26) when compared to the professional shoplifter who steals large amounts of merchandise daily to resell to a fence. The head of the Retail Bureau of the Washington, D.C. Board of Trade was recently quoted in a retail industry magazine.[3] He said the average citizen shoplifters actually steal more merchandise each year than the professional and for a simple reason—there are so many more of them. As a group these average citizen type shoplifters will do two things—shoplift from and sue stores. They will do both with success unless the merchants train employees to prevent shoplifting losses.

A large number of cases concern stores that have been sued civilly when a shoplifting situation was mishandled. These false arrest type cases were not filed by the professional or by the full-time amateur shoplifter, not at all. The stores in these cases were sued by the average citizen type shoplifter. Merchants must teach their employees how to handle cases involving the average citizen shoplifter. Because they are so numerous they constitute both a shoplifting and a civil liability threat to the company assets. These people will cost a store both merchandise and lost profit. Also, they will cost the store dollars, meaning money paid out to defend and settle civil actions.

The Full-Time Amateur Shoplifter

In most cities the police will circulate a list of known shoplifters around the retail community in a good faith attempt to help merchants with shoplifting loss-prevention. There is a problem with these lists. They do not contain the descriptions of average citizen shoplifters, because these people do not have criminal records. The average citizen shoplifter is not known to the police. The lists generally do not contain descriptions of the professional shoplifters, as boosters are rarely apprehended by merchants or their security staffs. The people on these lists are really the full-time amateur shoplifters, people who account for only about 20 percent of those caught shoplifting.

In conducting shoplifting awareness training programs, merchants should not rely totally on these known shoplifter lists. The typical shoplifter is much more likely to be the well-dressed woman shoplifting the compact or young adults. Finally, this group of full-time amateurs is not as likely to file a civil action against a store because they have prior criminal records—for shoplifting many times. In a civil action the fact that the plaintiff had a prior conviction for shoplifting will be brought to the attention of the jury. These full-time amateurs will shoplift from but will not sue stores. In others words, the civil liability threat is less with this category of shoplifters.

The Professional Shoplifter

The final type of shoplifter is the professional, or booster. They are small in number when compared to the average citizen and the full-time amateur shoplifters. The professional is typically a drug addict who has to shoplift daily to support a habit. They shoplift for one reason: money. They sell the goods to a fence (a buyer of stolen merchandise), who in turn sells the store's merchandise to the general public at about one-third or one-half its retail selling price.

Most often boosters work in pairs (a man and women), although sometimes two women team up to shoplift. They travel from city to city, shoplifting in malls and in freestanding stores as they go. Boosters are well dressed, because they wear clothing they have shoplifted. They shoplift enormous amounts of merchandise, because it is their job. Most professionals have a drug habit which must be supported by stealing merchandise and selling it to a fence. One former professional shoplifter[4] now conducts shoplifting loss-prevention training seminars for merchants all over the country. He made over $100,000 in his first year as a booster. Since the merchandise is sold to a fence for 25 percent of its retail value this means a booster has to shoplift over $400,000 at retail to net out at $100,000 per year. Boosters are in every major city. They are rarely caught by merchants and their security staffs. A booster would never consider filing a civil action against a store.

Merchants, especially security personnel, do not like to be told the professional shoplifter is rarely apprehended. This fact is documented by an article in a national security magazine which said less than 1 percent of the persons apprehended for shoplifting were professionals.[5] The reason so few professionals are stopped is that merchants are looking for the wrong people. Remember, the professional is well dressed and knows a lot about merchandise. Because professionals operate differently than both the average citizen and full-time amateur shoplifters, the information a merchant has about shoplifters does not apply to the booster.

Conclusion

This chapter is about the merchant's role in preventing shoplifting losses, both the loss of merchandise to shoplifters and the loss of dollars paid out in civil actions. It is important to understand that seven out of ten shoplifters will be the average citizen type. Training programs must take into account that many of the people caught shoplifting will also qualify as a credit account customer. These people have jobs, families, and otherwise good reputations. If the shoplifting situation is mishandled, these people will sue the store. It is important to understand that young adults and minors shoplift. They too can bring civil actions if mistreated by a store owner, manager, or security person. Since there is no typical shoplifter, the merchant must train employees to watch for people concealing merchandise in pockets, purses, shopping bags, and under clothing, because this is the way the average citizen type person shoplifts. Merchants must forget about shifty eyes or furtive gestures when it comes to identifying shoplifting suspects. Average citizen types, full-time amateurs and professionals all shoplift one way—with their hands.

Handling shoplifting situations is anything but a game of cops and robbers. Merchants are not peace officers, and most suspects detained for shoplifting are not criminals but average citizens. This chapter is about the legal tools available to merchants in handling shoplifting situations. Before the merchant becomes too wrapped up in the legal aspects of shoplifting, she or he should give some serious thought to the subject of *who shoplifts*. The court studies presented above are excellent sources of information—information not available to the retail community until the past five years. Study this material and develop a profile of who may be shoplifting in your stores. Shoplifting is not committed only by poor people—hardly. It is not confined to stores in high crime areas of a city. Shoplifting takes place in malls in the most affluent suburbs of America every day. Understanding who shoplifts is important for any store in formulating policies and procedures.

The Extent of Shoplifting Losses

An article in a recent retail trade magazine said the annual shoplifting loss was $22 billion.[6] The article went on to say that 90 percent of those arrested for shoplifting had cash, checks, or credit cards on their person which could have been used to pay for the merchandise shoplifted. Based on income level, 28 percent of those arrested were in the higher category, 45 percent in the middle, and 27 percent in the lower income group. There is no segment of the retail industry that escapes shoplifting losses. Consider some of the types of retail operations where people shoplift.

Stores generally with a full-time security staff to handle shoplifting:

Department
Discount
Food
Drug

Specialty stores, where the store manager and sales people handle shoplifting situations:

Children's apparel	Variety	Card
Women's apparel	Toy	Convenience
Men's apparel	Gift	Office supply
Family apparel	Novelty	Stationery
Sporting goods	Hobby	Home centers
Hardware	Music and record	Shoe

Groups indirectly involved with shoplifting:

Mall security staff personnel
Off-duty peace officers "moonlighting" in security for retail stores
Contract security guard service companies

The amount of merchandise shoplifted nationally is not the primary concern to a retail company.

Table 3–1. Shortage Report—the ABC Store, Anywhere, USA

Department	Shortage (%)
Cosmetics and beauty aids	3.91
Lawn and garden	1.38
Building supplies	2.34
Auto accessories	1.58
Bras and foundations	1.33
Women's dresses	3.19
Women's coats and suits	1.48
Lingerie, robes, and sleepware	1.35
Outerwear-sportswear	1.97
Costume jewelry	11.33
Jewelry and watches	2.48
Sports, leisure, beachwear	1.59
Handbags and accessories	2.59
Hardware	2.81
Toys and games	2.01
Sporting goods	1.85
Tapes and records	2.95
Photo supplies	2.06
Gifts and flowers	2.09
Candy	1.35
Tobacco	1.89
All other departments	1.61

Table 3–1 illustrates typical losses a department store is likely to suffer. This is much more important because it shows exactly where a store can be losing merchandise.

As a general rule, most retail companies will report theft losses as a percentage of sales. Let's examine a shoplifting loss from the viewpoint of making it up with additional sales. Suppose a clothing company is doing business at a 2.5 percent profit level and a $100 sweater is shoplifted. It will take almost $4,000 in additional sales to make up for the profit lost on that sweater. The owner or chief executive officer will be hard pressed to document that this lost profit can be made up with additional sales on a regular basis. Looking at the situation from a loss-prevention viewpoint, it can be said that preventing a $100 shoplifting loss is like finding $4,000 in additional sales. Store owners and retail executives realize they cannot consistently generate additional sales to cover shoplifting losses, so they train employees to prevent shoplifting in the first place.

Preventing Shoplifting by Sales Employee Vigilance

In department and discount stores only security staff members are authorized to prevent shoplifting by detaining persons after they have concealed merchandise and left without paying. By contrast, specialty stores do not employ full-time security people to patrol the selling floor looking for shoplifters. In specialty stores, the owner or manager has the responsibility of detaining persons who have concealed merchandise and left without paying. Most retail companies, whether it is a department store with a security staff or a specialty store without a security staff, do not authorize sales employees to stop shoplifting suspects. Sometimes sales employees feel they can make no contribution to the shoplifting loss-prevention effort. Nothing is further from the truth. The most effective deterrent to shoplifting is a well-trained and alert sales employee.

Consider the comments by a southern California department store manager about the value of sales employees in shoplifting loss prevention:

> Tags generally have been a good deterrent but *alert people are better than tags or anything else*. If a customer has not been recognized by a sales person saying "May I help you" or anything else the customer feels more anonymous and is more likely to try it [meaning attempt to shoplift]. (Riverside CA—The Press Enterprise, November 11, 1984).

The essential ingredient in all successful shoplifting efforts is *privacy*. A person needs privacy to conceal merchandise before walking out without paying. This is true whether one is talking about the average citizen, full-time amateur, or professional shoplifter. Nobody can conceal merchandise in a pocket or switch a price tag, if they are being watched. Store owners and retail executives sometimes complain they do not want to make security people out of sales employees. This is a valid comment. A suggestion is this—train sales employees to develop a *shoplifting alertness dimension* to their primary job of selling. Make sure they understand that by approaching a customer and asking, "May I help you," they are removing the privacy factor right off the bat. Make sure sales employees understand they are playing a vital role in preventing shoplifting by giving good customer service.

Sales employees are trained to greet customers and back off so the customer can browse. No sales person wants to be pushy or to be considered a pest. The point is simply this—sales employees cannot neglect customers, nor can they greet customers, back off, and forget about the possibility of shoplifting. They must be taught that the businessman, the teenager, the senior citizen, and even the woman pushing the baby stroller are all possible shoplifting candidates. Sales employees must be taught that the average citizen shoplifter does not use gimmicks to steal. Sales employees must understand that most people simply remove merchandise from a counter, rack, or shelf, conceal it, and walk out without paying. These people—especially the average citizen shoplifter—are successful because a sales employee gave them that needed privacy.

Merchants and their sales employees must remember the average citizen shoplifters are not true criminals. If these people are spoken to or otherwise acknowledged by eye contact, psychologically they are less likely to try and conceal merchandise. On the other hand, if no sales employee makes any kind of contact, two things happen. The customer is given the actual privacy plus a feeling of privacy. People feel more anonymous if they are not contacted, so they will be more likely to shoplift. Customer service techniques will vary from company to company. One technique all retailers agree on is greeting the customer and making eye contact. Speaking to and looking at a customer is the best way to remove that feeling of privacy as well as actual privacy.

Since most companies do not permit sales employees to detain shoplifting suspects, there are some basic facts a sales employee needs to know—about both shoplifting and their role in preventing it (see Table 3–2).

Generally sales employee turnover is high in most retail companies. For this reason owners and managers must provide sales employees with shoplifting loss-prevention information which is simple and basic. If nothing else, sales employees should understand that by giving good customer service they are playing an important role in shoplifting loss-prevention. Even though sales employees generally are not permitted to stop shoplifting suspects, the manager or owner should discuss shoplifting situations with them. Make it clear that the owner or manager will stop suspects, but that sales employees should always maintain an alertness to the possibility of shoplifting.

In my opinion specialty stores—ones without a security staff person on the selling floor—are especially vulnerable to shoplifting. During the years I prosecuted, rarely was a case filed by a specialty

Table 3–2. Facts Sales Employees Should Know to Prevent Shoplifting

1. Removing the privacy factor is critical. A shopper needs privacy to become a shoplifter. Any person must have privacy to conceal merchandise after removing it from a rack, counter, or shelf with his or her hands.
2. There is just one way in which people shoplift—with their hands. Train sales employees to watch a person's hands and think about where that person could conceal the merchandise if they were going to shoplift it.
3. The most effective way to remove the privacy factor is by greeting the customer and making eye contact.
4. Sales employees must learn to develop a shoplifting alertness dimension to their selling function. This shoplifting alertness dimension should center around facts about the average citizen shoplifter such as:
    a. They do not use gimmicks like booster boxes, booster coats, or putting merchandise between their legs. They simply conceal merchandise and walk out of stores, trying to act as normal as possible.
    b. These people are not true criminals. They shoplift because they get away with it. They get away with it in large part because sales employees give them that much needed privacy required to shoplift.
    c. A large percentage of the average citizen shoplifters make the decision to shoplift *after* entering the store. They decide to shoplift because they are being ignored by sales employees—because they have that privacy.
    d. Most of the average citizen type shoplifters will shoplift alone.
    e. Merchants only apprehend one out of ten persons who are shoplifting, and this estimate is probably low. Some estimate that only one in twenty persons who shoplift are caught.

store—a book, music, card, gift, sporting goods, apparel, hardware, toy, hobby, office supply, or home center store. Traditionally, retailers say they lose more merchandise to dishonest employees than to shoplifters. This sacred assumption has now been effectively challenged by Mr. Peter D. Berlin, the head of Shrinkage Reduction Services at Price Waterhouse. Using a hypothetical 137-outlet specialty apparel store chain, Mr. Berlin concludes that shoplifting accounted for $1,233,000 of the inventory shrinkage, while only $151,200 was attributable to internal theft. Retail and security executives in specialty stores should pay close attention to the formula Mr. Berlin developed and apply it to their own operations.

Specialty store owners and managers should develop training programs for sales employees. Some will counter by saying their stores use electronic article surveillance (EAS) systems to deter shoplifting. There is a twofold problem with EAS systems. First, not all of the merchandise is generally tagged with the warning devices. Second, employees may begin to rely totally on these EAS devices, forgetting all about giving good customer service. A store that uses these EAS systems may begin to neglect sales employee alertness training. EAS systems, assuming they are cost justified, are really only a support system for employees anyway. Since some retailers do not feel they are cost justified or cannot afford the systems, giving good customer service is the only preventive measure available.

Let's compare the role of a sales employee in preventing shoplifting with the role of a security staff member. The sales person prevents shoplifting by removing the privacy factor—by giving good customer service. By contrast, a security person never removes the privacy factor. Security people are taught to take the opposite approach. They roam the selling floor, generally in plainclothes, and watch for people concealing merchandise, switching price tags, or passing merchandise off to accomplices. They never approach a customer until that customer has obtained wrongful possession of merchandise and has left the store. Security people prevent shoplifting by stopping people *after* they have obtained wrongful possession of unpurchased merchandise. Sales employees prevent shoplifting by greeting customers and making eye contact *before* that person tries to obtain wrongful possession of goods.

It is important to remember that statistics on shoplifting apprehensions only show who gets caught. They are not reliable evidence of who shoplifts and gets away with it. For example, the security director of a department store chain with fifteen outlets in the Boston area was quoted as saying his security staff apprehends only one out of ten persons who are shoplifting.[7] One year his staff caught 4,000 persons. This means that 36,000 people shoplifted in those fifteen stores without being caught—even though the store had a full-time security staff. If stores with full-time security staffs catch only one out of ten, what about the number of people who shoplift and get away with it in specialty stores? If a specialty store owner or manager has not stopped a shoplifting suspect in the last thirty days, the chances are the shoplifting losses are high in that specialty store.

For years the retail community has talked about the *impulse* theory of shoplifting. Retailers say that people shoplift on impulse. The Pennsylvania study cited above said 57 percent of the first offenders interviewed admitted to shoplifting at least once a month before being caught. It went on to say the average person in the study shoplifted ninety-eight times before being charged with the offense. This appears to be much more than impulse. The study concluded that shoplifting was addictive among a large number of the average citizen types.

Consider an important statistic revealed in the Texas study among people who got caught. Sixty-eight percent of these first offenders said they made the decision to shoplift *after* they entered the store. This means these people thought they had sufficient privacy to conceal merchandise without being spotted. If 68 percent of those who got caught said they decided to shoplift after entering the store, a fair question is this—What percentage of those *who did not get caught* decided to shoplift after they entered the store? It is reasonable to assume that an equal number of successful shoplifters decided to steal after entering the store. This leads us to the question of why they got away with shoplifting? The answer is that the successful shoplifter was given that privacy necessary to conceal merchandise. Simply stated, privacy equals the opportunity to shoplift.

What does all this mean when it comes to sales employees and shoplifting loss-prevention training programs? It means the role of the sales employee is not always well-understood or well-defined. Although sales employees are legally authorized to detain shoplifting suspects, most stores do not allow it. In stores with a security staff, the sales employee is always told to call security when he or she sees a shoplifting situation taking place on the selling floor. In specialty stores the sales employee is normally instructed to notify the owner or manager when a person is seen concealing unpurchased merchandise.

This means shoplifting loss-prevention skills for sales employees are different than those needed by a store owner, manager, or security person.

The role of the sales employee is to remove the opportunity a person has to shoplift. This means removing the privacy factor by greeting customers and making eye contact. The role of the store owner, manager, or security person is to detain shoplifting suspects. Both are equally important. Shoplifting is an offense against the merchant's lawful possession of merchandise. Sales employees can deter people from obtaining wrongful possession of merchandise in the first place. Owners, managers, and security personnel prevent shoplifting by detaining people after they have obtained wrongful possession of merchandise and have left the store. By clearly defining the role of the sales employee, the retailer will make those sales employees more effective in preventing shoplifting cases.

Prosecution of Shoplifting Cases Deters Future Shoplifting Attempts

There is a famous California shoplifting case in which a female teenager was stopped for palming some costume jewelry. When the security employee asked why she failed to pay for the jewelry she told him her friends said the store was "an easy place to shoplift." Stores that do not prosecute quickly earn the reputation of being an easy mark when it comes to shoplifting. Remember, the studies have found shoplifting is not impulsive but habitual or addictive. Each time a person shoplifts without being caught it is easier psychologically for that person to come back to the store and shoplift again.

Most retailers feel that prosecution is a deterrent to future shoplifting attempts. I agree, especially when it comes to the average citizen shoplifter category. Many merchants tell the press they prosecute 100 percent of the people stopped for shoplifting in their stores. The facts of life are that stores prosecute some suspects and release others. In other words, many merchants follow a selective prosecution approach, because they do not take all their cases to court. The two studies which follow show many merchants use a selective prosecution policy.

Each year a national retail trade association publishes a survey of its members on a variety of loss-prevention subjects.[8] Consider their statement about prosecution of shoplifting cases—"Strong prosecution policies continued to be a major factor among retailers." The following statistics show these merchants prosecuted cases because they felt it was a deterrent to future shoplifting attempts. Note that they followed a selective prosecution approach, in that not all persons detained for shoplifting were taken to court.

	% OF CASES PROSECUTED	% OF CASES WHERE SUSPECT WAS RELEASED WITHOUT FORMAL PROSECUTION
1982	56	44
1983	66	33

Note: In both survey years these mass merchandise, discount, and specialty stores obtained convictions in 90% of the cases they prosecuted.

A study of west coast retailers also demonstrates that merchants use a selective prosecution approach.

TYPE OF STORE	% OF CASES PROSECUTED	% OF CASES NOT PROSECUTED
Discount	53	47
Food	42	58
Drug	26	74

Selective Prosecution Is Authorized by the Merchant's Privilege

Sometimes merchants think that all persons stopped for shoplifting must be arrested and prosecuted. This is not true. The legislatures of all states have passed laws known as detention statutes which have incorporated the *merchant's privilege* into their provisions. The privilege permits a merchant to detain a shoplifting suspect and investigate the facts without declaring the stopping to be a citizen's arrest. If the detaining merchant conducts an investigation of the facts, which is legally reasonable in manner, method, and time, the retailer has complied with the privilege. If the suspect later sues the store civilly, the merchant can plead compliance with the privilege as a defense to the civil action.

Most people think of these statutes as granting merchants conditional immunity from false imprisonment actions, which of course is what they do. There is another aspect of these statutes which permits the merchant to selectively prosecute. No detention statute in any state requires that all shoplifting suspects detained under the merchant's privilege be prosecuted. The privilege is one to detain, investigate the facts, and recover the unpurchased merchandise. A retail employee who detains a suspect under authority of the statutory privilege has a choice in the way he or she handles the shoplifting situation. The merchant can:

1. Detain, recover the unpurchased merchandise, warn, and release the suspect without involving the police at all; or
2. Detain, recover the unpurchased merchandise, call the police, and cause the person's formal arrest and prosecution.

In either situation a store owner, manager, or security person has prevented shoplifting by detaining suspects under authority of the merchant's privilege.

These detention statutes are the legal authority by which any retailer can adopt a selective prosecution approach. Under a selective prosecution policy, each retail company is free to prosecute as many or as few cases as the merchant feels will serve as a deterrent to future shoplifting attempts. As the west coast survey cited above clearly demonstrates, discount, drug, and food stores each prosecute a different percentage of persons detained for shoplifting. This flexibility in taking cases to court is the reason retailers should base their policy on the merchant's privilege. Understanding these detention statutes is an absolute necessity for any store in any state. This is true whether the store has a full-time security staff to handle shoplifting situations or whether shoplifting cases are handled by the store owner or manager. These statutes are the key to more effective prosecution with minimum risk of having a civil action filed against the store.

Merchants often complain that prosecution of shoplifting cases is expensive in terms of payroll costs. It may take two or three trips to court before a case is over. In most states a person found guilty of misdemeanor shoplifting has the right to appeal the case and have it tried all over again—this time in front of a jury. This means more trips to court. It means the merchandise cannot be returned to stock because it is prosecution evidence needed at the trial. Merchants look at payroll costs for the employee time spent in court and often conclude prosecution is not cost justified. By using a selective prosecution approach, the merchant can warn and release suspects in the inexpensive item cases, thus saving the payroll costs they complain about. In situations where the suspect shoplifted expensive merchandise, or shoplifted multiple items, the case could be prosecuted as a deterrent. These detention statutes, all containing the merchant's privilege, were passed to make it easier for merchants to stop suspects and recover unpurchased merchandise without being sued civilly. Always remember by using the merchant's privilege the store can prosecute as many or as few cases as it wishes. An effective prosecution policy generally means a selective prosecution policy.

The cost of prosecuting shoplifting cases is not always justified from another point of view. Court dockets in all cities are clogged with both civil and criminal cases. A businessman was convicted of shoplifting for eating a $.79 bag of nuts in a New York supermarket. He said he routinely spends $100 per week on groceries in that store. After the trial the Nassau County, New York authorities said the prosecution costs so far were $10,000 after the six-day jury trial.[9] Each shoplifting case must be dealt with on its merits but this case seems to be an excellent example of a store not considering a selective prosecution approach.

Merchants are fond of complaining about prosecutors and the court system as part of their shoplifting dilemma. They must remember that crimes against persons are always high priority with prosecutors, courts, and the general public. The Nassau County authorities were not overjoyed to spend $10,000 and tie up a judge for six days over a $.79 package of nuts, especially when the merchant had another legal option. All property offenses take lower priority in any court system and with the public. Shoplifting is an offense against property.

A merchant, like any other citizen who is the victim of crime, has the right to use the court system. The legislatures have given retailers a legal tool in these merchant's privilege statutes. Some court systems have a pretrial diversion program for first offender shoplifting cases. A well-thought-out shoplifting loss-prevention program will consider all options available, as well as how the courts and prosecutors view shoplifting cases. Many retail security people only show up around a prosecutor's office when they have a problem rather than maintaining any regular liaison. Do not get the reputation of the mystery security director in your city. Set up an appointment to discuss shoplifting and internal theft cases with the district attorney, because cost-effective prosecution is a two-way street.

The beauty of the merchant's privilege is that, besides giving the store a defense to a civil action if one is filed, it also authorizes selective prosecution of shoplifting cases. Store owners, managers, and security personnel must understand the provisions of these detention statutes because they are the ones who must make decisions based on evidence. Effectively handling shoplifting situations involves three factors: (1) deterrence of future shoplifting attempts, (2) cost effectiveness of taking cases to court, and (3) keeping the civil liability exposure to a minimum in each shoplifting situation. When store policy is based on these merchant's privilege statutes, a retailer in any state is using the best legal tool available.

Evidence-Gathering Skills Needed to Prosecute Shoplifting Cases Selectively

Without the ability to produce, investigate, and evaluate evidence found in the store, the merchant's privilege will be of little value. As the trade association survey above showed, merchants who did prosecute selectively obtained convictions in 90 percent of the cases filed. They also saved payroll costs because not all persons detained for possible shoplifting were taken to court. These stores had effective prosecution policies because their employees understood the evidence needed to obtain a conviction in court for shoplifting. They also knew when the evidence produced in the store told them a prosecutable case did not exist and should not be filed.

In all states, a merchant will present two types of evidence in court when seeking a conviction for shoplifting—eyewitness evidence and physical evidence. Eyewitness evidence is the observations made of the person concealing merchandise or switching a price tag. The physical evidence is the recovered merchandise and its price tag. This evidence is produced, reviewed, and evaluated in the store. If the retail employee decides to file the charge this evidence is presented some weeks later in court at the trial to establish the elements of the offense found in the shoplifting prosecution statute.

The fact that shoplifting prosecution statutes have different names from state to state often confuses retailers, especially those with stores in several states. A retail or security executive could comment:

> We have 31 stores located in four states. In Illinois the law is called Retail Theft, in Tennessee it is named Shoplifting/Concealment, in Texas they call the statute Theft and in Missouri the shoplifting prosecution statute is called Stealing. How can we write an effective policies and procedures manual when the laws are different in each state where we are doing business?

The answer is to teach employees to produce the two types of evidence—the eyewitness and physical evidence. In each state a merchant can obtain a conviction by presenting those two types of evidence in court. Training programs should teach employees how to produce evidence they will need later in court to prove the elements of the offense as found in the shoplifting prosecution statute. A person can be found guilty of retail theft in Illinois, of shoplifting/concealment in Tennessee, of theft in Texas, or of stealing in Missouri on the basis of eyewitness and physical evidence. Only evidence convicts a persons of shoplifting, not the name of the shoplifting prosecution statute.

There are five basic evidence-gathering skills a store manager, owner, or security person needs to acquire in order to prosecute shoplifting cases more effectively. Remember that shoplifting cases are won because a retail employee had the loss-prevention skills needed to gather evidence in the store.

1. *Logically and Systematically Gather Evidence.* When cases are lost in court, it is usually because of insufficient evidence. This means the store manager or security person overlooked some facts in the store, facts now needed as evidence at the trial. The ability to produce facts logically and systematically in the store that can be relied on at trial as evidence to prove the elements of the offense is a necessary skill.

2. *Make Decisions Based on Evidence Produced in the Store.* The essence of selective prosecution as authorized by the merchant's privilege is the ability to make decisions based on evidence found in the store. A retail employee must learn to judge when the evidence supports the decision to prosecute a case. Sometimes the evidence will reject the decision to prosecute and support the decision to release the suspect. Simply stated, not all persons detained should be prosecuted because there is not always sufficient evidence to justify filing the formal charge.

Sometimes merchants make decisions about how to handle shoplifting cases based on suspicion or opinion rather than on the basis of evidentiary facts (evidence). *Suspicion and opinion are not evidence in any court in any state.* No retail employee will be permitted to take the witness stand and testify, "Your Honor, *in my opinion* the woman is guilty of shoplifting." No retail employee will obtain a conviction by testifying that the suspect had shifty eyes or was exhibiting furtive gestures. Shifty eyes and furtive gestures are not facts that show the shoplifting suspect obtained wrongful possession of any merchandise and then left without paying for it. No retail employee can testify about suspicion and hope to win shoplifting cases. Retail employees must make the decision to detain shoplifting suspects and the decision to prosecute shoplifting cases on the basis of facts.

3. *Produce the Two Types of Evidence Needed to Obtain Convictions.* Merchants must learn how to produce the two types of store prosecution evidence needed to obtain a conviction for shoplifting in any state. Eyewitness evidence is produced by watching

the suspect on a sales floor conceal merchandise or switch a price tag or assist a companion who is concealing merchandise. The physical evidence is *always* produced after the suspect has first been detained on probable cause.

4. *Understand the Legal Meaning of the Suspect's Explanation.* Even veteran security people sometimes forget the suspect's explanation is also evidence: defense evidence. Merchants must become good listeners. They will see that store managers and security people produce evidence with their ears, as well as with their eyes. As all with even slight experience in dealing with shoplifting suspects know, very rarely will the person admit guilt. Most of the time people deny they are shoplifting. This is particularly true with the average citizen type shoplifting suspects. They will say, "I forgot to pay," or "I intended to pay," "I bought the merchandise at another store," "I bought it earlier at your store," or "I was returning the merchandise for a refund [or an exchange]." These explanations have specific legal meanings, both in the store and later in court if a case is filed. The merchant will be taught to understand these explanations for what they really are: defense evidence. The retail employee will see that in making the decision to prosecute the case, these explanations must be evaluated. In court a judge or jury will listen carefully to the defendant's explanation (called the *formal courtroom defense to the charge*) before making a decision about guilt. Retail employees must learn to do the same in their stores as part of the evidence-gathering process authorized by the merchant's privilege. Retail employees will see that these explanations are also important in making the decision not to prosecute. If a suspect is detained and claims to be returning merchandise, a store manager or security person should think twice before deciding to prosecute that case. The proper decision could well be to release the suspect because the explanation is true or because the merchant cannot prove it is false. The suspect's explanation is often the key piece of evidence used in deciding whether to prosecute the case or to release the suspect.

5. *Speak the Language of Evidence.* In France, people speak French. In court, people speak evidence. Judges, prosecutors, and defense lawyers all speak evidence. Thus store managers and retail security persons must learn the language of evidence in their stores while they are handling the case and deciding whether to file a formal charge. Merchants will be shown how to discuss a shoplifting case in their stores in terms of what the evidence will prove in court. If the store manager or security person is speaking the language of "suspicion" or "opinion," the chances of filing a nonprosecutable case and losing it are much greater.

Making a decision to warn and release the suspect or to prosecute the case is a process. First, the merchant must know what facts found in the store will become evidence in court to prove the elements of the offense. Next, the retail employee must discuss the case in language that will count in court if the case is filed. That discussion must be in the language of evidence rather than in the foreign tongues of suspicion or opinion. Training programs must teach employees to review and evaluate their cases by speaking the language of evidence.

Dealing with Shoplifting Situations from a Position of Strength

Retail merchants are not sworn law enforcement officials. They do not have the duty to prosecute any shoplifting suspect; they have the option of prosecuting or warning and releasing shoplifting suspects. Experience has taught merchants that prosecution can mean an employee is out of the store, often for several hours on two or three occasions, while the case is going through the court system. In many instances, the value of the recovered merchandise is small so prosecution seems not to be cost justified. But, if persons are not prosecuted—especially the average citizen shoplifter who steals habitually—the store is branded as an easy place to shoplift, and losses increase.

What counts is the ability to deal with *all shoplifting situations* from a position of strength. A position of strength means understanding the evidence and the options. It means having the ability to make decisions based on evidence. It means having the ability to systematically gather evidence before making a decision. A store owner, manager, or security person who understands the evidence and the prosecution options under the merchant's privilege is in the legal driver's seat. Strong cases can be prosecuted to deter future shoplifting attempts. Weak cases will be recognized as such along with the evidence reasons that make them weak cases. The ability to recognize the existence of a nonprosecutable case is an essential skill in shoplifting loss-prevention. Making the decision to release and understanding the evidence reasons why that decision is made is part of a selective prosecution approach.

There are three basic shoplifting situations a merchant in any state will encounter. The key to effectively handling all three depends entirely on the merchant's ability to understand evidence. Consider the three situations:

1. When a prosecutable case exists and should be filed.
2. When a prosecutable case exists but is not filed because the value of the recovered merchandise is small. Here the decision is to release without formal prosecution even though the case could have been taken to court.
3. When a nonprosecutable case exists because there is not sufficient evidence to establish the elements of the offense at trial. Here the suspect is released. Said another way, the decision made is not to prosecute.

The retailer's normal reaction when talking about training programs is wanting to know *how to prosecute shoplifting cases?* This overlooks the fact that some cases should not be prosecuted. It overlooks the fact many retail companies follow a selective prosecution approach. The other side of a selective prosecution policy is understanding when the evidence supports the decision *not to prosecute.* Making this decision not to prosecute is one which must be included in all training programs. As will be demonstrated in this book, when a merchant files a nonprosecutable case there is a strong possibility the store will be sued civilly after the wrongfully charged suspect is acquitted.

There are three basic situations where the evidence tells the merchant the proper decision is not to prosecute. First is the *returning merchandise* situation. The second situation is where the suspect is detained before passing the last cash register and explains that she or he "intended to pay" for the merchandise. The third situation where a nonprosecutable case exists is when the suspect claims ownership after being detained on probable cause. Selective prosecution is a valuable tool in shoplifting loss-prevention. It is effective only if merchants know when to prosecute and when to release suspects.

Reducing the Civil Liability Risk Factors When Handling Shoplifting Cases

This section will provide retailers with specific information about how to reduce the chances of having a civil action filed against the store by a shoplifting suspect. For several decades the insurance industry has taught its claims adjusters about the law of negligence as it applies to automobile accidents. It is time for the retail industry to teach itself about the law of false imprisonment, false arrest, malicious prosecution, slander, and assault and battery as it applies to handling shoplifting cases.

Reducing the civil liability risk factor in shoplifting cases is a neglected part of loss-prevention training programs. Generally merchants have concentrated on preventing shoplifting attempts by purchasing expensive electronic article surveillance and closed circuit television systems. They have spent some training time learning to prevent shoplifting losses by prosecuting cases as a deterrent to future shoplifting attempts. This is generally called *apprehension* to distinguish it from prevention. Actually, detaining is a prevention tool. If the person was not detained under the merchant's privilege the store will suffer a loss. Now the retail community is realizing the value of teaching itself a set of loss-prevention skills which center around reducing the civil liability risk. Preventing and prosecuting are not sufficient. It is the ability to prevent, prosecute, and do so with minimum exposure to false arrest type civil actions that make a shoplifting loss-prevention program complete.

There is good reason for merchants to learn about reducing the chances of having the store sued out of a mishandled shoplifting situation. In the fall of 1984, a university-based survey about civil law suits over crime and security was released.[10] The authors state that retailers are the most common defendants in successful security type civil law suits, suffering one-fourth of the major awards. Consider Table 3–3.

The study went on to break down the types of civil actions filed against retailers. The number one ranking was false arrest type cases filed by business invitees, which means customers. Table 3–4 is very informative.

Many of these false arrest actions are preventable. One loss-prevention goal for any store is to prevent the loss of dollars paid out in these false arrest civil actions filed as a result of mishandled shoplifting situations. Learning about the civil liability aspects of shoplifting is the way to reduce preventable losses.

Make no mistake about it—If a shoplifting situation is mishandled, a civil action can be the result. Remember, over 70 percent of the persons encountered are the average citizen type shoplifting suspect. They make excellent plaintiffs in civil actions against stores because they do not have prior criminal records. Instead, they have families, jobs, and otherwise good reputations in the community. If these average citizen types are improperly detained

Table 3–3. Primary Defendants in Security Lawsuits

Rank	Defendant	Number	Percent	Cumulative percent
1	Retailer	50	27	27
2	Other business	45	24	51
3	Residential landlord	33	18	69
4	Hotel/Motel	20	11	80
5	Common carrier	12	6	86
6	Municipality	6	3	89
8	State agency	4	2	91
8	Bank	4	2	93
8	All other	4	2	95
10	Office landlord	2	1	96
13.5	Public schools	1	—	—
13.5	Private schools	1	—	—
13.5	Private college/University	1	—	—
13.5	Hospital	1	—	—
13.5	Shopping center	1	—	—
13.5	Federal agency	1	—	—

and mistreated or if they are wrongfully prosecuted and acquitted, they will hire a lawyer and sue the store. The following data list the most common types of civil actions filed against stores and the reason the civil action can be filed.

TYPES OF CIVIL ACTIONS COMMONLY FILED	REASON THE CIVIL ACTION CAN BE FILED
False imprisonment False arrest Malicious prosecution	The retail employee handling the case made an evidence-gathering mistake
Slander Assault and battery	The retail employee handling the shoplifting case mistreated the suspect in a legally unreasonable manner.

As demonstrated, there are two basic types of mistakes merchants make which permit persons to sue the store civilly out of a mishandled shoplifting situation. First, the retail employee makes an evidence-gathering mistake. Second, the retail employee mistreats a suspect either at the time of the detention or while conducting an investigation of the facts. Both types of mistakes are preventable.

An example of an evidence-gathering mistake led a national department store's security personnel to wrongfully charge a woman with shoplifting. This thirty-five-year-old woman and her son were accused of leaving the department store without paying for a snowmobile suit. Outside the store she was handcuffed. The snowmobile suit was stripped off the boy while adult family members watched in terror. The woman said when they put the handcuffs on her she urinated. Before being arrested and charged with shoplifting she pointed out that the snowmobile suit was not new. She said it had been purchased some eleven days earlier. She said, "Anybody with eyes could see the suit was not new. It was filthy dirty and had a hole in it." The family

Table 3–4. Types of Civil Actions Filed against Retailers

Rank	Type of case	Number	Percent	Cumulative percent
1	False arrest—invitees	19	38	38
2	Inadequate security-invitees	13	26	64
3	Other improper security act	5	10	74
4	False arrest—employee	3	6	80
5	Improper investigation/polygraph	3	6	86
6	All other	7	14	100
	Total	43	100	100
	Missing cases	0		

members verified her explanation that the garment had been previously purchased.[11]

Notwithstanding her corroborated claim of ownership, a shoplifting charge was filed. After several trials she was found not guilty; then she filed the civil suit for false imprisonment and malicious prosecution. The jury awarded the mother $15,000 and her son $10,000 on the false imprisonment count. On the malicious prosecution count the jury awarded her $325,000. Under Michigan law the trial judge was required to triple all jury awards in malicious prosecution actions, making the total $975,000. When combined with the $25,000 awarded on the false imprisonment count, the total was a staggering 1 million dollars. The jury award was large, but the evidence-gathering mistake made by store security personnel was a common one. They filed a shoplifting case where the evidence to prove store ownership of merchandise was lacking. Under the merchant's privilege they had the choice of releasing or prosecuting the woman. If she had been released rather than prosecuted the only civil action possible was one for false imprisonment, an action to which the national department store had a defense under the merchant's privilege statute. This case is an excellent example of where a selective prosecution policy would have prevented loss—the loss of one million dollars.

Cases where stores have been sued civilly are clear on one point. Retail employees mistreat shoplifting suspects because store policy has not taught them how to systematically gather evidence to decide if a prosecutable case exists. Effectively handling shoplifting cases requires two sets of loss-prevention skills: (1) the ability to gather evidence, and (2) understanding how to avoid mistreatment of the suspect while gathering the evidence. The case law indicates that retail employees get in a hurry because they are afraid of making mistakes. Getting in a hurry is a mistake itself, one that leads to mistreatment which leads to civil liability exposure. Table 3–5 lists the most common ways a store owner, manager, or security person mistreats a shoplifting suspect. These mistakes are made either at the time the suspect is detained or while the retail employee is attempting to investigate the facts.

If retailers detain and make one of the following mistakes the courts will most likely find the employee's conduct violated the merchant's privilege statute. The store will then lose its immunity from false imprisonment liability granted by the statute and have to pay out money.

Training programs must consider one simple fact —Shoplifting is predictable. The studies show that 70 percent or more of the persons detained will be concealing merchandise in a pocket, purse, shopping bag, or on their person. There is no great mystery about how people shoplift. The merchant's training goal is to teach employees how to avoid evidence-gathering and mistreatment of suspect mistakes when they respond to these predictable shoplifting situations. Merchants must learn to play the percentages when setting policy and when conducting training programs. Concentrate on these cases involving concealment first. In about 30 percent of the shoplifting situations, people will shoplift by some combination of price tag switching, container stuffing, by shoplifting in fitting rooms, or by passing merchandise off to an accomplice. In these miscellaneous shoplifting situations, a merchant must also avoid making both the evidence-gathering and the mistreatment of the suspect type mistakes to minimize civil liability exposure.

Table 3–5. Mistreatment of Shoplifting Suspects

Use of excessive force.
Making an accusation of shoplifting.
Searching a suspect without first obtaining consent or permission.
Making threats of prosecution to obtain a civil liability release.
Making threats of prosecution to obtain a confession type statement.
Fingerprinting or photographing a shoplifting suspect.
Handcuffing a suspect who has not tried to escape or not become violent.
Conducting an investigation of the facts on the selling floor.

Suppose we are in a department store equipped with surveillance cameras and an electronic article surveillance system. Suppose further that this department store knows the value of employee alertness and has trained sales employees in high-shoplifting-incident departments. There will come a time—most likely each day—when the average citizen type has not been deterred by expensive cameras or antishoplifting tags. A security employee has just seen a well-dressed woman drop a $40 compact into her purse and head toward the mall exit. Suppose the setting is an apparel specialty store, one with an EAS system. The manager has just seen that same well-dressed woman remove a silk scarf from its place of display, tear off the price tag, throw the tag behind the counter, put the scarf around her neck, and head toward the mall exit. Neither the compact nor the scarf will activate the EAS system. Now the only way to prevent shoplifting is to detain that woman because she is in wrongful possession of unpurchased merchandise. In the department store situation, the compact is concealed. In the specialty store situation, the scarf is at all times visible to the store manager.

The retail employee cannot rely on devices to prevent shoplifting. The effectiveness of the store's shoplifting loss-prevention program boils down to one thing: the *specific information* about the dual legal aspects of shoplifting possessed by that retail employee. The crucial tool for that security person or store manager is knowledge of the merchant's privilege. The security person or store manager can detain on probable cause and either prosecute the case or recover the unpurchased merchandise, warn and release the suspect. Reducing the civil liability risk factor means reducing the chance of making a mistake when detaining this woman concealing a compact in her purse or when detaining the woman walking out of the specialty store with the scarf around her neck.

In a civil action, store policies and procedures are actually on trial. A jury decides if that department store security person or that specialty store manager used reasonable care in three respects: (1) in detaining the woman on probable cause, (2) whether the investigation of the facts after detaining her was legally reasonable in manner, method, and time, and (3) whether the decision to prosecute the case was based on evidence rather than on suspicion or opinion. If store policy is not specific in all three areas, mistakes will be made. Store policy and procedures must be based on the civil tort law requirement to use reasonable care before and during the encounter with a shoplifting suspect.

Some retail and security executives are reluctant to develop a loss-prevention manual that is specific. They like the guideline approach. Some even think that having no written loss-prevention manual at all is the best approach. These merchants feel that a person suing the store (the plaintiff) will subpoena the manual and then show that a security employee or a store manager violated company procedure. These retailers reason that it will be easier for a plaintiff to establish liability against the store by using its own loss-prevention manual against the company.

The problem with the guideline or the no manual at all approach is this: the civil tort law duty to use reasonable care when dealing with a shoplifting suspect exists whether the store has a loss-prevention manual or not. Stores that do not train employees about the specifics of the duty to use reasonable care will find employees violating the civil tort law. The result will be a civil action, one alleging false imprisonment, false arrest, slander, assault and battery, malicious prosecution, or some combination of them. As a practical matter, store manuals contain procedures that are well within the civil tort law as found in the merchant's privilege anyway. The fact that an employee did not follow procedures to the letter in a case will not be evidence by itself that the employee acted outside the scope of the merchant's privilege.

Employees in these no-manual or guideline-manual stores are more likely to mistreat a suspect or make an evidence-gathering mistake. Mistakes lead to civil actions. Training is the way to reduce the possibility of making mistakes. Training programs are evidence of a good faith effort by retailers to prevent the loss of goods and to comply with the merchant's privilege statutes. The security director or retail executive who testifies in court (or at a deposition) that her or his store conducts no regular shoplifting loss-prevention training programs is making it easier for a plaintiff to establish civil liability and recover damages. The absence of training programs is evidence that the store has made no effort to learn about the civil tort law duty to use reasonable care when handling a shoplifting case. Merchants can no longer afford the luxury of simply prosecuting shoplifters. They must learn to prosecute shoplifting cases and do so with minimum exposure to civil liability exposure. The civil liability risk will not go away by pretending it does not exist. The chance of being sued civilly is inherent in all attempts to handle shoplifting situations.

A final word about training programs is necessary: in my opinion, merchants spend too much time studying shoplifters. They love to show films about shoplifters, ones with shifty eyes or ones exhibiting furtive gestures. But shifty eyes and furtive gestures are not elements of the offense of shoplifting in any state. People shoplift with their hands, not with their eyes or their gestures. These films put the emphasis on the wrong persons. There is nothing in a film about shoplifters that will help a store owner, manager, or security person learn to gather evidence or to comply with the merchant's privilege. Sometimes these films even throw retail employees off the track about who shoplifts and how they do it.

Too often films show people using gimmicks to shoplift. A film may show a person using a booster box or putting merchandise between their legs. This film is interesting but misleading. Most people conceal merchandise in pockets, purses, shopping bags, and under clothing—they do not use gimmicks. These films are justified—in the budget anyway— under the banner of awareness. Awareness does have some value but the ability to gather evidence and comply with the merchant's privilege is the information that prevents shoplifting and reduces exposure to civil liability.

If films are of little value, merchants can validly ask what they should concentrate on in training programs. My answer is, concentrate on yourselves, on the way you respond to shoplifting situations. Examine the mistakes you make, and learn from them. Stores with a full-time security staff have a wealth of information right in their own file cabinets. Files about cases in which the store was sued civilly because of a mishandled shoplifting situation can be used in training programs. Stores can learn from past mistakes and not make the same ones in the future.

The Bottom Line

Retail and security executives like to talk about preventing shoplifting. Some accomplish this goal by spending large sums of money on closed circuit television systems and electronic article surveillance systems. Merchants who use these preventive devices justify the cost by saying: "We would rather prevent four thefts than apprehend one shoplifter." Retail and security executives who use these expensive systems do not claim to prevent shoplifting by stopping it entirely. They may be able to reduce the shoplifting shrinkage numbers, which, of course, is a loss-prevention goal. The fact remains that people shoplift every day in department and specialty apparel stores that have installed these devices.

Monitoring cameras and antishoplifting devices may be part of the answer for some stores. For other stores, these devices are too expensive to be cost justified. The problem with devices is the same as the problem with showing films about shoplifters with shifty eyes: both concentrate on the wrong people. Devices are aimed at deterring shoplifters rather than directed toward improving an employee's loss-prevention skills. My suggestion is to consider devices only as support systems. The first line of defense is always the alert sales employee. In some stores these sales employees will be backed up by a security staff. In specialty stores, the sales employee is backed up by a store manager who is authorized to detain shoplifting suspects and investigate the facts.

Store owners and retail executives are beginning to see the value of training employees to prevent shoplifting. Traditionally, the retail community has been content to ride along with the inventory shrinkage figures at about 2 percent of gross sales each year. They consider it a cost of doing business, passing it along to the customer. Now store owners and retail executives are beginning to see that time, effort, and money spent in training pays off at the bottom line. If somebody shoplifts the merchandise, it cannot be sold at a profit. If the merchandise is shoplifted, it takes a large amount of additional sales to make up for the profit lost on the shoplifted item. When sales projections are flat, this means preventing shoplifting is even more important for a retail company at the bottom line.

Preventing shoplifting losses means protecting profit margins. Merchants should teach sales employees one set of shoplifting loss-prevention skills centering around removing the privacy factor by giving good customer service. They should teach security personnel and store managers (in stores without a full-time security staff) another set of shoplifting loss-prevention skills, centering around gathering evidence and complying with the merchant's privilege statutes.

Forget about shoplifters with shifty eyes and forget about antishoplifting devices for a while. Instead, concentrate on developing a set of shoplifting loss-prevention skills, ones that can be used by store owners, managers, security personnel, and sales employees.

References

1. Study conducted by Larry A. Lloyd of the National Corrective Training Institute, Austin, Texas, September 4, 1981.
2. Study conducted by Lawrence A. Connor, director of Shoplifters Anonymous, Glen Mills, Pennsylvania, 1982.
3. "When You Least Expect It," *Chain Store Age Executive Magazine*, February 1985, p. 18.
4. Mr. W.R. (Dick) Deal was a professional shoplifter for many years. He was never stopped for shoplifting, nor was he ever with another professional shoplifter who was stopped. He now presents seminars for retail merchants, telling them how to identify boosters and how to prevent them from shoplifting.
5. *Security Management Magazine*, July 1978, p. 47. This magazine is published monthly by The American Society for Industrial Security.
6. "When You Least Expect It," *Chain Store Age Executive Magazine*, February 1985, p. 18.
7. "Shoplifters Pad Lists," *Kansas City Star*, November 20, 1981, p. 6.
8. National Mass Retail Institute—Security & Shrinkage, 3rd Annual Study of Inventory Shrinkage Control and Security Procedures in Retailing. Conducted by Arthur Young & Co., released January 15, 1982.

9. "The Nut Muncher Case," *Kansas City Star*, May 11, 1982, p. 7.
10. Lawrence W. Sherman and Jody Klein, "Major Lawsuits Over Crime and Security: Trends and Patterns 1958–1982," The Institute of Criminal Justice and Criminology, University of Maryland, College Park, MD, September 1984, p. 28.
11. Walker v. J.C. Penney and Co., *The Detroit Free Press*, September 5, 1984, p. 5A.

Chapter 4

Transportation-Distribution Theft and Loss Prevention*

LOUIS A. TYSKA, CPP

Finished products are most valuable and vulnerable to theft or misappropriation when in storage or inventory. Unfortunately, this is precisely when they are least controlled.

Manufacturing and marketing departments may go to extremes to plan, implement, and sell goods, but then we frequently turn out the lights and go home once the finished goods are in the inventory or transit system. Industry provides much control over these goods, but much more should be done to prevent theft and/or misappropriation.

The accountability for safeguarding goods in transit or storage lies completely with highest management for both the manufacturers and transporters of products. As goods pass through various modes of transportation and distribution, legal accountability changes hands and is frequently difficult to identify. It is precisely during these times that the thieves who prey upon the various transportation systems will take advantage of the weakness within the network of distribution.

The ultimate consumer pays the cost of transportation and distribution loss. These increased costs are based upon the need to file, process, and dispose of claims resulting from loss or damage in transit or storage. In addition, costs are involved in all reorders and reshipments as well as any dollar factors which may be directly charged to inconvenience and loss of timeliness in the marketplace.

Surveys show that the greatest loss a company can suffer is through actions of their own employees (up to 80 percent per various governmental reports). These internal thefts are committed by employees who have direct access and control of goods in transit. Adequate physical security is required, but improper employee screening and control leads to significantly greater losses. Effective control can be obtained by developing appropriate company procedures and implementing them. The attitude management assumes concerning security will dictate the success of the loss prevention program and its positive acceptance by employees.

Transportation and distribution management assumes accountability for goods in transit and responsibility for establishing standards of adequate physical security for the prevention of theft or damage. *The appointment of a security professional does not relieve management from its responsibility.* The role of security is related to the prevention of unauthorized entry, circulation, and exit of personnel and vehicles, perimeter fencing and entrance areas, the protective lighting and alarm systems, pass, badge and identification systems, and the prevention and detection of crime.

The establishment of physical security procedures proves difficult. Goods in transit of every type and size move through the intermodel system constantly. Each shipment must be identified and accounted for throughout its entire journey. The size and complexity of the large transportation facilities has made it vital that physical security depend on a comprehensive and continuing evaluation of protective measures and an aggressive program of auditing.

The appropriate degree of protection for any specific facility is based upon two factors: relative

---

*The Federal Government's various agencies and departments that have provided research and analysis for this chapter are acknowledged; specifically, such departments as Treasury-U.S. Customs, Transportation, Justice, and Commerce.

criticality and relative vulnerability. If a facility is both highly critical and highly vulnerable, then an extensive physical security program is a must.

Every facility is vulnerable in some degree to pilferage and theft. To determine the degree of vulnerability, the volume and value of goods handled, the past history of theft must be evaluated. If these factors rate high in incidences, then the relative criticality is high.

To determine relative vulnerability, the susceptibility of the facility to theft must be estimated. The degree depends on such variables as the type of facility involved, the value and volume of goods handled, accessibility to stolen goods' markets, physical layout and construction, and the loss prevention program in place if any. If the facility can be easily penetrated and confused by high volumes of rapidly moving goods, and the area marketplaces are highly motivated toward illegal practices, then the relative vulnerability is high.

Another complication is the multiplicity of managements using various parts of the specific facility simultaneously, for example, at a large metropolitan airport. Such facilities usually present a high degree of criticality as well as vulnerability. All managements using a facility of this nature must coordinate and cooperate closely with the specific facility management.

Conditions which greatly influence the scope of the physical and procedural security necessary for a specific facility include the following:

- Size
- Volume and value of the goods in transit
- Geographic location
- Economic and political conditions in the area
- Ability of public law enforcement agencies to respond
- Sophistication of the criminal element in the local area

The nature of a physical and procedural security program to be implemented in any facility should be no more involved than is called for by an analysis of the following:

- Criticality
- Vulnerability
- Impact of physical security measures on efficiency of operations
- Limitations imposed by the physical characteristics of the facility
- Cost required to implement a program versus losses

The security professional at each facility must continually evaluate and analyze, along with management, on a periodic basis. They must devise physical security measures consistent with continuing needs. Criticality and the vulnerability of a specific facility may vary from time to time as circumstances change.

Prevention Plan

Prevention of theft and pilferage demands commitment and determination on the parts of all employees and in particular security personnel. Physical security measures and an extensive security awareness program for every employee of the facility are essential. Theft prevention is an ongoing process which must be given dedicated attention, especially to high-value and highly vulnerable goods.

Our prevention plan begins with defining the three types of crimes committed against goods and materials in transit—pilferage, theft, and the organized crime influence on cargo loss from the intermodel transportation and distribution systems.

Pilferage

The protection of goods and materials while they are in the transportation system is a responsibility of management at all levels. Pilferage is one of the most annoying and costly causes of cargo theft. Prevention of pilferage is one of the primary functions of the physical and procedural security program. It is difficult to detect, hard to prove, and becomes costly if ignored. Cargo movements are made in all kinds of packaging, some is even moved without packaging—merely tagged and labeled. Small packages are more susceptible to pilferage, since they are more easily concealed and not as readily detected. Actual losses will depend on such variables as volume, accessibility, waiting and loading times, storage areas between carriers or modes of transportation, and a variety of elements which contribute to confusion. Since these factors vary between facilities as well as during different times of day, each must be considered separately in the physical security plan.

A pilferer is one who steals primarily to give in to the temptation of an unexpected, or sometimes arranged, opportunity, and has little fear of being caught. There is usually little or no planning or premeditation involved in pilferage. Pilferers generally act alone. They may take items for which they have no immediate need or use. They may take items for family or friends or for use in the home. The degree of risk involved in pilferage is slight unless very large numbers of persons are engaging in pilferage at the same time. Pilferage will occur whenever an individual feels the need or desire for a certain article and the opportunity to take it is provided by inadequate security measures. Though it involves unsystematic theft of small articles, pilferage is nevertheless very serious, and it may have a great cumulative effect if the stolen items have a high cash or potential resale value. There is always the possibility that casual pilferers, encouraged by successful theft, may turn to systematic theft.

Pilferers are normally employees of a large transportation facility or employees of other transportation segments whose interface requires them to be in the area. These persons are the most difficult to detect and apprehend.

Pilferage may occur anywhere in the transportation system where goods are being moved within a facility. Goods and materials left unprotected on carts, dollies, or other intraterminal vehicles for any period of time while awaiting the next move are extremely vulnerable. Such situations require adequate physical security measures since cargo accountability may become confused and difficult to determine.

One means of discouraging casual pilferage is to establish a parcel check system at all entry and exit points to a facility. Parcels carried into a facility will have to be logged in at an entry post upon entrance to the facility. Those leaving the facility must submit all parcels for inspection prior to departure. This will bring to everyone's attention that they may be detected if they do attempt the unauthorized removal of merchandise. Care must be taken to insure that personnel are not demoralized nor their rights violated by oppressive physical controls or unethical security practices.

An aggressive security awareness program can train employees to be diligent and to report pilferage. All employees must understand that pilferage is morally wrong, no matter how insignificant the value of the item taken.

Supervisory personnel should set a proper example and maintain a desirable moral climate for employees by establishing an air of concern for the merchandise.

In establishing any deterrant to pilferage, security personnel must not lose sight of the fact that most employees are honest and disapprove of thievery of any kind. Mutual respect between security personnel and employees of the particular facility must be accomplished if the facility is to be pilferage free. Any security measures which infringe on the human rights or dignity of others will jeopardize, rather than enhance, the overall protection of the facility.

Theft

Theft in the transportation system challenges both management and security personnel. It is the transportation system management's responsibility, working with local law enforcement agencies through its security professional, to reduce or eliminate cargo theft where possible. To determine the extent of theft at any given facility, there is a need to pinpoint the actual amount of loss which is occurring. This is not an easy task. Cargo accountability records and methods are generally designed to single out thefts by category. Details of such losses may not be disclosed until some time later, usually upon receipt by the ultimate consignee or customer. The specific location may be far removed from the actual point of theft and the management accountable for the delivery may not have been the accountable management at the time of the loss.

Facilities located in and around large metropolitan centers are the most vulnerable for thievery. More goods and materials are being moved, terminals are more crowded, and more people are involved, adding to confusion. Markets for criminal redistribution are more readily available.

Thieves differ from pilferers in that they steal for profit. Thieves steal according to preconceived plans and they steal any and all goods in transit. They are selective only according to their *fence* requests. A thief can work alone but generally has accomplices. It is not unusual to find an accomplice who is an employee of the facility and in a good position to give relevant information to the thief. Some accomplices may even be in an advantageous position to locate or administratively control targeted goods or cause them to be removed from high security areas. The specific theft may be a one-time occurrence based on inside information pertaining to a particular shipment, or may extend over long periods, even years.

Like pilferage, theft occurs anywhere but, unlike pilferage, theft is planned and organized. It can occur in more unsuspected places, involving switching trailers, misrouting railroad cars, or taking possession of a cargo in the name of a legitimate freight forwarder. The opportunity for theft in the transportation system begins in the shippers installation and continues throughout the system and into the warehouse of the consignee after receipt. Employees of a shipper or of a consignee may be working to defraud. They may be part of an organized theft group. Dishonest documentation can result in the transportation system getting blamed for the loss. Actually the goods were never sent, or they were stolen after delivery. One common practice is called *short-loading*. Control of goods in transit is one of the greatest challenges to the transportation industry.

Thieves can infiltrate into transportation ranks as legitimate employees. They can strike at a time of their choosing. Management can do none of these. Management can be ready, however, and can take measures to prevent theft and make it unprofitable and unappealing. Management protects goods by realizing that they are vulnerable and by studying and knowing the problems facing thieves and counteracting them.

Thieves' first problem is to identify and locate the cargo to be stolen. Then they must learn the system. They most assuredly will have to check storage areas and documentation. Countermeasures should include the control and movement of personnel and sensitive information through awareness programs. People have a tendency to talk too much and to the wrong people. If losses are occurring in a facility or at a certain place in a facility, management should carefully observe the behavior of employees; unthinking or disloyal employees will give away goods in transit, sometimes unknowingly.

Thieves, after locating the desired goods, must gain access to them and then take possession. To gain possession they may be required to survey such security factors as physical safeguards or guard procedures for weaknesses, attempt to bribe guards, alter or forge shipping documents or passes, or create a disturbance to divert the attention of security personnel while the actual theft is taking place.

The next problem is the removal of the cargo. Thieves may have to remove it to a vehicle or vehicles, falsify documents, or alter the exterior by painting a vehicle for this purpose. The final problem is to dispose of the stolen goods. Thieves need an outlet for stolen goods and must divert them back into the legitimate market for profit. They most probably will sell the goods through fences, pawn brokers, flea markets and discount houses. Frequently thieves steal on order and know in advance who will receive the stolen goods.

The knowledge that the thief has the problem of locating, taking possession, moving and selling stolen goods offers management the opportunity for constructive preventive measures. Thorough investigation and discovery of the means used to accomplish a theft offers a chance to prevent it from recurring. In addition, thieves must work with other people to dispose of their loot. The primary concern of thieves in selecting a target is its monetary value. Since they steal for profit, thieves will look for items from which they can quickly realize the greatest financial gain. This means that they must already have or be able to quickly find a ready market for the stolen merchandise—such items of high value as pharmaceuticals, metals, or electronic components. A thief may, if the profit is substantial, select a target of great size and weight.

There are many ways by which stolen cargo or items may be removed from transportation facilities:

- Terminal operations are extremely vulnerable to systematic theft. It is here that facility personnel and truck drivers have direct contact with each other, and a readily available means of conveyance which offer opportunity for collusion. Although most truck drivers and employees are honest, some do become victim to temptation caused by poorly controlled goods in transit. For instance, a receiving clerk might certify the receipt of property which the truck driver actually disposed of prior to arrival at the facility, or in some instances at the consignee's location. A facility employee, or a thief acting as an employee, can provide a truck driver with cargo and assist in concealing it aboard the truck for unauthorized removal from the facility. Employees can assist a truck driver in removing property by executing a false invoice which may appear to be legitimate when inspected by security personnel.

  The driver's responsibility starts when the truck is loaded and the cargo is receipted for; it ceases when the truck is unloaded at destination and the receipt is signed. The distribution and security of the load on the truck are also the responsibility of the driver and the management in custody of the cargo.

- Railway employees assigned to switching duties at a facility can operate in a similar manner. Since a railway car normally cannot be directed to a desired location so that stolen cargo can be removed, additional confederates will usually be required to transfer the stolen goods from the railway car, at some point or siding outside the facility, into some other means of transportation for removal to the ultimate destination. This increase in the number of persons involved will reduce profits and increase the chances for discovery and apprehension.

  The carrier's responsibility commences when the loaded and sealed railroad car is coupled to

the engine. The carrier's responsibility ceases when the railroad car is spotted in the consignee's yard or unloaded at destination.
- Trash disposal and salvage activities offer excellent opportunities to the thief to gain access to valuable material. Property may be hidden in waste material to be removed by a confederate who has the duty to remove trash from the facility.
- There are many other methods which may be employed by a thief to remove cargo from transportation facilities, such as moving items to a wrong location within the facility to be picked up by a confederate, intentionally misrouting cargo to a location where it can be picked up by a confederate, collusion with security personnel, failing to keep goods under surveillance, or removal by custodial or vendor vehicles.
- Thieves do not need to be employees nor be in collusion with employees; they can pose as employees or have false credentials from a nonexistent transportation company. There are many methods for thieves to pass themselves off as legitimate persons employed by legitimate companies in the transportation field. This is particularly true in the overcrowded terminal facilities in the large metropolitan areas.

Organized Crime

Just how much theft of goods from the transportation modes can be attributed to individual thieves and how much to organized crime is not accurately known. The U.S. Department of Justice believes that the vast majority of thefts, both in quantity and in dollar value, are the result of organized crime.

Just as the bookie operates within the framework of organized crime, so does the thief of goods in transit. Since organized crime provides a major network for the disposal of stolen goods, in effect, organized crime has become a theft-to-order operation with the transportation system as the victim. Some complaints have been registered but usually are settled by claims being paid, unhappy consignees, and large shipments or stolen merchandise being offered for sale at unbelievable bargains hundreds of miles from the scene of the actual thefts. Someone must pay for these costs, and it is usually passed on to the public. Legitimate businesses must pay the very large bill for such indirect costs as:

- Spiraling insurance costs or growing deductible coverage that stem from organized crime's unique competitive methods.
- Increased tax burdens resulting from nonexistent records in organized crime enterprises.
- Customer drop-off when legitimate enterprises are forced to raise prices to compensate for theft harrassment.

Organized crime is a real and costly hazard to the security of goods in the transportation system. In comparison to the hazards of the pilferer and the unorganized thief, organized crime presents infinitely more complicated problems.

These problems are varied and consist of every way that the human mind can conceive of to steal, swindle, defraud, and separate a rightful owner from their property—particularly while in the custody of others. Some recommended preventive measures for assisting in reducing the risks associated with organized crime thefts are the following:

- Engage a trustworthy security professional who works directly for the senior manager of a transportation facility or other levels of management as appropriate.
- Know your supervisors and make them aware of the need for consistent security awareness.
- Investigate all losses promptly and thoroughly and report to the appropriate authorities.
- When losses occur and tend to portray a pattern, maintain extra vigilance of employees and, if the situation dictates, make appropriate personnel changes.
- Provide periodic detailed inspections of the security measures in effect and the transportation operations procedures.

Organized crime activities consist of highly organized, tightly controlled management which will take advantage of any weakness in the transportation system. Can you answer these basic questions?

- What type of organization is required to systematically pinpoint the location of goods and to set them up for theft?
- How is it that a shipment of merchandise is found on the shelves of a retail store thousands of miles from the location where the theft occurred?
- What organizational structure is required to transport and ship stolen goods?
- What organizational patterns are used to effect the sale of such stolen goods in many locations after a theft?

While pondering these questions, remember that when one person is apprehended, the organization continues without interruption.

Dishonest employees, whether working alone or

in collusion with other employees or nonemployees, are restricted in their methods of operation only by their own imagination and the existing opportunities to avoid detection. In order to minimize the opportunities to commit thefts while avoiding detection, the following sequence of facts concerning thefts, existing opportunities to avoid detection, and elimination of those opportunities within the reasonable cost/risk planning range must be evaluated:

1. Losses experienced (total, including cost of dispositions)
2. Distribution of losses by type:
   a. Pilferage (concealable on person or in vehicle)
   b. Theft (bypassing existing controls)
   c. Damage (accidental, deliberate, elements)
   d. Error (over, short, missing documents)
   e. Unexplained (no evidence to substantiate method causing loss)

Cargo Package and Movement Controls

The responsibilities of the shipper, the carrier, and the consignee must be clearly established if cargo is to be adequately protected in transit. Custody varies according to the size of the shipment and the means of transportation.

For shipments of less than the total capacity of a truck, railroad car, aircraft, or ship, the carrier assumes responsibility when the agent acknowledges receipt at the shipping point and is relieved when delivery is made to the consignee.

The responsibility of the carrier for large shipments depends upon the mode of transportation used. If transportation is by truck, the carrier assumes custody when the vehicle is loaded. The truck carrier's responsibility ceases when the truck is unloaded at its destination. The railroad carrier's responsibility begins when the loaded and sealed car is coupled to a locomotive. He or she relinquishes custody when the car is spotted in the consignee's yard or unloaded at its destination.

Air and water carriers assume responsibility when an agent takes possession of the cargo from a shipper. This responsibility ends when the cargo is removed from the vessel or aircraft and is delivered to a freight forwarder, another carrier for transshipment, the consignee, or the agent.

A shipper load and count (SL & C) movement in any of the above modes relieves the carrier of liability for shortages, provided an accurate seal record is maintained.

Trucking Operations

Inclement weather and the possibility of accident are ever-present dangers to cargo in transit. It is incumbent upon the security officer to develop preplanned procedures for real operation, which should be part of the security plan.

The integrity of those responsible for line haul should be unquestionable. Only responsible, screened employees should be assigned as drivers and helpers to transport high-value cargo. The equipment used must be in excellent condition to avoid breakdowns that require emergency security protection. This is often difficult, sometimes impossible, and always expensive to provide. Modifications to equipment are often desirable. They can include additional locking and alarm devices, removal of outside door hardware, and installation of oversize fuel tanks (where permissible).

Employees should receive strict instructions concerning procedures to be followed during rest periods or meal stops while in transit. The vehicle should be locked and parked where it can be observed. The nature of the cargo carried should not be discussed with anyone. The driver should not deviate from the preplanned route. In the event of equipment failure, the driver should notify the nearest terminal immediately. If it is necessary to move high-value cargo over a weekend or on a holiday, it should be delivered to a terminal where maximum security can be provided.

To further protect vehicles on the road, large, brilliant-hued numerals should be painted on the roof and sides. This will help identify the equipment if it is stolen or hijacked and a helicopter search is initiated. The frequency of these types of thefts have diminished recently but the threat or vulnerability to hijacking still exists, making the need for rooftop markings necessary.

In states where the use of double bottoms is permitted, the most valuable cargo should be loaded in the lead van.

Seals

Seals are invaluable to the protection of cargo in transit, but they are only as effective as the controls maintained over them. The responsibility for seal control must be vested in a specific individual who will maintain a record of all seals issued and to whom. Unissued seals should be stored in a locked container with limited access.

All cargo under seal entering a facility should be

checked and the seal numbers recorded. If a seal is found to be broken or removed, or if the number does not agree with the one on the shipping document, an immediate item-by-item inventory of the shipment must be made.

Transfer Points

Transfer points, where cargo is transshipped from one mode of transportation to another or between two carriers of the same mode, are frequently areas where large losses can be expected. Cargo is often left unprotected, particularly in remote parts of the facility or yard.

Shipping documents are apt to be carelessly handled at these points and shipments can be intentionally misrouted into the hands of the criminal.

To reduce loss by theft or misrouting at a transfer point, management must maintain strict accountability of all cargo and demand that a carrier who receives cargo from another would accept custody for it in its original condition. Shipping documents must be examined carefully to guard against falsification or tampering. Employees of carriers using the facility must be able to prove their identity.

In-Transit Storage

While cargo is in the custody of the transportation system, it must frequently be stored for varying periods of time. During these intervals it is highly susceptible to loss, theft, or damage. A few elementary precautions will help reduce the claims filed against the responsible carrier:

Individual shipments of cargo should be kept intact and stacked as a unit. This will prevent accidental separation and loss of a portion of a shipment.

Fixed position lights in a storage area should be diffused to eliminate deep shadows. Each guard should be equipped with a strong, high-beam flashlight. Stacks should be arranged in accord with existing lighting to avoid creation of deep shadows.

Greater emphasis should be placed on personnel control than upon structural or mechanical protection in those areas where cargo is stored for short periods of time. The increased traffic generated in this situation, and consequent increased vulnerability of cargo to pilferage and theft, require more alertness by guard patrols and more careful checking of the credentials of all individuals who are given access to the area. If possible, the storage area should be broken into sub-areas, with cargo handlers and truckers limited to the one sub-area compatible with their credentials and business purpose.

Protected cargo, a term common to shippers of Department of Defense material, consists of items which require special handling because of their value or sensitive nature. The Department of Defense places protected cargo in three categories:

1. **Sensitive.** Small arms, ammunition, and explosives which have a ready use during civil disturbances, terrorist acts and other types of domestic unrest and which, if in the hands of militant or revolutionary organizations, present a definite threat to public safety.
2. **Pilferable.** Items vulnerable to theft and having a ready sale potential in illicit markets, such as alcoholic beverages.
3. **Controlled.** Items which require additional control and security in accordance with published regulations and statutes, including: money, negotiable instruments, narcotics, registered mail, precious metal alloys, ethyl alcohol, and drug abuse items.

High-value pilferable (protected) cargo should be kept separate from other material and provided a greater degree of security. This material is best stored in a crib or security cage.

High-value, low-volume cargo, such as fissile radioactive materials, jewelry, furs, and securities require special handling and special security measures during storage. Cargo of this nature should pass from accountable officer to accountable officer. It should kept be under close surveillance constantly and always be stored in an exclusion area. When placed in an exclusion area it should be registered, including time, date, accountable party, and witness. The same procedure should be followed when the cargo departs from the facility.

Controlled Areas

A controlled area is any area whose access is governed by special restrictions and controls. In establishing controlled areas, consideration must be given to preserving the facility's cargo-moving goals as well as to its past loss and theft record.

All high-value cargo should be under surveillance or in a controlled area unless it is locked or sealed in the vehicle that transports it. Carrier vehicles, particularly trucks, trailers, and railway cars, containing high-value cargo, must be guarded or protected in a controlled area until they are released to authorized personnel for movement.

To be fully effective, a controlled area should be under surveillance by physical or electronic methods and movement within the area controlled. A barricade providing limited access does not, in itself, constitute a controlled area.

The transportation industry provides a public service. As a result, some of its operations must be open to the general public. The general offices, personnel office, and freight receiving offices may need to be outside the controlled area. Where practical, consideration should be given to the installation of convex mirrors in warehouse storage areas so that supervisory and guard personnel can have additional surveillance capability.

A controlled area can extend over many acres and include vehicle marshalling yards, docks, warehouses, and service or supply buildings. Such an area is a first line of defense for the protection of cargo in the transportation system.

Limited Area

Within the controlled area a limited area can be established. This will provide a higher degree of security. A different pass, issued to fewer people, should be necessary for entry to a limited area. Sorting, recoopering of crates, and storage may be accomplished here.

Exclusion Area

An exclusion area can be located inside the limited area. Again, a different pass should be required and the number of people granted access strictly limited. The exclusion area is used only for handling high-value, low-volume cargo. The crib, vault, or cage that comprises the exclusion area should be kept locked or under surveillance at all times.

Access points to any controlled area, regardless of the degree of security involved, should be locked at all times or under physical or electronic surveillance. Strict control of the keys or combinations to locks is essential. Management should make periodic checks to determine the integrity of controlled areas in addition to any checks conducted by security representatives.

Package Control

A good package control system is an invaluable aid in helping to prevent or minimize pilferage and theft. No packages, except those with proper authorization, should be permitted to be brought into controlled areas without inspection or accountability.

A simple system should be established to control movement of packages, materials, and property into and out of the facility. Limitations as to types of property authorized, persons allowed to move authorized property, and approved points of entrance and exit should be included in the facility physical security plan.

A standard package checking system may be used at the entrance gate to a controlled area for the convenience of employees and visitors. When practicable, all outgoing packages should be inspected except those properly authorized for removal. When 100 percent inspection is impracticable, frequent unannounced checks should be conducted at random times.

Vehicle Control

Strict control of all vehicles entering or leaving a controlled area should be maintained. Parking lots should be located outside of all cargo exchange and/or controlled areas. The only vehicles entering or leaving a controlled area should be bona fide cargo carrying or handling equipment or emergency vehicles. There are generally three types of vehicles working within the controlled areas: facility vehicles, primarily small trucks, cargo handling vehicles, cargo loading vehicles, etc.; cargo pick-up and delivery and freight forwarder vehicles; and the inhouse cargo carrier vehicles. Of primary concern to the security staff are the cargo pick-up and delivery vehicles and freight forwarder vehicles. These vehicles should be both checked in and checked out, with adequate records being maintained to insure that they are the authorized vehicles for specific cargo movements. The facility vehicles generally remain on the facility but should be properly recorded if they are required to leave the facility. The cargo carrier vehicles should be inspected and documented when arriving or departing. It is essential to maintain accurate records of all cargo carrying vehicles entering or leaving controlled areas.

A close inspection of all trucks entering or leaving a facility should be a general requirement. An orderly system should be established to limit and control the movement of trucks and other conveyances within controlled areas.

All trucks and conveyances entering a controlled area should be required to pass through a control gate manned by security. Truck drivers, helpers, passengers, and vehicle contents should be carefully screened. The security check at truck entrances should cover both incoming and outgoing trucks and should include the following:

- Appropriate entries on a truck register, including vehicle registration, name of truck firm, description of load, and date and time of entrance and departure.
- Identification of driver and helper, including proof of affiliation with the company owning the truck or conveyance.
- A license check of the vehicle operator.
- Examination of the truck or conveyance.

Passes or some form of documentation should be issued to truck drivers and helpers who have been identified and registered. This documentation should permit only limited access to specific loading and unloading areas.

Incoming traffic should be kept to the minimum essential for the efficient operation of the facility, and escorts should be provided if vehicles are permitted access to controlled or restricted areas.

For trucks with loads that are impractical to examine, door seals may be used at the entrance gate. These seals will be opened by a designated representative at the receiving end. Likewise, the truck doors may be resealed for exit or for other stops within the facility.

Loading and unloading operations should be strictly supervised to assure that unauthorized materials or persons do not enter or leave the facility via trucks or other conveyances. This would also apply to all service and/or support vehicles who, under most circumstances, have regular access to a given facility.

Warehousing and Storage

Storage is not a normal function of the transportation system for goods in transit. It does, however, become necessary at times to provide at least temporary warehousing. The necessary storage of goods and material is more applicable to the manufacturers or shippers as well as the variety of consignees who receive the goods. It is management's responsibility in all instances to assure adequate protection for goods at all times. This includes terminal storage at either end of the route or, in some instances, enroute.

Storage areas may include a warehouse, shed, open areas, or any portion of a facility which is designated for holding purposes. All of the security considerations and principles that apply to the prevention of loss and theft of goods in the transportation system apply equally to warehouse areas.

Open Storage

Open storage is normally used only for those goods not subject to damage by weather conditions, and bulky, nonperishable items that are not sensitive to being easily pilfered. It should never include high-value, low-volume items.

When property is stored in open areas, it should be properly stacked and placed within but not near fenced-in areas. All off-loaded cargos must be properly stacked or stored regardless of the time of receipt.

It is imperative that complete individual shipments of cargo be kept intact. This helps prevent the inadvertent separation and eventual loss of a portion of one shipment in the system of accountability.

Stacks of goods in storage should be a minimum of 50 feet from the perimeter fencing. They should be as symmetrical as possible, and the aisles between stacks or lines should be wide and straight. These arrangements not only provide for visibility by the security staff, but also allow for the proper movement of vehicles and cargo hauling equipment.

Fixed position lights in a storage area should be of a diffused type to eliminate deep shadows.

Covered Storage

The same principles of even stacking and adequate aisle space recommended for use in open storage are applicable to covered storage. Goods in storage should be placed to conform with existing lighting in order to keep deeply shadowed areas to a minimum.

In areas where goods and materials are constantly being moved in and out, or otherwise stored for very short periods, more emphasis should be placed on personnel movement control and security than upon structural or mechanical protection. This is especially so during working hours in order to prevent pilferage or theft by unauthorized personnel or workers.

Highly theft-prone types of goods and material in

storage should be kept separate from other cargoes and given a greater degree of security. Sensitive items are those that can be pilfered or stolen and disposed of quickly. They include such items in a separate high value storage area to provide a much higher degree of physical and procedural security protection. In instances where a separate building is not available or warranted by the requirement for sensitive storage, areas that have previously been described as limited or exclusive areas should be designated and utilized.

High Value, Low-Volume Storage

All high-value, low-volume items, like jewelry, gems, furs, and securities require special handling and special security measures when in storage. Goods of this nature should pass from one accountable person to another. They should be held under close observation at all times and should always be stored in an exclusion area. All goods placed in the exclusion area should be registered and logged for time, date, and recipient. The goods should also be logged out upon removal, with similar notations made. It is essential to maintain accurate records of all persons who enter and leave an exclusion area.

Shipping and Receiving Areas and Personnel

Control of shipping and receiving areas and personnel at a plant, warehouse, or distribution center can do much to prevent cargo theft. A number of practical measures are available to shippers and receivers for improving security of cargo-handling areas and controlling personnel to minimize opportunities for theft. To provide better physical security, shippers and receivers may find it to their advantage to adopt the following precautions in areas where cargo is moved to or from carrier equipment:

1. Maintain perimeter controls. Mark off a perimeter area a suitable distance, at least 20 feet, from the dock edge and from the wall of the office where carrier personnel report. Place signs reading "Restricted Area—Authorized Personnel Only" along the perimeter line facing the dock and office. Make sure that shipping and receiving areas are well lighted. Use lighting with a foot candlepower level of 50–60, if possible. Include floodlights to light the interior of railcars and truck vans.
2. Keep all cargo doors closed when not loading or unloading.
3. Do not allow cargo being loaded or unloaded to remain in the operating area between dock front and perimeter line or, generally, in close proximity to railcars or trucks.
4. Provide a secure room in the shipping and receiving area for control of sensitive or high-risk cargo during the shipping or receiving process. Limit access to this room and exercise tight control over movement of such cargo to the carrier or from the carrier to storage.
5. Maintain strict control and accountability for all keys to locked areas, security rooms, and containers.
6. Store seals securely within the office area. Limit issuance of seals to a select few employees. Maintain accurate records of all seals issued.

Strong measures of physical security are essential, and shippers and receivers run serious risk of cargo theft if they neglect them (see Appendix 22a). Close personnel control and supervision are even more important. They provide the key to high-level security for shipping and receiving operations. Good personnel control begins with preemployment screening. Employers should make the screening process as thorough as possible. It should include a check with previous employers and other available sources. After hiring, employers may find it advisable to take the following precautions:

1. Require identification badges with photographs for every employee.
2. Maintain an up-to-date signature file or other verification system for all employees authorized to sign receipts and other shipping documents.
3. Require employees to enter and leave the premises through a single personnel door or gate. Prohibit access of employees' vehicles to the cargo area.
4. Maintain controls such as special passes for employees who leave the facility during duty hours.
5. Limit to a select few the employees authorized to process receipts for shipments of sensitive and high-value items.
6. Require identification badges for all persons visiting the facility.
7. Indoctrinate all personnel to challenge persons moving about the facility who are not accompanied by an employee.
8. Remind all employees periodically of the penalties for theft, including loss of jobs, and the

possible impact of serious theft from the company.

Good supervision will do much to improve cargo security at the shipper's and receiver's premises (see Appendix 22b). The shipper and receiver may find it useful to instruct company supervisors to observe the following practices:

1. Make their presence conspicuous in the shipping and receiving area, continually overseeing cargo-handling operations.
2. Make frequent checks of the quantities of inbound and outbound cargo being handled. Make occasional unannounced spot audits, especially of loading operations.
3. Rotate cargo-handling personnel among different carriers, where feasible.
4. Rotate cargo checkers and handlers on different work cycles, where possible.
5. Prevent nonemployees from assisting in shipping and receiving operations.
6. Report any suspicious activity to the security office.

Documentation

Much cargo loss can be traced to poor documentation practices attributable to the volume of paperwork necessary to control both large and small shipments. Nevertheless, documentation is one of management's most important methods for controlling cargo handling and combating the increasing sophistication of thieves and pilferers. To reduce cargo theft, shippers and receivers should be familiar with the security hazards associated with documentation, and the ways in which documentation practices can be improved to increase cargo security.

The purpose of documentation is simple. Documentation gives the characteristics of the cargo, tells how, when, and where a shipment is to move, provides accountability, and forms a basis for carrying out the financing of individual shipments. Documentation, while essential to the movement of cargo, tends by its very nature to expose cargo to risk of theft. It creates serious security hazards which can be minimized only by stringent control measures.

Accessibility

Many persons have almost unlimited access to the detailed information required to document cargo shipments. The origin, carrier, route, destination, description of commodities, weight cube, value, and time of shipment can be obtained from various shipping documents. The large number of document-processing points associated with each shipment magnifies the problem. Because of easy access to detailed information, shipments can easily be set up for theft and pilferage by company or carrier employees or even by outsiders.

Errors

Errors in documentation constitute a security hazard because they lead to cargo theft. The large number of subsidiary documents which duplicate data found on source documents repeatedly contain errors in transcription. Consequently, shipments are misrouted, delayed, or frustrated. The cargo often ends up lying unattended and unaccounted for in terminals, piers, or warehouses where it becomes a target for thieves, because the longer cargo is delayed at any given point along its route the more susceptible it is to theft.

Falsification

Company and carrier employees can falsify documents and steal cargo without the theft being readily detected. This is particularly true when large volumes of cargo are being shipped and supervisors are lax in their duties. The spot audit is a useful safeguard against falsified documents.

Late Submission

Failure of shippers and receivers to forward documents promptly to the necessary parties delays the processing of cargo, and cargo delayed in transit is extremely vulnerable to theft.

There are a number of practical measures that shippers and receivers may find it advisable to take if they wish to reduce theft and pilferage caused by poor documentation practices:

1. Maintain a continuing review of the company's documentation procedures and change them where necessary to improve cargo security.
2. Analyze records of incidents and losses to determine causes and direct necessary corrective action.

3. Limit access to documentation. Maintain strict controls over the storage and distribution of invoices, shipping orders, manifests, and other vital materials.
4. Avoid wide dissemination of documents. Do not make excess copies of documents. Discourage transmission of information by telephone or telegram, particularly concerning high-value items.
5. Transcribe information carefully from source documents to subsidiary documents, and check the transcription closely.
6. Forward shipping documents promptly, especially those needed at foreign destinations.
7. Limit the number of persons having knowledge of cargo shipments.
8. Maintain strict control of information on the shipment's routing, time of dispatch, time of arrival, and carrier.
9. Give security personnel advance notice of cargo requiring surveillance and protection. Inform the carrier when special protective services are required on its part.
10. Ensure that documentation procedures provide for a clear audit trail of all cargo shipped and received. Require cargo checkers to use self-inking identification stamps to facilitate audit by overcoming the problem of illegible signatures on receipts. Maintain close control of such stamps.
11. Inspect cargo immediately upon receipt. Note shortages on receipt documents and notify carrier. If theft is suspected, notify the proper law enforcement authorities.
12. Enclose a packing list in each shipping container or package, if practicable, to facilitate prompt and complete survey if pilferage occurs. Do not put the packing list on the outside of the container or package.
13. Select documentation employees carefully, ensure thorough training, including instruction in the security hazards of documentation, and require close supervision.

Receipt of Shipments

The warehouse is particularly vulnerable to theft and pilferage. In addition to security of the physical area and effective control of personnel, however, other safeguards can improve security. An inbound register should be maintained, recording carrier name, commodity, quantity, time of arrival, time of departure, driver's name, truck van or railcar number, and any note of discrepancies. In addition to the inbound register, the receiver may find it advantageous to maintain special control procedures for trucks:

1. Refuse to spot a truck at the dock until receiving personnel are available. A receiving clerk or warehouse worker should meet the driver and request the bills of lading or way bills. He should examine these documents against the purchase order to assure legibility, authenticity, and completeness. All discrepancies should be corrected immediately or the cargo refused.
2. Check that driver has a valid gate pass and retain it until ready to sign the release. Note the date and hour of release on the pass so that gate security personnel can ascertain normal time lags in travel from the receiving area to the gate.
3. Examine door seals, if called for by the documentation. If the numbers of the seals are at variance with those on the documents, notify the supervisor immediately so that she or he can notify the shipper and the receiver's security office.
4. Assure that the receiving clerk safeguards the documents through the checking, unloading, and storage process, taking care not to deface or lose them. Require the receiving clerk's signature on the documentation, with date and time information.
5. If a full truckload of the same cargo is moved to a location other than the receiving dock for unloading, do not break the seals until the truck is ready for unloading. Have an employee of the receiver facility accompany the truck while moving from the receiving area to the unloading point.
6. If a truck cannot be fully unloaded before the close of business, close all doors with suitable locking devices until the next work day.
7. Assure that the supervisor checks the interior of the truck after unloading and before the receiving clerk goes to the next vehicle. Then require the driver to depart immediately.

Similarly, the receiver may wish to maintain special control procedures for railcars:

1. Assure that the documents are legible, complete and authentic before issuing the cargo receipt. Require immediate correction of any discrepancies.
2. Inspect the seals on loaded railcars and check against the numbers recorded on the documents. Where possible, assign an employee of the traffic section to check the seals as the cars are spotted

by the rail carrier. Immediately notify the carrier of discrepancies.
3. Where discrepancies are found in the seals, make a physical check of the contents of the railcar, with a representative of the rail carrier present. Notify the receiver's security office.
4. If a railcar is spotted in a holding area within the facility and later moved to the unloading area by the receiver's equipment, lock the railcar while in the holding area, using the receiver's locking devices.
5. Assure that the supervisor checks the interior of the railcar after unloading and before the receiving clerk goes to the next railcar. Then close the railcar doors immediately.

The security of sensitive or high-value cargo or other goods requiring storage in security rooms calls for additional safeguards. They should be moved directly from truck or railcar to the security room, and the receiving clerk should require a signed receipt from the security room employees who receive the cargo.

Through lack of supervision, control, and checks during the cargo receiving process, the thief or pilferer gains opportunity. A large receiving facility is a complex and active place with much movement of merchandise, personnel, and equipment. Unless the devices of the thief are known and adequate controls are maintained, theft is bound to occur. An informed knowledge of the methods of the thief will help the receiver to adopt and apply more effective controls. Some examples follow.

The Partially Emptied Carton

Opening a carton, removing part of the contents, filling the void with some type of waste material, and resealing the carton hides a theft unless someone handling the carton notices the weight differential. The carton, however, is not always resealed. Therefore, the receiving clerk should immediately check the contents and reseal any open cartons. A strange sound from a carton should also alert the receiving clerk. Substitutes such as rocks and bottles that can create sound have been found. If possible, make frequent spot checks at receiving platforms by opening a small percentage of cartons not otherwise subject to loss or damage notations.

Lunch and Break Periods

Leaving the receiving area unguarded and cargo doors open during lunch and break periods is an invitation to the thief to load checked cargo back on the truck. Cargo doors should be closed and at least one employee should remain on duty in the receiving area during these periods.

Pallet Patterns

After pallet pattern and count have been established, a thief who knows pallet patterns can easily change the pattern by putting one less carton per layer on the pallet. Visible evidence of this change is minimal, and several cartons can be extracted from a full truck or carload. These cartons are difficult to see when slipped into dark confines of the truck, van or railcar. The receiver should periodically check the pallet load count and inspect the inside of the van or railcar before signing the gate pass or closing the car doors.

The Falsified Document

Large quantities of cargo are lost through theft in receiving activities by employees who falsify the quantity received and leave merchandise in the truck or railcar to be hauled out of the terminal. A receiving clerk in collusion with a driver or a rail employee can steal much cargo simply by leaving the quantity he desires in the truck or railcar and certifying on the documentation that the full quantity was received. Two simple procedures will do much to control this problem:

1. Require each receiving employee to call the supervisor for a final check of the truck or railcar to assure that no merchandise is left in the vehicle. If dock lights are insufficient to illuminate the interior of vans and railcars, the supervisor should use a flashlight or lantern. He or she should assure that merchandise is not concealed under trash or in the cab of the truck. In damage-free railcars, the supervisor should check behind the bulkhead doors.
2. Take a 100 percent inventory periodically of the stored receipt counts of each receiving clerk for one day's receipts.

Trash

Dropping valuable items into trash bins and later retrieving them from the dump is a common practice. In some cases, the employee has an agreement will the dump attendants. Alert observations of employees by supervisors will do much to prevent this type of theft, but also it is desirable to apply trash collection and removal controls, such as establishing timetables for placing trash in bins. Crushing or shredding the contents of trash receptacles prior to removal from the premises will discourage thieves from depositing stolen items in refuse.

Trailers

Unattended trailers and trailers remaining overnight at a receiving point offer an excellent means for the illegal removal of merchandise. Failure of supervisors to make at least one check of each trailer (during unloading and when empty before releasing it) can result in the loss of an entire trailer load. Simply by signing the documents and the cargo receipt, a receiving employee may be able to steal an entire trailer without bringing it to the receiving area. When supervisors insist on checking each carrier vehicle before release, this type of theft can be prevented. When trailers are left unattended by receiving personnel, the driver has a blank check for loading cargo. Lock or seal empty and unattended trailers remaining overnight, weekends or holidays to prevent them from being filled with stolen goods.

Pallet Exchange

If collusion exists between the receiving employee and the driver, empty pallets can be quickly loaded into the van to conceal cargo. Falsifying the cargo receipt showing that the entire quantity was received will cover up the theft. Close supervision is the only answer.

Serious Oversight

A few cartons used to support the end of an unloading conveyor can be conveniently left on the truck. If noticed by receiving personnel, the driver claims that it was an oversight. Checking carrier vehicles before release can prevent this type of theft.

Seal Switches

If seals are left in a readily accessible place, carrier personnel can steal them, break the existing seal, and substitute a stolen one after going to another area and loading illicit cargo before going to the perimeter gate. There are cases where a receiving employee has furnished to the driver in collusion a second seal, the number of which has been entered on the gate pass. This seal is placed on the door of the vehicle after illicit cargo has been loaded elsewhere.

Packaging, Pallets, and Containers

Shipping goods in the right kind of package or container is essential to good cargo security. Good shipping practice requires that packaging be suitable to protect cargo against all hazards of transportation, and theft has become a serious hazard. Inadequate packaging is not only a major reason why shipments arrive damaged, but weak or damaged packaging invites pilferage by making it easy to steal the goods. Most often goods are stolen from broken packing cases in small quantities either in the warehouse or during cargo-handling operations.

There are a number of precautions that shippers and receivers may take to curb losses by increasing difficulty and risk for the thief or pilferer:

1. Select shipping containers strong enough to protect the load from damage and to hold together without breaking open under stress or rough handling.
2. Select shipping containers that are so difficult to breach and close so tightly that the thief must destroy the case to get at its contents.
3. Ensure that the shipping container has effective closing devices. Make sure that all flaps of fiberboard containers are fully closed and reinforced with 3-inch paper tape.
4. Reinforce heavy shipments with strap, normally applied girthwise.
5. Examine wooden containers to assure that all nails are driven home and reinforced strapping used.
6. Avoid using second-hand packaging materials where possible. Marks of previous nails and straps and obliterated addresses make it difficult to determine on inspection whether pilferage has occurred.
7. Adopt the unitized load principle through use of pallets and van containers to the greatest extent feasible.

The object of the unit load principle is to give greater speed, security, flexibility, and economy to cargo movement by consolidating small packages into unitized loads of optimum size for the use of mechanical cargo-handling equipment. The main applications of the unit load principle are the pallet and the van container.

The pallet affords security to cargo by assembling small packages in units held to the pallet by heavy strapping and various forms of covering. Consolidation into a palletized unit reduces the danger of loss or pilferage of individual packages. To reach a package, the thief ordinarily must cut the strapping or slash the covering or otherwise leave visible evdience of having disturbed the load. There are several variant forms of palletization:

1. Consolidation of packages on a standard wood, plastic, or fiberboard pallet, secured by strapping.
2. Consolidation of packages into a fiberboard container sealed and strapped to a skid.
3. Consolidation of packages onto a standard pallet, with the load secured by transparent shrink wrapping—a heavy plastic coating formed to the shape of the load by application of heat and by consequent shrinkage upon cooling.

The van container normally offers an even greater measure of security to cargo. Its security attributes are substantial because:

1. It usually is constructed of steel, aluminum, or plywood; it is difficult to breach without leaving visible marks.
2. It can be padlocked and sealed, affording security equal to that of truck trailers and railcars. If seals of good quality and design are used, their removal is difficult without signs of tampering.
3. Entire pallets can be accommodated in the van container. This materially increases the degree of protection against pilferage afforded to the goods.
4. Containers are adapted to intermodal movement and can be transported door-to-door with a minimum of delay and opening for inspection of the contents. Each opening thus eliminated reduces opportunities for theft.

Experience has shown that theft or pilferage is most likely when containers are open at the shipping and receiving dock. This calls for the same security procedures as to personnel, documentation and physical controls as in the case of trucks. Pilferage while containers are en route is most likely if the container has been damaged or is not properly maintained. Proper attention to sealing, with use of reliable, tamper-proof seals and maintenance of accurate seal records, including noting seal numbers and the reasons for any breaking of seals, will do much to reduce the incidence of pilferage en route. Good maintenance practices are the first line of defense against such pilferage, and the shipper should be careful to reject a damaged or defective container if it is offered for the transportation of cargo.

Liability and Claims

General

In arranging for the movement of cargo, the shipper or receiver must be prepared to accept the fact that on occasion even good security measures may fail, and theft or pilferage may occur. It is at this point, after the harm is done, that liability and claims procedures come into play. Knowledge of the rules of carrier liability and of claim procedures will prove indispensable to the shipper or receiver if she or he is to obtain indemnity. The fact of loss, however, should give shippers and receivers an incentive to go beyond indemnification and consider how to prevent recurrence.

From the standpoint of cargo security, the most important aspect of liability and claims is claims prevention. Every safeguard the shipper or receiver adopts in an act of claims prevention. Moreover, the shipper or receiver can obtain professional help in taking preventive measures on a systematic basis. Many insurance companies have claims prevention specialists who can survey a shipper or receiver's operations and recommend ways in which greater security can be attained. Use of consultative services of this kind is a prudent course of action even for the shipper or receiver with relatively good loss experience.

Carrier Liability for Loss to Cargo

When loss occurs through theft or pilferage, liability of the shipper or receiver depends on variable elements. Since there are no uniform rules of carrier liability, the presence and amount of the shipper's indemnity depend in large measure on the mode of transport used to ship cargo. Each mode of transport operates under differing rules with wide variations in limits of liability and the defenses available to the carrier. For the informed shipper or receiver, these variations among modes of transport determine whether the receiver relies on the carrier's liability or obtains additional protection through cargo insurance.

The present monetary limits of carrier liability should be known by the user and confirmed with the particular carrier or modes of transportation to be used.

Surface transportation in foreign countries usually is subject to local law. Most countries in Western Europe, however, are parties to international conventions which limit the liability of highway carriers.

In air transportation the shipper may obtain an increase in the carrier's limit of liability by paying an additional charge. The same is usually possible in water transportation by agreement between the shipper and carrier.

These monetary limits on carrier liability offer only partial assurance of indemnity to the shipper in view of the numerous defenses available to the carrier. Commonly, the fault of the shipper or owner of the goods is a valid defense. Thus, for example, defective packing frequently exonerates the carrier. The low limits of carrier liability and uncertainty as to recovery have led shippers as a common practice to obtain cargo insurance on goods they ship by air or water carrier. For shipments moving by rail or motor carrier, the need to obtain such insurance may be less compelling, but the shipper may find it prudent to consult an underwriter, especially if the shipment is of high value or would be unduly exposed to theft or pilferage. Additional assistance may be obtained from the carrier's claims prevention section.

Claims Procedures

When loss from theft or pilferage occurs, the shipper or receiver should take care to observe the requisite claims procedures, especially as to notification of loss to the carrier, the formal filing of claims, and, if need be, the institution of suit. As a matter of law or under the contract of carriage, these actions ordinarily must be taken within specified time limits (see Appendix 4a). Failure to comply may foreclose the shipper or receiver's right to indemnity, regardless of the merits of his claim. (C.F.R. 49 Part 1005 Appendix 2–1a).

Claims Prevention

From a practical standpoint, claims prevention is the most important element of liability and claims. The shipper or receiver must have a reasonable assurance of indemnity if goods are pilfered or stolen, but it is much more beneficial to reduce theft or pilferage to a minimum. The best way is through systematic claims prevention measures. Many of the measures recommended elsewhere in this booklet are in actuality basic claims prevention measures, for they aim to remove opportunities for theft or pilferage and thus reduce the incidence of claims. These recommendations deserve the most serious consideration.

It is desirable to go about a claims prevention program in a systematic way. Even if loss experience is relatively low, the shipper or receiver usually will find it an advantage to make a thorough survey of the operations from beginning to end with the specific objective of locating weak spots. Such a survey should cover all parts of the operation: documentation, communications, procedures, personnel, and physical facilities. One method could be to trace a variety of individual shipments step by step. Another might be a concentrated examination of each single element such as documentation or physical security.

Some shippers or receivers have the capability of making such surveys with their own staff. If a shipper or receiver lacks this capability or wishes to have the objectivity of an outsider, professional help is available through an insurance underwriter or broker.

Many will be able to conduct security surveys with their own staff. Others will prefer to call in specialists from the particular insurance company or association with which they are affiliated.

A single survey, while useful, probably would not suffice in most instances. The prudent course might be to repeat at suitable intervals. In this respect the shipper or receiver could be guided by loss experience. It would be a sound practice, however, to order a security survey whenever an important new operation is being set up, as, for example, when a firm is entering the export business for the first time.

In addition to such surveys, shippers and receivers may find it useful to review their claim files periodically to see if there is a pattern with respect to the number and types of claims resulting from the operations of other shippers and receivers with whom they do business. Examples would be repeated instances of faulty packing or count discrepancies.

Cargo Security Prevention Standards

Buildings

All terminal buildings housing cargo should be constructed of a material that will deter unlawful entry.

Ground floor windows should be steel barred. Bars should be spaced at intervals of not more than 6 inches and set in a steel frame which is securely affixed to the structure. All windows of buildings which adjoin other structures should also be barred.

Delivery and receiving doors should be constructed of a material that will deter unlawful entry and should be equipped with a self-locking device that will engage immediately when the door is closed. All delivery and receiving doors should remain closed and locked when not actually in use.

Pedestrian doors should be capable of being locked with a substantial lock. Workforce facilities such as rest rooms, locker rooms, and eating or lounging areas should be separated from the area in which cargo is stored.

Maximum security cribs, constructed of a substantial material which deters entry on all four sides, overhead, and from the floor, and provides adequate space for storage of high-value cargo, should be incorporated in each terminal building. They should be constructed in an area that is visible to management or security personnel or under frequent surveillance by the security patrol. (Only designated personnel should be allowed to enter this area; it may be desirable to secure it with a double lock requiring keys held by two different persons.)

The office in which delivery and pick-up orders are processed should be an area to which only authorized personnel have access. Room arrangements should be such that documents being processed are not available to or observable by unauthorized persons.

Fencing and Gates

Whenever possible, perimeter fencing should be provided around terminals in which merchandise is stored. The number of entrances and exits should be held to a minimum.

Fencing should normally be of the chain-link type, maximum 2-inch mesh, at 9 gauge, and no less than 6 feet in height surmounted by an additional 2 feet of barbed wire. Where installed in soft earth, the bottom of the fencing should be anchored below grade and then backfilled. Where installed over concrete, the bottom of the fence should be no more than 2 inches above the surface, but high enough that, when sagging, it does not touch the surface. The top of the fence should be surmounted by at least three strands of barbed wire occupying no less than 2 feet vertically, but positioned at a 45-degree angle to the vertical.

Gates should be constructed of the same chain link fence material, be surmounted by an additional 2 feet of barbed wire, be within 2 inches of hard ground or paving, and be capable of being locked.

The fence line should be maintained in good condition and provisions made to avoid bumping and distortion of the fence by motor vehicles and other equipment. Fence lines should be kept free of shrubbery and other objects impeding the line of sight.

Gatehouses

Gatehouses should be self-contained units, equipped with at least two modes of communications for assured redundancy. These may be commercial telephone, radio, private telephone, or alarms. The exist gatehouse should be set back from the gate so that exiting vehicles can be stopped and examined on terminal property without the gate being opened.

Adequate lighting must be provided in the area of the gatehouse so that documents and identifying features, and contents of incoming and outgoing vehicles, can be examined by the guard.

The area around the gatehouse should be free of encumbrances that restrict the guard's line of vision. Procedural signs advising drivers and visitors of the conditions of entry should be prominently displayed on the exterior of all entry gatehouses, preferably where they can be read by drivers before turning off the public street or road into the approach to the gate.

When warranted, photo recorders should be installed in gatehouses to provide a photo record of persons and their documents entering the terminal facility.

Locks and Key Control

All padlocks should be of a single standard type for control purposes. The use of other than the approved type is then easily determined. When possible, the base of the padlock should indicate the company name. These padlocks should have multiple pin tumblers (at least six), interchangeable cores, a minimum tension pull resistance of 6,000 pounds on the shank portion, and shrouded shackle.

The use of nonstandard padlocks should be prohibited. All lockers, gear boxes, and cooper and/or carpenter shanties should be locked only with padlocks as described above.

No gates or exit areas should be secured by padlocks with the use of chains. All gates and exits should have proper latches which are secured to the surface with nonexposed bolts. The area surrounding the lock, particularly on the exposed side, should have metal backing to prevent accessibility.

The use of electrical switch locks is recommended in gate and exil areas. These permit use of a microswitch to record and indicate the opening of a given lock on a panel or central control indicator.

Key control should be rigid. For every key given out, a signature should be required and a card file maintained which would indicate the history of each key. Duplicate keys should be secured under absolute control. The distribution of submaster, master or grandmaster keys should be highly restricted. The key control of all equipment, particularly that which could be used in the commission of a theft, such as hi-los, stackers and yard tractors, should be very strict. No keys should be left in any equipment overnight or when not assigned. This also should apply to equipment held in the maintenance or repair shops.

Alarm and Communications Systems

All terminal buildings in which merchandise in the trasportation scheme is moving should be equipped with an intrusion detection system. However, alarm systems should be considered as an *adjunct* to fencing, lighting, or guard forces, not in lieu of.

Circuit boards, lines, and control panels for alarm systems should be placed in such locations and constructed in such a manner that they are protected from vandalism or deliberate attempts to disrupt or destroy their usefulness.

There are various recognized alarm systems; the type that meets the needs of a particular facility can best be determined by a qualified security specialist.

The three most popular alarm systems are:

1. The *audible alarm*, which, when activated, draws attention to the facility or a portion thereof.
2. The *silent alarm*, which is designed to alert private guard forces or municipal police departments. When considering this system, managers should determine the time required for the guard force to respond to the alert.
3. The *visible alarm*, which increases light intensity in a given area by utilizing additional lights, by explosive flares, or by rockets.

Lighting

Adequate lighting should be provided between dusk and dawn at all entrances and exits, along all boundary lines, around all storage structures, and in all parking areas.

The primary power source at a facility is usually a local public utility. However, to protect against a public power failure, an alternate source of power adequate to provide lights at key control points should be provided.

A gasoline-driven generator that starts automatically upon the failure of outside power is the preferred alternate, although battery power may also be used.

Tables 2–1 and 4–2 provide standards for area coverage and lighting intensity.

Guard Requirements

Standards for the guard force should include complete background investigations and physical and mental examinations to meet established standards.

The guard should be physically capable of vigorous physical training in self-defense.

Uniforms should be distinctive and complete. Uniforms and equipment should meet high standards of cleanliness and maintenance.

Prior to any duty assignment, basic training must be given to all guards. Training should include such techniques as patrol (both mounted and dismounted), report writing, log and recordkeeping, use of security equipment, fire and safety regulations, self-defense, crowd and riot control, and firearms instruction.

Under most conditions connected with transportation and distribution systems, guards should not be allowed to remain too long in a given post; they should be rotated periodically. However, under certain conditions, keeping the guard at one post could be an asset, especially when personal recognition of people is desired. Examples are personnel entrances to an office building, tradesworkers' gates, monitoring stations, areas where recognition and protocal are required, and administrative posts.

Vehicle Control

Movements of all vehicles within the terminal should be strictly controlled. The controls must be a stan-

Table 4–1. Lighting Area Coverage

Type of area	Type of Lighting	Width of lighted strip (in feet) Inside Fence	Outside Fence
Isolated perimeter	Glare	25	200
Isolated perimeter	Controlled	10	70
Semiisolated perimeter	Controlled	10	70
Nonisolated perimeter	Controlled	20–30 (Total width from structure)	30–40
Vehicle entrance	Controlled	50	50
Pedestrian entrance	Controlled	25	25
Railroad entrances	Controlled	50	50
Vital structures	Controlled	50	

Table 4–2. Lighting Intensity

Location	Foot Candles on Horizontal Plane at Ground Level
Vehicular and pedestrian entrances	2.0
Vital structure and other sensitive areas	2.0
Unattended outdoor parking area	1.0

dard operating procedure and prominently posted.

All trucks and conveyances entering the facility should be examined by gate guards. Guard checks should include registration of truck, name of driver and helper, license check of vehicle operator, and name of driver or helper. If possible, a photo record of the driver and documents should be made before entering the terminal. Strict supervision of loading and unloading operations should be incorporated into procedures. No vehicle should be opened except when actually being loaded, unloaded, or inspected.

Other procedures for vehicle control and operations should include a record on all inbound and outbound trailers and containers and, when not being moved, their storage locations, and a report of any seal discrepancies noted.

Parking lots for personal vehicles of employees and visitors should be located outside and separated from freight handling areas. Only cargo-carrying or handling vehicles should be permitted in controlled areas.

It is advisable to separate entry and exit roadways and gates if the facility permits. Flow of traffic, then, is essentially one-way.

Special Problems—High-Value Cargo

Transportation of merchandise such as fissile radioactive materials, jewelry, furs, optical goods, cameras, electronic articles, whisky, and ammunition present special security considerations. Precautions include scheduling arrival of goods at hours which permit prompt pickup, use of secure areas, continuous guards, selected reliable drivers, delivery in vehicles with two-way radio equipment, a checkpoint system, convoying and *shotgun* guards, use of antihijack equipment (locked doors, no steps, high cab) and coordination with federal and local police agencies.

All protective measures for movement of high-value merchandise must be set out in a standard operating procedure.

Only that equipment which is sound should be utilized in transport of high-value cargo.

When regular accounts requiring regular handling of high-value cargo are acquired, modifications to equipment should be made. This could include removal of running boards, addition of extra locking devices, alarm equipment, and oversize gas tanks (where permissible). Routing of vehicles should vary continuously where practicable.

Key control for all equipment undergoing maintenance should be stressed. Often, a breakdown in key control takes place when equipment is turned over to the maintenance department for service. When equipment breaks down while on the road, a preplanned procedure should be effected to safeguard high-value cargo.

The shipper can be advised as to how to avoid security pitfalls, whether in packaging or advertising. He or she must carefully control information regarding shipments, destinations, times of arrival and departure and routes to be traveled among employees, and on labels and packaging. Further, the consignee must plan for immediate pickup at destinations to avoid high-value cargo being needlessly exposed.

Appendix 4a
Cargo Security Checklist

The checklist below may be used for many different types of facilities. It permits each facility manager to select those elements pertaining to an establishment and location in making one's own security survey.

Barriers

1. Is the perimeter of the facility or activity defined by a fence or other type physical barrier?

2. If a fence or gate is used, does it meet the minimum specifications?
   a. Is the top guard strung with barbed wire and angled outward and upward at a 45 degree angle?
   b. Is it at least 10 feet total height?
   c. Is it located so that it is not adjacent to mounds, piers, docks, or any other aid to surmounting it?
3. If building walls, floors, and roofs form a part of the perimeter barrier, do they provide security equivalent at least to that provided by chain link fence? Are all openings properly secured?
4. If a masonry wall or building forms a part of the perimeter barrier, does it meet minimum specifications of perimeter fencing?
5. If a river, lake, or other body of water forms any part of the perimeter barrier, are security measures equal to the deterrence of the 10-foot fence provided?
6. Are there openings such as culverts, tunnels, manholes for sewers and utility access, and sidewalk elevators which permit access to the facility properly secured?
7. List number, location, and physical characteristics of perimeter entrances.
8. Are all portals in perimeter barriers guarded, secured, or under constant surveillance?
9. Are all perimeter entrances equipped with secure locking devices and are they always locked when not in active use?
10. Are gate and/or other perimeter entrances which are not in active use frequently inspected by guards or management personnel?
11. Is the security officer responsible for security of keys to perimeter entrances? If not, which individual is responsible?
12. Are keys to perimeter entrances issued to other than facility personnel, such as clearing, trash removal, vending machine service personnel?
13. Are all normally used pedestrian and vehicle gates effectively and adequately lighted so as to ensure:
    a. proper identification of individuals and examining of credentials
    b. that interiors of vehicles are clearly lighted
    c. that glare from luminaries is not in guard's eyes
14. Are appropriate signs setting forth the provisions for entry conspicuously posted at all principal entrances?
15. Are clear zones maintained for the largest vehicles on both sides of the perimeter barrier? If clear zone requirements cannot be met, what additional security measures have been implemented?
16. Are automobiles permitted to park against or too close to perimeter barrier?
17. What is frequency of checks made by maintenance crews of condition of perimeter barriers?
18. Do guards patrol perimeter areas?
19. Are reports of inadequate perimeter security immediately acted upon and the necessary repairs effected?
20. Are perimeters protected by intrusion alarm devices?
21. Does any new construction require installation of additional perimeter barriers or additional perimeter lighting?

Lighting

1. Is the perimeter of the installation protected by adequate lighting?
2. Are the cones of illumination from lamps directed downward and away from the facility proper and away from guard personnel?
3. Are lights mounted to provide a strip of light both inside and outside the fence?
4. Are lights checked periodically for proper operation and inoperative lamps replaced immediately?
5. Do light beams overlap to provide coverage in case a bulb burns out?
6. Is additional lighting provided at vulnerable or sensitive areas?
7. Are gate guard boxes provided with proper illumination?
8. Are light finishes or stripes used on lower parts of buildings and structures to aid guard observation?
9. Does the facility have a dependable auxiliary source of power?
10. Is there alternate power for the lighting system independent of the plant lighting or the power system?
11. Is the power supply for lights adequately protected? How?
12. Is the standby or emergency equipment tested periodically?
13. Is emergency equipment designed to go into operation automatically when needed?
14. Is wiring tested and inspected periodically to ensure proper operation?
15. Are multiple circuits used? If so, are proper switching arrangements provided?
16. Is wiring for protective lighting securely

mounted?
- a. Is it in tamper-resistant conduits?
- b. Is it mounted underground?
- c. If above ground, is it high enough to reduce possibility of tampering?

17. Are switches and control properly located, controlled, and protected?
    - a. Are they weatherproof and tamper resistant?
    - b. Are they readily accessible to security personnel?
    - c. Are they located so that they are inaccessible from outside the perimeter barrier?
    - d. Is there a centrally located switch to control protective lighting? Is it vulnerable?
18. Are the lighting system designed and location recorded so that repairs can be made rapidly in an emergency?
19. Is adequate lighting for guard use provided on indoor routes?
20. Are materials and equipment in shipping and storage areas properly arranged to permit adequate lighting?
21. If bodies of water form a part of the perimeter, does the lighting conform to other perimeter lighting standards?

Alarms

1. Is an alarm system used in the facility?
   - a. Does the system indicate an alert only within the facility?
   - b. Does it signal in a central station outside the facility?
   - c. Is it connected to facility guard headquarters?
   - d. Is it connected directly to an enforcement headquarters outside the facility proper? Is it a private protection service? Police station? Fire station?
2. Is there any inherent weakness in the system itself?
3. Is the system supported by properly trained, alert guards?
4. Is the alarm system for operating areas turned off during working hours?
5. Is the system tested prior to activating it for nonoperational periods?
6. Is the alarm system inspected regularly?
7. Is the system tamper resistant? Weatherproof?
8. Is an alternate alarm system provided for use in the event of failure of the primary system?
9. Is an alternate or independent source of power available for use in the event of power failure?
10. Is the emergency power source designed to cut in and operate automatically?
11. Is the alarm system properly maintained by trained personnel?
12. Are periodic tests conducted frequently to determine the adequacy of response to alarm signals?
13. Are records kept of all alarm signals received to include time, date, location, action taken, and cause for alarm?

Communications

1. Is the security communication system adequate?
2. What means of communications are used?
   - a. Telephone
     - (1) Is it a commercial switchboard system? Independent switchboard?
     - (2) Is it restricted for guard use only?
     - (3) Are switchboards adequately guarded?
     - (4) Are there enough call boxes and are they conveniently located?
     - (5) Are open wires, terminal boxes, and cables frequently inspected for damage, wear, sabotage, and wire-tapping?
     - (6) Are personnel cautioned about discussing cargo movements over the telephone?
   - b. Radio
     - (1) Is proper radio procedure practiced?
     - (2) Is an effective routine code being used?
     - (3) Is proper authentication required?
     - (4) Is the equipment maintained properly?
   - c. Messenger—Is the messenger always available?
   - d. Teletype—Is an operator available at all times?
   - e. Public address
     - (1) Does it work?
     - (2) Can it be heard?
   - f. Visual signals
     - (1) Do all guards know the signals?
     - (2) Can they be seen?
3. Is security communications equipment in use capable of transmitting instructions to all key posts simultaneously?
4. Does the equipment in use allow a guard to communicate with guard headquarters with minimum delay?
5. Is there more than one system of security communications available for exclusive use of security personnel?
6. Does one of these systems have an alternate or

independent source of power?
7. Has the communications center been provided with adequate physical security safeguards?

Personnel Identification and Control

1. Is an identification card or badge used to identify all personnel within the confines of the controlled areas?
2. Is the identification medium designed to provide the desired degree of security?
3. Does the identification and control system include arrangements for the following:
   a. Protection of the meaning of coded or printed components of badges and passes
   b. Designation of the various areas requiring special control measures to which the badge holder may be authorized entrance
   c. Strict control of identification data
   d. Clear explanation and description of the identification data used
   e. A clear statement of the authorization and limitations placed upon the bearer
   f. Details of where, when, and how badges shall be worn
   g. Procedures to be followed in case of loss or damage to identification media
   h. Procedure for recovery and invalidation
4. If a badge exchange system is used for any restricted area, does the system provide for:
   a. Comparison of badge, pass, and personnel
   b. Physical exchange of restricted area badge for general authorization badge at time of entrance and exit
   c. Logging a record of each badge exchanged
   d. Inventory of badges issued by security personnel at the start and completion of tours of duty
   e. Location of personnel who have not checked out of the area at the close of each tour of duty
   f. Security of badges not in use
5. Are messengers who are required to traverse areas of varying degrees of security provided with special identification
6. Are the prescribed standards for access to exclusion areas supplemented with arrangements for the following:
   a. At least one representative of management or security is in the area at all times when work is in progress
   b. No other persons are permitted to enter the area until one representative of management or security has entered
   c. A representative of management or security remains until all others have departed
7. Are personnel, who require infrequent access to a critical area and who have not been issued regular security identification for the area, treated as visitors thereto, and issued either
   a. a visitor's badge or pass
   b. a special pass
8. Are all personnel required to wear the security identification badge while on duty?
9. Do guards at control points compare badges to bearers both upon entry and upon exit?
10. Is supervision of personnel charged with checking identification badges sufficient to ensure continuing effectiveness of identification and control system?
11. Are badges recorded and controlled by rigid accountability procedures?
12. Are lost badges with one bearing a different number or one that is otherwise not identical to the lost one?
13. Are procedures relative to lost, damaged, and/or forgotten badges adequate?
14. Are temporary badges used?
15. Are lists of lost badges posted at guard control points?
16. Are badges of such design and appearance as to enable guards and other personnel to recognize quickly and positively the authorizations and limitations applicable to the bearers?
17. How long ago were currently used badges originally issued?
18. Do existing procedures ensure the return of identification badges upon termination of employment?
19. Are badges similar or identical to employee badges issued to outside contractor employees working within the installation?
20. Have local regulations governing identification and control been revised in any material respect since first established?
21. Are all phases of the system under supervision and control of a security officer?
22. Is an effective visitor escort procedure established?
23. Are visitors required to conspicuously display identification on outer garments at all times while on installation?
24. When visitors leave the installation, are they required to turn in their identification badges, and is the departure time in each case recorded on the visitor's register?
25. What procedures are invoked when visitor

identification badges are not turned in prior to departure of the visitor?
26. Is there a central receptionist?
    a. If yes, specify functions.
    b. Are functions performed under the supervision of a security officer?
27. Are receptionists (or guards) stationed at different focal points to maintain visitor control?
28. Are there special procedures applicable to visitors requiring access to cargo handling documents?
29. Are special visitors, e.g., vendors, trades workers, ultility workers or special equipment workers issued a special distinctive type of visitor badge?
30. What measures are employed, other than the issuance of identification badges, to control the movements of personnel from other transportation companies working within the perimeter of the facility?
31. Does the system used for identification of truck drivers and helpers conform with security regulations?
32. Is the security officer the single responsible official for all aspects of visitor control?

Package and Material Control

1. Is there standard procedure on control of packages and materials?
2. Are all guards conversant with the package control measures?
3. Are notices on restriction and control procedures prominently displayed at each active entrance and exit?
4. Is there a checkroom where employees and visitors can leave their packages?
   a. Is an adequate receipt system in effect?
   b. Are packages inspected in the owner's presence before a receipt is issued?
   c. Is access to the checkroom restricted to authorized personnel only?
   d. Is a policy established for disposition of items left beyond a specified period?
5. Are spot checks of persons and vehicles conducted and, if so, are frequency and scope thereof indicated?
   a. Regular search
   b. Spot search
   c. Special search
6. Are detection devices used?
   a. X-ray or other similar device
   b. Metal detector
   c. Other; evaluate effectiveness
7. Is a property removal slip, signed by an authorizing official, required when property is being removed from the facility?
8. Are removal slips available in the security office for signature by officials authorizing property removals?
9. Are property removal slips surrendered to guards at exit points?
10. Are special rules established for package and material handling?
    a. Is package and material pass used to exempt bearer from search?
       (1) Are time, date, bearer's name, using agency, and description of the contents properly recorded thereon?
       (2) Are preparation and issue rigidly controlled?
       (3) Is it serially numbered?
       (4) Does it provide for signature of validating officials?
       (5) Is signature card readily available to guards for comparison?
    b. Is a trustworthy and identified courier used at all times?
11. Are special clothing issued for wear in the facility to prevent the introduction or removal of unauthorized items?
12. Is an effective procedure used for control and search of special vehicles?
    a. Emergency vehicles
    b. VIP vehicles
    c. Special courier vehicles
    d. Vendor's vehicles
    e. Vehicles with loads which are impracticable to search.
13. Is there close coordination between security headquarters and the activities that handle cargo movements?
14. Are new employees given appropriate instructions relative to the handling and safeguarding of cargo?

Vehicle Control

1. Are vehicles which are allowed regular access to the facility registered with the security officer?
2. Have definite procedures been established for the registration of private cars, and are they issued in writing?
3. Do the vehicle registration requirements apply also to motor vehicles owned or operated by employees of any individual, firm, corporation,

or contractor whose business activities require frequent use of vehicles on the facility?
4. Is annual or more frequent registration required?
5. What information is incorporated in registration application forms?
6. Do the prescribed prerequisites for registration include a valid state registration for the vehicle and a valid state operator's license?
7. Is mechanical inspection of vehicles and/or proof of financial responsibility required as a prerequisite of authority to operate a vehicle within the facility?
8. Are decalcomania or metal permit tags affixed to all vehicles authorized to operate within the facility?
9. Do registration permits bear a permanently affixed serial number and numerical designation of year of registration?
10. Do the regulatory control criteria for registration include:
    a. Prohibition against transfer of registration permit tags for use with a vehicle other than the one for which originally issued
    b. Replacement of lost permit tags at the registrant's expense
    c. Return of tags to the security officer when the vehicle is no longer authorized entry into facility
    d. Destruction of invalidated decalcomania or metal tags
11. What is the nature and scope of registration records maintained by the security officer?
12. Do the gate guards make periodic checks to insure that vehicles are operated on the premises only by properly licensed persons?
13. Is a specified system used to control the movement of commercial trucks and other goods conveyances into and out of the installation area?
14. Are loading and unloading platforms located outside the operating areas, separated one from the other, and controlled by guard-supervised entrances?
15. Are all trucks and other conveyances required to enter through service gates staffed by guards?
16. If trucks are permitted direct access to operating areas, are truck drivers and vehicle contents carefully examined?
17. Does the check at entrances cover both incoming and outgoing vehicles?
18. Are truck registers maintained?
19. Are registers maintained on all company vehicles entering and leaving the facility?
20. Are escorts provided when vehicles are permitted access to operating or controlled areas?
21. Does the supervision of loading and unloading operations ensure that unauthorized goods or people do not enter or leave the installation via trucks or other conveyances?
22. Are company trip tickets examined?
23. Is a temporary tag issued to visitor's vehicles?
24. Are automobiles allowed to be parked within operating or controlled areas?
25. Are parking lots provided?
26. Are interior parking areas located away from sensitive points?
27. Are interior parking areas fenced so that occupants of automobiles must pass through a pedestrian gate when entering or leaving the working area?
28. Are separate parking areas provided for visitors' vehicles?
29. What is the extent of guard surveillance over interior parking areas?
30. Are there restrictions against employees entering private vehicle parking areas during duty hours?
31. Are automobiles allowed to park so close to buildings or structures that they would be a fire hazard or obstruct fire fighters?
32. Are automobiles permitted to be parked close to controlled area fences?
33. Are parking facilities adequate?

Lock Security

1. Has a key control officer been appointed?
2. Are the locks and keys to all buildings and entrances controlled by a key control officer?
3. Does the key control officer have overall responsibility for issuance and replacement of locks and keys?
4. Are keys issued only to authorized personnel?
5. Are keys issued to other than facility personnel?
6. Is the removal of keys from the premises prohibited?
7. Are keys not in use secured in a locked, fireproof cabinet?
8. Are current records maintained indicating:
    a. Clear record of person to whom key is issued
    b. Time of issue and return of keys
    c. Buildings and/or entrances for which keys are issued
    d. Number and identification of keys issued

e. Location and number of master keys
   f. Location and number of duplicate keys
   g. Location of locks and keys held in reserve
9. Is a current key control directive in effect and understood?
10. Are locks changed immediately upon loss or theft of keys?
11. Are inventories and inspections conducted by the key control officer to ensure compliance with directives? How often?
12. If master keys are used, are they devoid of marking identifying them as such?
13. Are losses or thefts of keys promptly investigated by the key control personnel?
14. Must all requests for reproduction or duplication of keys be approved by the key control officer?
15. Are locks on inactive gates and storage facilities under seal? Are they checked periodically by guard personnel?
16. Are locks rotated within the installation at least semiannually?
17. Where applicable, is the manufacturer's serial number on combination locks obliterated?
18. Are measures in effect to prevent the unauthorized removal of locks on open cabinets, gates, or buildings?

Guard Forces

1. Is a guard force provided? Is it responsive to management authority?
2. Indicate authorized and actual strength, broken down by positions.
3. Have there been changes since the last survey in either the authorized or actual guard force strength?
4. Is present guard force strength commensurate with the degree of security protection required?
5. Is the use of guard forces reviewed periodically to assure effective and economical use?
6. Is supervisory responsibility for guard force operations vested in the security officer?
7. Is a guard headquarters provided?
8. Does the guard headquarters contain control equipment and instruments of all alarm warning and guard communications systems?
9. Are guards familiar with the communications equipment used?
10. Does the guard headquarters have direct communication with local municipal fire and police headquarters?
11. Do members of the guard force meet the minimum qualifications standards?
12. Are guards armed while on duty? If so, with what type of weapon?
13. Are the weapons kept in arms racks and adequately secured when not in use?
14. Are ammunition supplies properly secured and issued only for authorized purposes?
15. Is each member of the guard force required to complete a course of basic training and take periodic courses of in-service or advanced training?
16. Are the subjects included in the various training courses adequate? Does the training cover:
    a. Hand-to-hand combat
    b. Care and use of weapons
    c. Common forms of pilferage, theft, and sabotage activity
    d. Types of bombs and explosives
    e. Location of hazardous materials and processes
    f. Location and use of fire protective equipment, including sprinkler control valves
    g. Location and operation of all important steam and gas valves and main electrical switches
    h. Conditions which may cause fire and explosions
    i. Location and use of first aid equipment
    j. Duties in the event of fire, explosion, natural disaster, civil disturbance, blackout or air raid
    k. Use of communication systems
    l. Proper methods of search
    m. Observation and description
    n. Patrol work
    o. Supervision of visitors
    p. Preparation of written reports
    q. General and special guard orders
    r. Authority to use force, conduct searches, and make arrests
17. Are periodic examinations conducted to ensure maintenance of guard training standards?
18. Are activities of the guard force in consonance with established policy?
19. Is supervision of the guard force adequate?
20. Are general and special orders properly posted?
21. Are guard orders reviewed at least semiannually to ensure applicability?
22. Are periodic inspections and examinations conducted to determine the degree of understanding and compliance with all guard orders?
23. Do physical, functional, or other changes at

the installation indicate the necessity for, or feasibility of:
   a. establishing additional guard posts
   b. discontinuing any existing posts or patrols
24. Is two-way radio equipment installed on guard patrol cars?
25. Are duties other than those related to security performed by guard personnel?
26. Are guard patrol cars equipped with spotlights?
27. Does each guard on patrol duty carry a flashlight?
28. Do guards record or report their presence at key points in the installation by means of:
   a. portable watch clocks
   b. central watch clock stations
   c. telephones
   e. two-way radio equipment
29. Are guard assignments and patrol routes varied at frequent intervals to obviate an established routine?

Recommendations

Any recommendations for correction of observed deficiencies must be reasonable in cost. Each recommendation must be carefully weighed to ensure that the additional security obtained is worth the expense involved. In laying out the long-range security programs involving extensive construction, hiring of additional guards, installation of illumination and other protective devices, as well as specifying immediate remedial actions, primary consideration must be given to dollar value received versus dollar value spent. Whenever possible, maximum use will be made of existing facilities, supplies and equipment by renovation in order to reduce expenditures. It is also possible to improve security by improving discipline in administrative procedures. This step is probably quickest to produce results and most cost-effective.

Appendix 4b
Personnel Security Checklist

This checklist has been prepared as an action guide for transportation industry management to use in upgrading personnel security measures. The DOs and DON'Ts contained herein should not be considered to be all-inclusive but should be viewed as steps from which to build a more effective personnel security program.

Management Policy and Response

DOs

1. Recognize that employees are participants in a substantial majority of theft and pilferage losses.
2. Promulgate company policy regarding personnel security measures.
   a. Assign authority and responsibility for execution of the personnel security program to officials within the organization.
   b. Involve personnel security considerations in the decisionmaking process of the organization.
   c. Provide support and cooperation of all levels of management of the personnel security program.
3. Integrate personnel security measures into the existing employment system—including a firm commitment by management to elements of the program.
   a. Identify weaknesses inherent in the present employment process that might allow employment of applicants with questionable backgrounds.
   b. Examine procedures being used by security-conscious employers for screening and investigating their applicants.
   c. Implement procedures designed to upgrade employment practices regarding personnel

security. Specific managerial prerogatives are discussed later.
   d. Adopt measures to assure periodic review of the personnel security program.

DON'Ts

1. Don't evade the problem of employee theft and pilferage.
2. Don't respond to the problem with *lip service* policy and ineffective procedures.
3. Don't ignore the economic advantages of incorporating effective, but relatively inexpensive, personnel security measures into the existing employment process through the exercise of certain management prerogatives.

Employment Application Forms

DOs

1. Require submission of a detailed employment application by all prospective employees—including applicants for clerical and maintenance positions as well as applicants for cargo-handling positions.
2. Design the application form to include information that will be helpful in judging the applicant in terms of honesty, integrity, and reliability. The following information should be required by the application forms:
3. a. Gaps in employment continuity
   b. Frequent job shifts
   c. Complete employment history, including:
      (1) Reasons for leaving
      (2) Sufficient data with which to make contact with former employers and supervisors
      (3) Salary information
      (4) Brief statement of duties and responsibilities
   d. Educational background, including specific information regarding schools attended, dates, etc.
   e. All names used by the applicant
   f. Type of military discharge
   g. Citizenship
   h. Present residence and prior residence information for the past ten years
   i. Affiliations and organizations
   j. Selective Service classification
   k. Personal references
   l. Bonding history
   m. Criminal history—when permitted by law, indictments, arrest, and conviction data should be obtained
   n. Conditions to which applicant agrees by signing the application form including:
      (1) Misrepresentations on the form shall be considered acts of dishonesty.
      (2) Permission is granted to the employer or agent to investigate the applicant's background, including a credit check.
      (3) The application for employment in no way obligates the employer to hire the applicant.
4. Carefully review the employment application form for accuracy and completeness prior to consideration for processing.
5. Consider the use of a separate form for obtaining security-related information, e.g., fingerprints, driving record, criminal history. Such a procedure makes the applicant aware of the organization's interest in employing personnel with a high degree of honesty, integrity, and reliability.

DON'Ts

1. Don't employ applicants prior to submission of a detailed employment application form.
2. Don't use a standard type of application form that makes little or no provision for obtaining security-related information.
3. Don't accept applications that are inaccurate or incomplete.

Fingerprinting and Photographing of Applicants

DOs

1. Include the fingerprinting of applicants and the taking of identification photographs in the pre-employment process.
2. Make arrangements with the local police department for fingerprinting and a local photographer for taking applicant I.D. photos if in-house facilities for such procedures are not feasible.

DON'T

1. Don't fail to recognize the discouragement to undesirables and the deterrence to thieves or

criminals provided by fingerprinting and photographing requirements.

Interviews

DOs

1. Make provisions for a personal interview of all applicants to be conducted by trained interviewers.
2. Design the interview session to assure that the following security-related elements are completed:
   a. Verification of information submitted on the employment application form.
   b. Clarification of details regarding questionable or derogatory information detected on the application form or during the initial pre-employment background investigation.
   c. Obtaining additional information not contained in the application.
   d. Obtaining information from the applicant which will help to appraise personality, character, motivation, honesty, integrity, and reliability—and to judge appearance and personal characteristics face-to-face.
   e. Informing the applicant about the company, including security policies and procedures.

DON'Ts

1. Don't allow interviews to be conducted solely by department managers or line supervisors.
2. Don't expose the applicant to a brief, non-comprehensive type of interview that is void of security considerations.

Confirmation of Personnel Data

DOs

1. Confirm significant data contained on the employee application form and contact references.
2. Conduct thorough background investigations—go beyond the basic confirmation of factual data. Include searches for information regarding the applicant's character, integrity, honesty, and reliability. Such information should include, but not be limited to:
   a. Any deliberate misrepresentation, falsification, or omission of material facts.
   b. Any criminal, infamous, dishonest, immoral, or notoriously disgraceful conduct, habitual use of intoxicants to excess, or drug addition.
   c. Conviction of crimes of violence, including assault with a deadly weapon.
   d. Any facts which furnish reason to believe that the individual may be subjected to coercion, influence, or pressure which may cause her or him to act contrary to the best interests of the company.
   e. Any previous dismissal from employment for delinquency or misconduct, theft, or embezzlement.
3. Select the most effective method of conducting background investigations within the existing capabilities of the organization. Methods available include:
   a. Mail verification
   b. Telephone interviews
   c. Contracts with outside firms, ranging in cost from $10 to $70 each, depending on the service required.
4. Search applications being processed against local security and/or trade association indexes to ascertain whether any derogatory information is on file regarding:
   a. The applicant
   b. The names of applicant's references
   c. The name of applicant's spouse
   d. The identity of applicant's friends or relatives working in the industry
5. Organize local trade indexes to minimize the chances of hiring applicants already determined to possess undesirable characteristics by another member of the transportation industry.

DON'Ts

1. Don't adopt procedures that allow the employment of applicants without completion of a thorough background investigation.
2. Don't let the cost or the time required to conduct background investigations be the sole factor in rejecting such procedures.
3. Don't restrict background investigations to mere verifications of factual data.
4. Don't limit the scope of background investigations by subscribing to only one method. Tailor procedures to the need and to the resources available locally.

Appendix 4c
Physical Security Checklist

1. Do you have a written security plan which is updated at least once a year?
2. Are the functions of security clearly defined?
3. Do you have a preplanned procedure for internal and external notification in case of theft or emergency?
4. Do you have fixed management responsibility for theft or pilferage?
5. Are your records stored and safeguarded properly?
6. Does your shipping and receiving facility have adequate lighting?
7. Is your entire shipping and receiving area fenced or enclosed?
8. Are your entrance and exit gates secured properly or are they secured only by the use of chains and padlocks?
9. Do you have a sufficient number of competent checkers for your needs?
10. Do you have a sufficient number of supervisors?
11. Do you check your facility during the evening and early morning hours to determine the caliber of supervision during those times?
12. Is there a pass system for logging vehicles in and out?
13. Do you have a security crib for placement of high-value cargo?
14. Where called for, are interior or exterior alarms used?
15. Is the monitoring of your alarms provided on a continual basis?
16. Do you have padlocks issued to equipment operators for use when vehicles are left unattended?
17. When assigning keys to individuals, are signatures obtained?
18. Do you issue master keys only where operationally essential?
19. Are periodic checks conducted on key control and locking devices?
20. Are spare keys and locking devices kept under tight controls?
21. Do you check to see if equipment parked or left unattended is left with keys in the ignition?
22. Is there a predesignated, high-value holding area for cargo held in shipping and receiving areas?
23. Do you periodically examine the refuse removal and cleaning service equipment and personnel when they leave your facility?
24. Have you screened the vendor companies with which you are doing business?
25. Are you conducting a preemployment investigation on all prospective applicants?
26. Are you taking fingerprints and/or photographs of all new employees?
27. Are casual employees included in your preemployment screening process?
28. Is there a probationary period for new employees?
29. Are your personnel records safeguarded?
30. Have you displayed reward posters with proper wording and considerations?
31. Are you engaged in a security education program?
32. Is your total management team security conscious?
33. Are you providing any type of in-service training for dealing with security problems?
34. Is your paper documentation controlled?
35. Are you tallying cargo properly as it goes on and off equipment?
36. Are you using seals and recording the numbers?
37. Are seal numbers being verified at destination?
38. Do you keep your seals under tight security controls?
39. Are seals being affixed by responsible management personnel?
40. Is there a report made to security for trailer or container seal discrepancies if found?
41. Are files maintained which indicate the number of times an employee is involved in a theft or shortage?
42. If dealing with interline activities, are controls maintained on their movements and equipment checked?

43. Do you restrict your employees, including management, from parking near cargo operations?
44. Are controlled parking areas provided for employee's cars?
45. Do you prevent unauthorized employees access to areas where cargo or paper flow dealing with cargo is being handled?
46. Are unauthorized personnel and unidentified visitors allowed in your shipping and receiving areas?
47. Are employees prohibited from visiting their vehicles during working hours?

Chapter 5
Bank Security

R.E. ANDERSON

Bank security is "a positive program of action designed to protect bank employees, the public, and bank assets against crime and other hazards."[1]

This is an excellent definition of what bank security *should* be. Unfortunately, the state of the art of bank security is not as healthy or as positive as many bankers would like you to believe. Security training is a serious problem. Although bank security training is required by Title 12 of the U.S. Code, many financial institutions expose their new employees to risk and danger without first having provided them with any basic security training.

The Bank Protection Act of 1968: Public Law 90-389

The United States Congress passed the Bank Protection Act in 1968 as a direct result of increased criminal attacks on financial institutions. The Act required the various regulatory agencies (Federal Reserve System, Comptroller of the Currency, Federal Deposit Insurance Corporation, and the Federal Home Loan Bank Board), to establish minimum standards for the installation, operation, and maintenance of security devices and procedures.

The regulations enacted by the four regulatory agencies were practically identical and all codified into various parts of Title 12 of the U.S. Code (12 CFR). A copy of the Bank Protection Act of 1968 and Section 21 of Title 12 of the U.S. Code (12 CFR 21) are included in the Appendix of this chapter for information and reference.

The Act was targeted at the external crimes of robbery, burglary, and larceny and should *not* be viewed as a broad and comprehensive program designed to impact on all bank crimes and risks. The act really never considered kidnap/extortion crimes or the two crimes that result in the greatest financial losses to banks: internal frauds and embezzlements; and stolen and forged check frauds.

Because the act tended to focus public attention on external attacks on financial institutions, there is some concern that it has had the effect of laying a big smoke screen over a rather sensitive and embarrassing major security problem and risk: internal frauds and embezzlements. Many of these crimes are never reported to the FBI or the regulatory agencies and these statistics are grossly underreported, representing only the tip of the iceberg.

Efforts to amend and improve the Bank Protection Act are almost always opposed by the banking industry. The Attorney General and the U.S. Department of Justice once petitioned the regulatory agencies to make some much needed changes but the banking industry was able to persuade the regulators that no changes were necessary. This presents a most unusual dichotomy because many expert bank security officers recognize a need for some improvements in the bank security regulations. Obviously, the banking industry has been able to exert much more influence on the regulatory agencies than the bank security officers have, and this does not bode well for the future safety and protection of bank employees, the public, and bank assets.

Current Risks Affecting Financial Institutions

Financial institutions have both internal and external risks and exposure to criminal attacks. The greatest danger to human life occurs in the external crimes of kidnapping and robbery. About 300 bank robberies occur to every kidnapping. The greatest financial losses are sustained as a result of external check frauds and internal embezzlements.

Before discussing recent crime trends and their effect on financial institutions, a few insights into different bank crime statistics may prove helpful in interpreting the data and the risks more realistically. With the exception of check frauds, external bank crime statistics are much more accurate and reliable than internal bank crime statistics. There are many reasons for this disparity in crime reporting, but the principal reason for the underreporting of internal crime is that many financial institutions are reluctant to make a full and open disclosure of the true scope and extent of internal crime. To do so would cause an even greater loss of public confidence in banks and bankers. Image is a large factor in the reluctance to report all these crimes.

Kidnap/Extortion

Often viewed as the most dangerous crime facing financial institutions, kidnap/extortion criminals frequently victimize the banker's family, not just the banker alone. Known as Hobbs Act crimes, about 250 cases occur each year in financial institutions. Violence is employed in about 12 percent of these crimes each year or in about 30 incidents.

The probability of an individual bank being victimized by this unusually vicious crime is not very great statistically. But when kidnappings have occurred in banks, the results have often been very traumatic, and the emotional and psychological scars have frequently been of greater concern than the loss of funds paid for ransom.

Robbery

Bank robbery, more than any other, was the crime that really precipitated the passage of the Bank Protection Act. Paradoxically, the most dramatic increases in the crime of bank robbery occurred after the act was passed to curtail these very crimes. Since the passage of the Act in 1968, bank robberies have more than tripled. From 1977 to 1979, bank robberies surged up 54 percent. Figure 5–1 depicts the very serious trends of bank robberies from 1960 to 1979.

A few banks have been successful in controlling bank robberies and reducing risk and losses while the great majority of banks were experiencing significant increases. Most of the credit for the success in deterring bank robberies has been attributed to the installation of bullet-resistant barriers in high-crime banking office locations. A 1979 study by Virginia Commonwealth University of people protection barriers in banks produced these principal findings:

- Less than 3 percent of all bank robberies are attempted in banks protected by bullet-resistant barriers.
- Bullet-resistant barriers act as effective deterrents to bank robberies according to the robbers themselves.
- Most customers (about 92 percent) feel that other banks could install bullet-resistant barriers to deter robberies and protect personnel.
- Most employees agree that working in a protected environment provides them more safety and greater peace of mind.[2]

For the recovery of funds stolen during bank robberies, tear gas/dye pacs have proven to be very effective. Although used primarily as asset recovery devices, they have caused the arrest of several thousand bank robbers and shown a deterrent effect as well.

In another research project undertaken by Virginia Commonwealth University in 1980, the following findings and conclusions were cited:

- Banks that use tear gas/dye pacs recover 71 percent of the currency stolen during bank robberies. Banks that do not use these devices recover less than 18 percent of their stolen funds.
- Experience has shown that tear gas/dye pacs provide valuable assistance to law enforcement agencies in the apprehension of bank robbers.
- The reluctance of many banks to install these devices because of their fear of injury and liability is not supported by any valid empirical evidence. In fact, the evidence, nationwide, supports the conclusion that risk of injury and liability is minimal and almost nonexistent.[3]

Burglary

The most serious type of burglary attack faced by financial institutions involves the compromise of an alarm system followed by a successful penetration of a vault containing safe deposit boxes. The liability and loss can reach into many millions of dollars. Fortunately, only a relative few burglaries of this type occur each year in the United States. In practically all cases, the alarm systems did not reflect the current state of the art and should have been upgraded years before.

Burglary attacks on night depositories are more common than vault attacks. With thousands of additional automatic teller machines being installed, these devices are expected to compete with the after-hour depositories for burglary honors in the

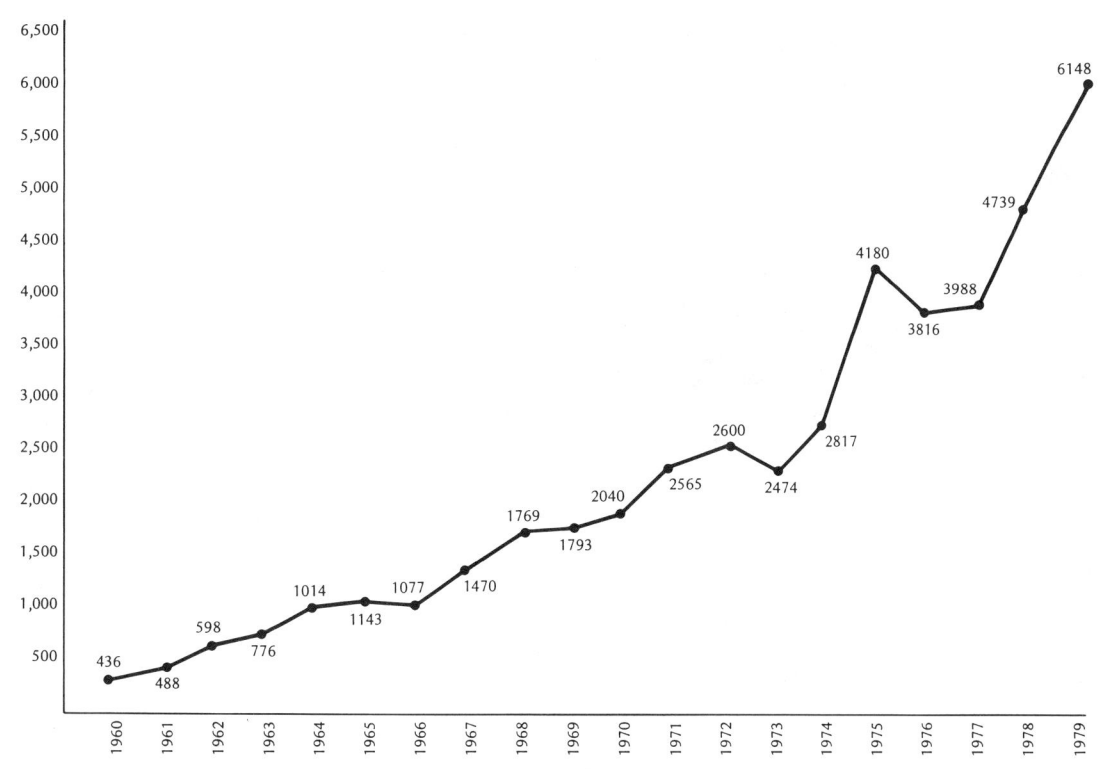

Figure 5–1. Graph depicting bank robbery trends from 1960 to 1979.

future. The need for internal line security (ILS) and for hi-line coded response alarm protection for these devices should be considered a necessity in the future. Burglary crime trends from 1977 to 1987 disclose an average of 435 attacks on banks each year.

Stolen and Forged Checks

Informed estimates place check fraud losses between $2 to $3 billion annually for financial institutions. These crime statistics are not as reliable as the bank statistics for kidnap, extortion, robbery, or burglary. The losses sustained from check frauds far exceed the total of all losses from robberies, burglaries, and extortions put together. Only internal fraud and embezzlement losses compete with check fraud losses as the cause of greatest criminal losses to financial institutions.

Larceny

During the past 20 years, the FBI has reported an average of 275 larcenies each year under the Federal Bank Robbery and Incidental Crimes Statute. The data is obviously underreported, and knowledgeable bank security officers do not believe for a minute that there are only a few hundred larcenies committed each year in all of the financial institutions in the United States.

The photograph in Figure 5–2 was not posed. This external larceny was captured on film by 16mm bank surveillance cameras that were operating on sequence taking one photo every 15 to 18 seconds. Although tellers are not supposed to leave their stations unattended and unprotected, the best solution appears to be in designing teller counters that protect more effectively against this type of larceny. Tellers are only human and will always be somewhat vulnerable. Well designed teller counters always remain in place.

Internal Fraud and Embezzlement

Internal bank crimes committed by officers and employees have not received the attention or notoriety that other crimes have received. There is considerable evidence that these crimes are seriously underreported and represent only the tip of the iceberg.

96 INSTITUTIONAL AND INDUSTRIAL SECURITY

Figure 5-2. Photogaph taken by surveillance camera during robbery.

Bank fraud and embezzlement crime trends from 1964 to 1976 are shown in Figure 5-3. Data for the period of 1977 to 1979 is not shown since a new form of reporting now uses a data base of cases where investigation was completed, as opposed to the amount of crimes actually reported during a given year. The upward trend of these *reported* crimes is clearly discernable. Although the actual number of these crimes may be a matter of controversy, the rather clear upward trend since 1964 should be quite apparent even to nonbelievers.

The alarm and concern over bank robberies has

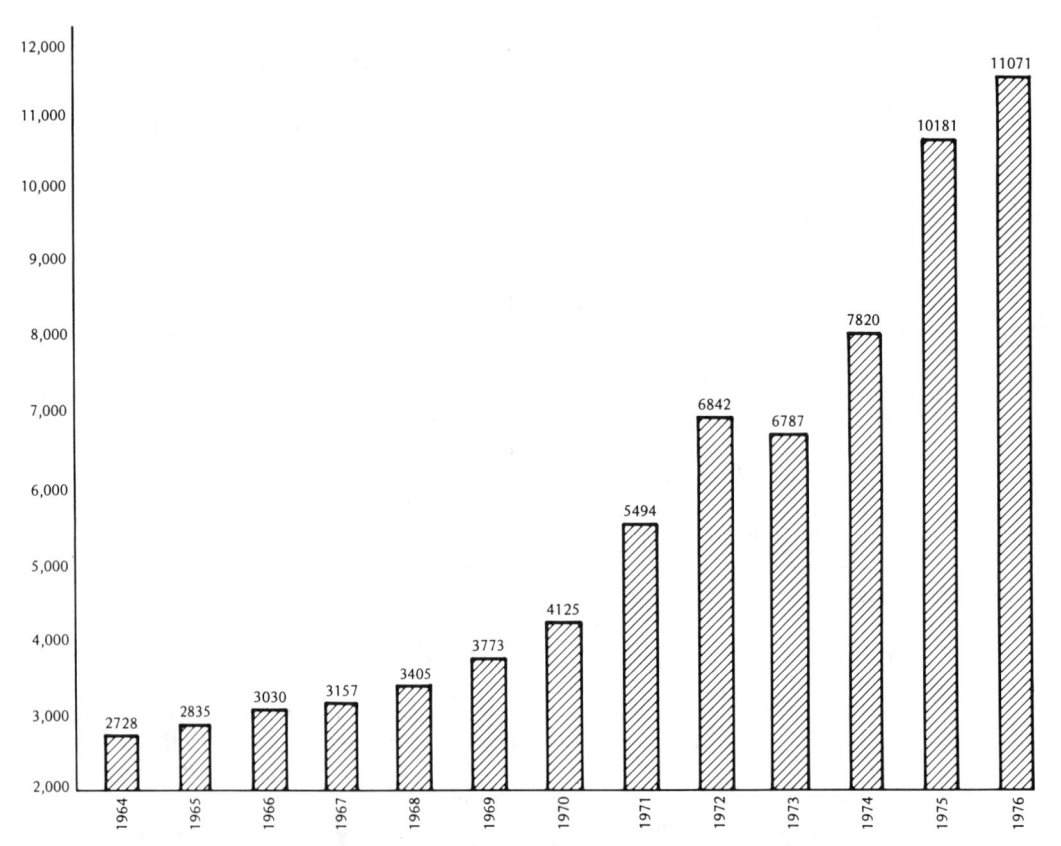

Figure 5-3. Graph showing bank fraud and embezzlement crime trends from 1964 to 1976.

tended to draw attention away from internal fraud and embezzlement crimes; yet during 1979, financial institutions *reported* losses of almost $145 million due to bank fraud and embezzlement crimes. Although bank robberies are certainly a most serious crime, the internal frauds and embezzlements may eventually cause much more damage to the character and reputation of banks than bank robberies ever will. A full and open disclosure of the true scope and extent of internal bank fraud and embezzlement would be very much in the public interest. Will it ever happen? Probably not.

Role of the Bank Security Officer

> Within 30 days after the opening of a new bank, the board of directors of each bank shall designate an officer or other employee of the bank who shall be charged, subject to the supervision by the bank's board of directors, with responsibility for the installation, maintenance, and operation of security devices and the development and administration of a security program which equals or exceeds the standards prescribed by this part.

As a result of the Bank Protection Act and Title 12 of the U.S. Code, many thousands of bank security officers were created practically overnight. What really happened? About 90 percent of all the financial institutions simply assigned the title of Bank Security Officer to a bank officer as an additional duty in addition to other duties.

The role of the parttime bank security officer was created in 1969 and is the rule rather than the exception, even today. Although parttime security officers may be justified in small banks, it was never the intent of the U.S. Congress that large banks with more than 100 banking offices would refuse to hire at least one fulltime professional to do the job properly. A survey by Bank Administration Institute's security commission a few years ago established that six of the largest banks in the United States had not yet hired at least one fulltime professional security officer.

In discussing the role of the bank security officer, it is important to keep in mind that thousands of security officers are bankers with numerous other duties and very little prior experience in safety, security, or investigations. Many complain that they simply do not have enough time to do the job properly. Fulltime security officers are very much in the minority and are usually found in the large commercial banks and very large savings and loan associations. Neither the American Bankers Association (ABA) nor the Bank Administration Institute (BAI) could estimate how many fulltime bank security officers there were in August, 1980, although this would appear to be a very important factor in upgrading bank security. Knowledgeable fulltime bank security officers estimate that there are 8 to 10 parttime bank security officers for every one fulltime security officer. The effect of having so many parttimers is what might logically be expected. The role of many bank security officers has been diffused, fragmented, inadequately defined, and poorly supervised.

In many banks, the role of the security officer involves the bare minimum mandatory requirements set forth in 12 CFR. Some have not yet met the minimum standards that were originally established in 1969. We might refer to these security activities as *bare bones* programs. Conversely, in some of the medium-sized and larger banks with fulltime security officers, the role of the bank security officer involves a broad range of activities that may include:

- Protecting people against external crime
- Protecting property against external crime
- Compliance with security statutes, regulations, and the bank's official security program
- Administering or coordinating security training
- Managing physical security devices
- Emergency planning and operations
- Liaison and coordination with law enforcement officials
- Conducting investigations
- Security planning and research activities
- Computer security
- General security management and administration

In between the bare bones role and the rather extensive role cited above, a review of other programs will reveal just about every other type of bank security program. There is no real industry-wide set of standards and goals for bank security, and many bankers and regulators seem to prefer it that way.

In order for bank security to make any reasonable progress in overall risk reduction during the 1990s it will be necessary to:

- Recruit and retain highly qualified security professionals in many more medium- and large-sized financial institutions.
- Establish security standards and goals for all financial institutions.
- Get board of directors and top management much

more actively involved in bank security programs.
- Improve security training significantly.

Security Devices

When used properly, individual security devices function as integral parts of an overall system to prevent and reduce crime in financial institutions. Unfortunately, many banks purchase security devices on the basis of crisis management rather than any rational, coordinated plan. In order to minimize liability, bank security officers should only use Underwriter's Laboratories (UL)-listed and certified security devices.

Bullet-Resistant Barrier Research Report

The Security Division of United Virginia Bankshares in cooperation with Virginia Commonwealth University has analyzed the robbery deterrent value and cost-effectiveness of bullet-resistant barriers in providing safety to bank employees. Customers and tellers at many different banking offices in Virginia participated in the study. More than 150 customers participated in the study.

Some of the principal research findings disclose that:

- A high percentage of bank customers have a positive attitude toward the presence of bullet-resistant barriers in their banks.
- Contrary to popular myth, most customers (about 75 percent) do *not* perceive bullet resistant barriers as devices that make banking less personal than it should be.
- Ninety percent of bank customers see a distinct advantage in banking at an office protected by bullet-resistant barriers.
- About 85 percent of the customers say it is *not* harder to conduct business in a bank equipped with bullet-resistant barriers.
- About 92 percent of the customers feel that other banks should install bullet-resistant barriers to deter robberies and protect personnel.
- Ninety-one percent of employees agree that working in a bullet-resistant protected environment provides them more safety and greater peace of mind.
- Bullet-resistant barriers are much more cost-effective than utilizing armed guards. The savings range from $250,000 to $500,000 for each banking office.
- Bullet-resistant barriers do deter robberies, according to the robbers themselves.

Automatic Teller Machines (ATMs)

These machines which dispense cash and accept deposits are expected to increase in numbers significantly during the next four years. Informed bankers project a total of 100,000 to 225,000 ATMs in service by 1995, compared to about 15,000 in service in 1980. Bank security officers will spend a lot of their time and resources on ATMs during the 1990s in order to solve some of the safety and security problems already associated with ATMs, such as:

- Bank security officers are frequently left out of ATM planning and operations until some serious crime or incident occurs.
- The safety and protection of customers and bank personnel servicing the ATMs has been frequently ignored in the planning and design stages.
- Camera surveillance needs for fraud prevention and investigation are often overlooked or ignored until after the installation has been completed.
- Bank marketing and operations officers tend to ignore Title 12 of the U.S. Code (12 CFR) that specifies that ATMs are security devices and that the bank security officer is responsible for "the installation, maintenance, and operation of security devices."[5]
- Sometimes these ATMs malfunction and the losses can be significant. An automatic teller machine at a bank in Kentucky once shelled out $15,405 to a customer who probably just meant to withdraw $25.

After-Hour Depositories

Sometimes called *night depositories*, these devices assist customers in making deposits when the banking office is normally closed to the public. Recent security areas of concern with after hour depositories include:

- Night depositories tend to be attacked by burglars much more often than vaults and safes.
- Some of the night depository alarms need upgrading in order to provide internal line security and improved protection to the head and chute where most of the attacks are centering.

- Unprotected night deposit funds left exposed during banking hours have resulted in higher losses when bank robberies have occurred. The average amount of funds taken during bank robberies in 1979 increased 28 percent over 1978 and night deposit funds contributed to these higher losses.

Alarms

The two primary bank alarms are the robbery alarm and the burglary alarm. Secondary alarms that interface with the primary alarms would include alarms for automatic teller machines, night depositories, safes outside of a vault, and sometimes, premises alarm protection. Since vaults and safes will only delay a career burglar, the electronic alarm systems are unquestionably the most vital component of a bank's protection system. Here are a few of the more significant alarm problems noted in recent incidents:

- Practically all of the successful major bank vault burglaries in the last decade were accomplished by defeating old vintage burglary alarms that should have been upgraded years ago.
- Banks have erroneously treated alarm systems like vaults. They install them and think they will last 20 or 30 years along with the vault. Electronic alarm systems require periodic upgrading if they are going to remain relatively secure from compromise.
- Bank false alarm rates are so high that many police agencies now charge a fee for each false alarm. Others are removing bank alarms from police consoles.

Alarm Activation Devices

Effective use of these devices during emergencies is the real payoff. Unfortunately, the actual activation of these devices during 1979 should be a cause of concern to all responsible bankers. Here's the sad story from 1979:

- Robbery alarms were *not* activated in 19 percent of the bank robberies.
- Surveillance cameras were *not* activated in 21 percent of the bank robberies.
- "Bait" money was *not* given out in 37 percent of the bank robberies.
- Tear gas/dye pacs that were readily available were *not* given out in 51 percent of the bank robberies.

Time-Delay Locks

From the time the main vault or safe is opened in the morning until it is closed at night, banks become vulnerable to robbery attacks directed at reserve head teller currency or reserve vault currency that could result in very large losses. Federal law recognized this risk and requires that vaults be opened as late as possible in the morning and closed as early as possible in the evening (12 CFR 21.4 (b) (5)).

In an effort to minimize the risk when the vault is open, some banks have installed 15 minute time-delay locks on the lockers containing reserve head teller currency and reserve vault currency. Since hardly any robbers were willing to wait around for the time-delay locks to run down and permit access, these devices can serve as valuable deterrents to vault robberies. Large, prominent signs are often used to indicate that time-delay locks are in operation and employees cannot gain immediate access to these funds (see Figure 5-4).

Safe Deposit Boxes

Quite often, safe deposit boxes are not perceived as important security devices that can involve significant liability. Frequently, bank security officers are not even involved in safe deposit box decisionmaking until some type of security incident occurs. Many banks have not yet established minimum standards for safe deposit boxes and some banks simply purchase the cheapest boxes available. Recent successful vault attacks are directing attention to the vulnerability of safe deposit boxes and concerned banks are reexamining the quality of protection being provided to safe deposit customers.

Some of the security concerns in the current use of safe deposit boxes include:

- A relatively new lock puller device that permits a lock to be pulled in less than a minute in order to gain access to a safe deposit box. In the hands of a dishonest bank vault attendant or a professional

```
-12"- | TIME DELAY LOCKS IN OPERATION.
      | BANK EMPLOYEES CANNOT OPEN LOCKS
      | FOR 15 MINUTES.
        -36"-
```

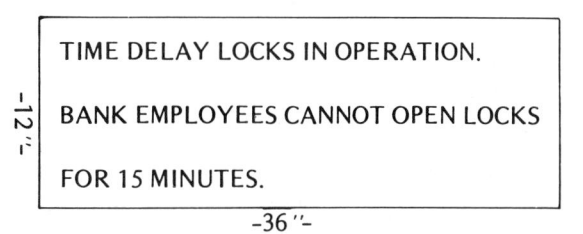

Figure 5-4. Sign to indicate presence of time delay locks.

burglar, this device could be used to open all the safe deposit boxes in a vault in a very short time. The cost of a lock puller is about $50, and almost anyone can obtain one by simply paying the purchase price. It should be noted that the manufacturer has advised bankers that it is their responsibility to prevent the device from being used by bank employees to commit an internal attack on safe deposit boxes.
- Loss of key and lock controls often results because—existing locks and keys are sometimes duplicated in the same vault. It is recommended that contracts for new safe deposit boxes contain a clause that no existing locks or keys are to be duplicated unless a waiver is obtained from the bank security officer.
- Available vault space for safe deposit boxes is often not used effectively, causing a loss of rental income to the bank. Boxes are not added according to any prearranged plan that will ensure good space planning.
- Banks that serve a transient population and have a high turnover in safe deposit box rentals incur a considerable expense in changing safe deposit locks unless they do it themselves. UL certified key changeable locks can be reset and changed by bank personnel in the presence of the new customer who receives the new keys in a sealed container.

Surveillance Systems (Cameras and CCTV Systems)

Surveillance systems are a good example of a *catch-22* at work in financial institutions. At the present time, surveillance systems are not required to be installed. However, if a surveillance system is installed, 12 CFR requires that the system meet the minimum standards set forth in Appendix A of Title 12 of the U.S. Code. *Nothing* in the present regulations requires a bank to install surveillance cameras even if that bank is robbed a dozen times. Most law enforcement officials would prefer regulations that would make surveillance systems mandatory in financial institutions that have sustained multiple bank robberies.

Some of the more common problems that banks have and are experiencing with surveillance systems would include the following:

- Bank personnel frequently do not use the cameras when they are needed most. In 1979, available cameras were used in only 74 percent of the bank robberies. This means that about 1,600 cameras costing about $3 million remained idle during the most critical time of their existence.
- Most banks opted for 35mm cameras positioned near the doors in the late 1960s and early 1970s. Film for these cameras is expensive and they usually do *not* operate on continuous sequence taking pictures automatically. Bank personnel must activate these cameras during emergencies; and the data discloses that they frequently fail to do so. Sixteen millimeter cameras using continuous sequence photography are not totally dependent on bank personnel and offer a much higher probability of successful use. The film cost is also considerably less than 35mm film.
- Many cameras located over doors do not provide the flexibility needed for recording check fraud crimes as well as robberies where most of the action occurs—at the teller line. Cameras positioned behind the teller line are often much more successful in obtaining photos of criminals than cameras located over the doors. This is especially true when 16mm cameras are located behind the line using automatic sequence photography.
- The rapid increase in the price of silver and the cost of film is causing sequence camera photography costs to escalate. Although closed circuit television systems have a much greater initial cost than cameras, the increased cost of film is making CCTV systems look much more cost-effective over the long term to many banks. It is projected that CCTV systems will be used more frequently than cameras in the latter half of the 1980s.

Most law enforcement officials are strong advocates of using surveillance systems in financial institutions. They say that use of these systems has resulted in the apprehension of thousands of bank robbers and bad check passers and that they also serve to deter crime by their very presence.

Vaults and Safes

Some financial institutions are not large enough to contain a vault and may use a safe in which to store the bank's valuables. The actual distinction between a safe and a vault is related to size rather than to the degree of protection provided. A vault is designed for the safekeeping of valuables, as is a safe; however, a vault is large enough to permit entrance and movement *within* by one or more persons.

Minimum standards for both safes and vaults are set forth in Appendix A of 12 CFR, a copy of which may be found at the end of this chapter. In addition to the minimum standards required by 12 CFR, bank security officers can reduce liability by ensuring that safes also meet or exceed Underwriters Laboratory (UL) standard 687 and that combination locks meet UL standard 768.

Although vault doors now carry UL classifications according to the amount of time necessary to attack them successfully, vaults themselves do not yet have UL standards but are rated by the Insurance Service Office (ISO). Most bank vaults would be rated somewhere between 5R and 10R with 5R representing compliance with the *minimum* standards set forth in Appendix A of 12 CFR.

Only a few years ago, Underwriter's Laboratories agreed to test various vault doors and classify them based on the net amount of time required to attack them successfully with certain designated tools and equipment. As a result, bank vault doors are now classified for the first time according to penetration time as follows:

Class I   ½ hour
Class II   1 hour
Class III   2 hours

It is strongly recommended that only UL-certified vault doors be used by banks in order to minimize risk and liability.

Neither vaults nor safes are impregnable and they cannot resist prolonged and sustained attacks. Nor were they designed to do so. For this reason, the bank's burglar alarm system becomes critically important and must be upgraded from time to time in order to prevent defeat, circumvention, and compromise. Bank vaults without modern alarm systems using random, interrogate-response line security that is digitally coded and AA-UL-certified are like battleships and carriers without radar or electronic countermeasures—very vulnerable.

Security Training

Title 12 provides for "training, and periodic retraining, of employees in their responsibilities under the security program, including the proper use of security devices and proper employee conduct during and after a robbery, in accordance with the procedures listed in Appendix B of this part."[6]

In-house security training is the real payoff for safety and protection because this training is provided to those employees who are actually exposed to risk and danger on a day-to-day basis. Yet, it is not at all unusual to go into a bank and find tellers working on the line who have had absolutely no basic security training. However, this author has been unable to find one instance where a regulatory agency has imposed a civil penalty on any financial institution under Section 5 of the Bank Protection Act for violating this regulation.

A former Attorney General of the United States and many others have identified training as one of the areas of Title 12 of the U.S. Code that requires revision in order to quantify some reasonable minimum security training standards. The present training requirement is so vague and general that it is practically meaningless. However, improvements in 12 CFR, important as they may be, will have little meaning unless the regulatory agencies demonstrate some willingness to impose sanctions on those banks that have developed a poor track record of compliance.

Since the regulatory agencies have not done much about upgrading security training since 1968, it is doubtful they will show any more concern for security training in the 1990s than they did in the 80s. In all probability, serious efforts to improve bank security training will probably require the concerted action of the U.S. Department of Justice and the U.S. Congress. The regulators simply cannot be relied upon to do the job properly.

Security training is probably the single most important subject in an overall bank security plan/program. In the view of many professional bank security officers, if we can significantly upgrade bank security training, most of the rest of the program will fall into place and also be improved. Security training is vitally important if bank security programs are going to be improved in the 90s. Reasonable minimum standards and goals for security training for banks with different risks and exposures are urgently needed and banks should waste no time in getting on with the job.

Guidelines for Establishing Minimum Training Standards

The present general and extremely vague training requirement in 12 CFR simply has not done the job. There is a very clear and present need for more specific training standards to improve overall bank security programs in the United States. There has been much resistance to any additional security training in the past and probably will be in the future, nevertheless, standards must be established and enforced—and soon!

For our purposes, it may be helpful if we can

outline the concept and general principles that might form the foundation of any viable and effective minimum training standards for bank security in the 1980s:

- Not all banks are subject to the same degree of risk and exposure to criminal attack; therefore, all banks should not have the same minimum training standards.
- Banks with higher risk and exposure to crime should have higher standards of minimum training than banks with lower risk.
- Both training in protection against robbery and violent crime, and training in fraud and property crime, should be included in any minimum training standards.
- Training in robbery and violent crime prevention and protection should receive a higher priority and more attention and resources than fraud and asset protection training.
- The exact amount of training to be provided by each bank should be related to the crime rate of that particular community or some equivalent measure of risk and exposure. The standards should be adjusted as risk and exposure varies upward and downward from time to time. Unfortunately, crime tends to increase more than decrease.

The practical effect of these minimum training standards would translate into the following illustration:

- Banks in New York City; Los Angeles; San Francisco; Flint, Michigan; New Orleans, and Miami would have very high training standards.
- Banks in Atlanta, Cleveland, Dallas/Ft. Worth and San Diego would have high training standards.
- Banks in Cincinnati, Honolulu, Louisville, Salt Lake City, and Raleigh/Durham would have moderate training standards.
- Banks in Sioux Falls, Milwaukee, Cedar Rapids, Lincoln, and Syracuse would have training standards just above minimum levels.
- Banks in communities with very low crime rates would have minimum or *bare bones* training standards.

Current Bank Security Training Organizations

There is no apparent shortage of organizations that conduct bank security training; however, the quality and effectiveness of these various training programs varies considerably since the faculties are constantly changing in many of these organizations. A partial list of organizations that conduct bank security training programs at various locations throughout the United States would include:

- American Bankers Association, Washington, D.C.
- Bank Administration Institute, Park Ridge, Illinois
- The Mosler Safe Company, Hamilton, Ohio
- R.E.A. Associates, Dallas, Texas
- American Society of Industrial Security, Washington, D.C.
- International Association of Chiefs of Police, Gaithersburg, Maryland
- Various universities and consultants

Critical Problems in Bank Security

If bank security is to be significantly improved in the 1990s, some of the critical problems left over from the 80s will have to be resolved first. Here is just a partial list of some of the more serious issues and problems that retarded the professional growth and development of bank security in the 1980s:

The members of the board of directors of far too many banks have shown little or no interest in their bank security program. This is especially true of banks operating in relatively high crime environments. Some of these directors would not even recognize their bank security officer if they were in the same elevator. Directors and other members of top management will have to demonstrate much greater interest, awareness, and concern for bank security.

The various regulatory agencies have very few security professionals on their bank examining staffs. The examiners cannot make very much of a contribution to improving bank security if they know little about it. There is an important need to add some highly qualified security professionals to the regional staffs of the four regulatory agencies.

The regulatory agencies have demonstrated no willingness to really enforce the will and intent of the U.S. Congress as expressed in the Bank Protection Act and Title 12 of the U.S. Code (12 CFR). Violations have been numerous and often flagrant with little or no penalties imposed. Banks that provide no basic security training to tellers are in clear violation of 12 CFR, and yet the bank examiners have rarely taken any action to enforce the regulations. Bank security will not receive the attention of bank executives in some banks in the 80s until they are persuaded that the regulatory agencies mean business and intend to enforce the provisions of 12 CFR.

The federal system of criminal justice and particularly the role of the 94 U.S. attorneys is not conducive to bank crime deterrence nor to speedy trials and swift justice. The present practice of the federal system picking and choosing which cases they will accept and leaving the rest for local law enforcement is creating serious problems of unequal justice and disparity in the treatment of offenders. Bank security in the 1990s will be retarded unless the feds and the locals can get their acts together and agree on *who* is going to do *what* for the bank security officers.

Many banks have been reluctant to hire and retain at least one highly qualified fulltime bank security officer. The industry is proliferated by parttime bank security officers with little or no experience or credibility in the current state of the art of bank security—a highly specialized, technical, and complex subject in contemporary banking. There is an important need for more fulltime highly qualified bank security officers in many of the large and medium-sized financial institutions in the 1990s. Progress will depend on it.

There is almost a complete lack of standards and goals relating to bank security programs throughout the entire banking industry. At the present time, no minimum training standards exist in 95 percent of the banks. Tellers in banks that have been robbed ten times still don't know how much training they should receive to meet minimum standards. Most banks have no standards with respect to the use of Class I, II, or III vault doors. All too often, the tendency is simply to purchase the cheapest available piece of security equipment without any standard that recognizes the varying degrees of risk involved. There is a definite need for more thorough and comprehensive standards and goals if bank security is to achieve reasonable progress in the 1990s.

Most banks have been asset-protection oriented in the 80s. People protection programs have frequently not received the same priority or attention given to property protection programs. There is a critical need to bring people protection programs into better relative balance with asset protection programs in the 1990s. This is particularly relevant in a high robbery risk banking office that has already sustained multiple robberies. Some bank personnel have refused to work in these high risk locations. Unless more attention and resources are allocated to people protection programs in the 90s, additional employee grievances should be expected.

Computer security is not receiving the attention it requires in many financial institutions. Banks are overrelying on auditing EDP programs *after* they hire data-processing personnel. Very little prevention is being applied to the front end of the hiring process itself to ensure that suitable EDP personnel are hired for this risk sensitive function. Computer frauds are often internal people-related frauds and risk reduction in the 1990s will require more adequate hiring and clearance procedures in order to prevent more computer frauds. Detecting computer frauds after they have occurred is helpful, but *prevention* should receive a higher priority in the 90s than it received in the 80s. Does it make much sense to keep adding EDP auditors to the staff when the bank doesn't even take finger-prints from or perform an employment or credit inquiry on newly-hired EDP personnel?

Many banks have failed to upgrade their vault burglary alarm systems since they were first installed many years ago. These banks are particularly vulnerable and would have significant liability in the event of a successful safe deposit box vault burglary. There is a definite need to improve vault protection in a great many banks by adopting minimum standards for line security that would upgrade many of the older, less secure burglary alarms existing today in many financial institutions.

The great majority of banks still do not fingerprint their employees even though they may do so legally pursuant to Section 19 of the FDIC Act. Internal bank fraud and embezzlement losses amounted to $103 million in cases investigated by the FBI during the first six months of 1987. Although fingerprinting is certainly not a panacea for internal crime, it certainly can eliminate some unsuitable employees and even prevent some of these losses in the future.

The quality of bank security training and audiovisual training aids varies widely throughout the industry. Some minimum standards are urgently needed to control some of the inferior programs and films that have invaded bank security in recent years. Agencies without any bank security experience or expertise have produced films and training programs simply for profit without regard to any standards of excellence, safety, or even compliance with Title 12 of the U.S. Code (12 CFR). ABA and BAI will have to do a better job of screening out some of these poor quality products in the 1990s if bank security is to advance and improve. The least that ABA and BAI should do is *not* to advertise some of these unsuitable training films in their publications for profit.

Worthless Checks

Experts estimate that worthless checks amount to $12 million a day—$40 billion a year. Commercial business establishments sustain the majority of these losses—banks account for an estimated $2–$3 billion of the loss.

Between 75 and 90 percent of all worthless checks have a serial number between 101–150. Many banks pay more attention to identification of the check passer than they do to the *check* itself. Although identification is often helpful, it is also very easy to obtain fraudulent identification. Don't overlook the check itself as a means of detecting check frauds. A fast examination of the check itself is important and can often prevent a check fraud loss. For instance, keep in mind that a U.S. government treasury check is the only *unperforated check*. All other valid checks will have at least one perforated side.

The ink is another clue to the validity of the check. Magnetic ink is 100 percent dull and has no glare or shine; therefore, if you tilt a check towards the light and get a glare or reflection from the magnetic ink, you have a firm indication of a counterfeit or forged check.

Many commercial business establishments that sustain check fraud losses blame the banks who issue starter kits for not exercising reasonable control. Some banks have lost the good will and business of commercial customers because of loose new account opening activities (i.e., "Don't send my new checks to my residence. I'll stop by the bank and pick them up.").

Drive-up tellers are the most frequent target of many check fraud criminals. Inexperienced tellers should not be assigned to this window because they will not be as familiar with fraudulent checks as an experienced teller.

The *average* bank customer uses about 200 checks each year. In most banks, tellers can tell fairly well how long a customer has been with the bank (personal checking accounts) simply by reading the serial number (i.e., serial number 1,000 would usually indicate a customer of 4 to 5 years).

No financial institution can be immune from fraudulent checks but each can follow these precautions to reduce this crime to the minimum.

References

1. *Bank Security Manual*, Bank Administration Institute, Park Ridge, Illinois, 1973, p. 1.
2. *Safety and Protection Against Bank Robberies*, United Virginia Bank/Virginia Commonwealth University, Richmond, Virginia, James C. Willis, June 1979, pg. iii.
3. *TRAP Report*, United Virginia Bank/Virginia Commonwealth University, 1980.
4. *Minimum Security Devices and Procedures*, Part 21 (12 CFR) of Title 12, U.S. Code Section 21.2 for National Banks, 12 CFR 216.2 for Federal Banks, and 12 CFR 326.2 for FDIC banks.
5. Title 12, U.S. Code (12 CFR-21), Section 21.2.
6. Title 12, U.S. Code Section 21.4 (b) (9) for National Banks, 216.4 (b) (9) for Federal Banks, and 326.4 (b) (9) for FDIC regulated banks.

Chapter 6
Hotel Security

An Overview
NORMAN D. BATES, J.D., CPP

The Role of Security in a Hotel

General Responsibilities

The major purpose of security in a hotel is to provide services to guests and the hotel employees. This statement expresses more than mere philosophy, since the vast majority of the calls for security assistance are service-oriented. *Service* in this context may be defined as any function which is not directly related to the prevention of crime or the apprehension of criminals. Specifically, the type of tasks and services required of security include, but are not limited to, the following:

Medical Assistance

A large portion of service calls, as much as 25 percent, are associated with providing some level of medical assistance. This will range from simply referring the guest to the house physician for medication up to actually performing CPR on a victim of heart failure. The latter is not an uncommon occurrence, even in the urban setting where one would expect a fairly rapid response from the local ambulance service.

General Guest Assistance

Under this catagory there are a variety of services to include such tasks as unlocking the guest's room when he has forgotten his key, taking a report of a lost item or delivering a recovered item to the guest.

Physical Security

Physical security is the protection of tangible assets including the building and all of its contents through the use of security hardware and systems. The type of hardware or technology now being used in hotels includes locking systems, closed-circuit television cameras, video recording equipment, intrusion and panic alarms, fencing, walls, lighting, fire alarms and any other barrier or system which tends to protect the property and its contents and occupants. Although security departments have varying degrees of involvement in the selection and installation of such equipment, they will all generally be responsible for the monitoring and maintenance of the systems.

Fire Prevention/Fire Watch

It is clearly an understatement to say that fire is a concern in a hotel. The major responsibility for fire prevention and control has frequently been placed with security. Prevention entails the inspection of equipment such as fire extinguishers, hoses, smoke and heat detectors for signs of wear, tampering, or some other condition that could indicate malfunction.

Fire prevention also includes training. Although security may have the functions of prevention and control, all hotel employees, regardless of position, share in the prevention and control of fires. Security often will have the responsibility for the training of other employees. Training in most instances will be limited to some knowledge of the causes of fire, uses of the different fire extinguishers, and the employees' role during an alarm or actual fire.

The fire watch is probably the oldest function that security has performed and unfortunately represents the stereotype perception of what hotel security is all about. The image of the "night watchman" is that of the retired person who wanders the empty halls during the late night hours carrying a bulky, black

105

leather-bound clock. This is not an accurate representation of the nature and responsibility of the hotel security function. In a modern hotel or motel, the late or overnight shift still has the task of patrolling the building at regular intervals with the objective of catching a fire in its early stages. The onslaught of new technology in fire detection systems has facilitated the task, but the responsibility remains with the security guard. The monitors used to discover fire are now so sensitive that smoke may be detected before physically perceptable by a human being. This is not to suggest that security never need leave the office; rather, the provision of the early warning system enhances the overall safety of the hotel.

Key Control

Security must be responsible for maintaining the control of keys and locks in any hotel or motel. Unfortunately, this is not always the situation. This task is often assigned to the engineering or maintenance department and commonly, a lack of coordination exists between security and the department which has the keys.

Key control can be divided into two areas: guest room keys and administrative keys. Guest rooms in hotels and motels require a large variety of keying systems; from traditional cylinder locks with brass keys to the more sophisticated systems using a small computer, which records all entries to the room and facilitates the rekeying of a room lock within seconds. The type of system found on any property will depend on how old the property is and whether a renovation has recently occurred. It is a common practice to install more sophisticated systems in new properties and to retrofit older hotels with the more traditional type of lock (due to the higher cost associated with a retrofit of a system that requires extensive wiring).

The second area of key control involves administrative keys. These include master guest room keys, individual offices, storerooms, and locker keys. This category of key control is extremely important due to the increased accessibility allowed to employees, especially those with master keys.

A number of inadequate security cases have developed based on the premise that there was either poor or lacking key control within the hotel. Liability for the lack of controls will be discussed later in the chapter.

As stated above, security must be assigned the overall task of maintaining key control, even if it is limited to recordkeeping and coordination, for security has the responsibility to investigate crime on at least a preliminary basis until the local police have assumed jurisdiction.

Arguably, key control is a guest service due to the impact on the guests. Ultimately, security is designed to provide a safe environment for the guests and key control enhances that safe environment.

Loss Prevention

Many public sector and experienced security professionals know that financial and labor resources directed at the prevention of crime are more cost-effective than those resources used to apprehend, prosecute, and recover stolen property (assuming that it is a recoverable loss). Furthermore, it makes good business sense to *avoid* losses rather than waste corporate assets trying to *recover* after the fact.

The major emphasis of security has got to be to prevent losses before they occur. In a hotel, there are numerous opportunities to prevent crime and other potential losses.

Loss prevention requires *proactive* thinking. A security manager cannot afford to wait until a loss has occurred; he/she must anticipate the probability of a loss and plan for it. The common problem associated with hiring retired law enforcement officers is that they are generally reactive when dealing with security problems. It is obvious that this apprehension and prosecution approach to crime is not working, since crime has continued to rise in the past twenty years.

A security manager must become familiar with all aspects of the hotel business in order to develop a sense for what could go wrong. Therefore, both security manager and staff should interact with other departments on a frequent basis and not limit the interaction to negative situations such as interviewing someone suspected of wrongdoing. A hotel's general manager has the ultimate responsibility to establish priorities for the security department, as for any other department. Top priority must be the *prevention* of crime and other losses and to discourage activity that focuses too narrowly on the catching of crooks.

Other more substantial risks of loss exist, including the reputation of the hotel, guest accidents, alcohol liability related incidents, activity by employees which include the theft of business information, workers' compensation claims, and so on. The hotel manager who limits security to the old traditional role of fire watcher and prostitute chaser is grossly underutilizing a major resource.

Training

What role does security have in training—a major one! A security manager has the responsibility to train security employees in a variety of issues, including first aid, fire prevention, law, self-defense, public relations, and patrol practices. Security also has a role to play in training other employees.

In a typical hotel, the total number of security personnel will amount to about 1–2 percent of the total number of employees. Yet, security is generally expected to provide twenty-four hour protection for the property. Any security manager who expects to do the job without help from other employees is absolutely wrong. From the first day on the job, all hotel employees should be instructed that they too have a role in maintaining the security of the hotel and its guests and employees. It is in the best interests of the employee, in fact in the pecuniary interests, to assist in the security effort. This does not mean that untrained, nonsecurity employees should try to apprehend a suspicious character or to render first aid on their own. The nonsecurity employee could assist by way of a phone call or information essential to the overall success of the security program.

Why train? Trained employees will perform better and display greater enthusiasm for the job. Training keeps morale and productivity up while reducing the chance that the task is improperly performed. Within security, improperly performed tasks can quickly lead to serious liability problems. So, in addition to a better staff, adequate training reduces risks.

The security manager enlists the aid of nonsecurity employees by being creative. There are several possible approaches. First, if the hotel has an employee or orientation program, the security manager should participate in the welcoming of the new employees and give a short talk on the nature of security's role. She or he should encourage questions from the group and ask for their support, for example, by calling the security office whenever they observe unusual behavior by guests, visitors, or other unknown persons. He or she can meet with individual departments on a regular basis to inform them of specific security problems or concerns. For example, by meeting with the housekeeping department to discuss the theft of guest property or prostitution activity, they may help to reduce either or both of these problems by working with security. The security manager should talk informally with employees at lunch and in other appropriate settings to encourage open communication. Educational seminars on such topics as security in the home, self protection on public transportation systems, and self-awareness for prevention of assault also help. The mechanisms used by the security manager in developing the necessary rapport with the employees are limited only to the imagination and the level of support from management.

Arrests/Prosecution of Criminal Offenders

Chasing suspects, appearing in criminal court, and all the other activities related to prosecuting criminals should be the last resort for most hotel security departments. Even the safest of hotels will experience crime. Recognizing this as fact, a security manager must be prepared to effectively deal with the problem. Proper preparation includes the training of the security staff in both legal aspects as well as the procedure for arrest. It is important to establish a good working relationship with both the local police and district court. Not all cases will be prosecuted with the assistance of the police. Before arrest and prosecution is pursued as an option, the security manager must consider the corporate policy and obtain the verbal and written approval of senior management and, where possible, legal counsel.

Department Organization

The structure, size, and reporting level of a security department will drastically affect the department's efficiency and success within the hotel operation.

The appropriate size of a department is probably the most difficult task to determine, particularly when a new department is being formed for either a new property or when a proprietary department is being established to replace contract services. Unfortunately, there are no national standards to provide guidelines for what would be considered adequate levels of protection. A security manager must look at a variety of sources within the state for some indication of what would be acceptable for that hotel. Some of these sources include:

- Relevant case law—perhaps a hotel or similar institution was found civilly liable for inadequate security because the staffing was considered to be below acceptable levels
- Any statutory regulation by either the state or local government which mandates; for example, a minimum number of officers performing the fire watch
- A look at your local competitors to see what level

of protection they have determined necessary for their properties

Some factors that must be taken into account, however, are the size of the properties reviewed, number of employees, geographical location, crime rate for the neighborhood, and any other factor which could affect the demands placed on security in that particular hotel. A security manager can take advantage of the information sources found in the various professional security organizations, by interacting with other security managers in similiar properties, and by reading trade publications related to the industry.

The structure of the department is fairly standardized, in that there is usually a manager or director, assistant manager (although smaller properties frequently argue that they cannot afford this position), shift supervisors, and line officers. It is essential that the security manager have at least one person as an assistant or senior supervisor who can assume the responsibility for the department in case of absence. Further, some form of supervisory coverage during the hours the manager is off duty must be maintained or the hotel will risk liability for *inadequate supervision*, a recently developed theory of tort liability based on the premise that the officers were performing their duties without any supervision or system of supervision.

The final issue for security department organization is the reporting level of the department. This issue asks the question: To whom does security report in the organization? Security managers are typically department heads with equivalent status of any other hotel department head. Unfortunately, there are great inconsistancies in the reporting level for this position. Some of these variations include reporting to the general manager, assistant general manager, rooms division manager, comptroller, and personnel manager.

A security manager's objective should be to report to the highest level in the organization that is practical. Although it is preferable to report to the general manager, the general manager may not be accessible frequently enough, particularly in a large hotel, to provide the assistance and feedback that the security manager will require. The reason why the security function should report to the highest level of management is primarily due to the sensitive nature of the information generated within security, and the possibility of criminal activity by employees at the middle or upper levels of management. A security manager is placed in an extremely difficult position whenever faced with the problem of theft or other illegal activity which is allegedly occurring at the department head or higher level. She or he must be assured of total cooperation and confidentiality during the course of an investigation. Some more serious offenses, such as drug sales, may be investigated by the local police. However, with the fiscal restraints faced by most police departments today, a security manager is generally left alone to resolve the company's problems.

Security's Reputation

A hotel's reputation as a well-run, clean, and comfortable establishment is essential for its success. The same argument can be made for the reputation of the security department and certainly, the hotel's reputation will depend somewhat upon the success of the security program. That program's success is tied directly to the reputation and image that the department holds in the opinion of the hotel's employees, its guests, and the general public.

How does one establish a good reputation for the security department and its program of loss prevention?

In order to be considered capable and effective from the employees' perspective, the members of the security department and the manager in particular, must project a positive image. Image is affected by a variety of factors, including outward appearance—grooming, demeanor and attire, personality—friendly and caring but professional at all times, consistency in treatment of employees in situations requiring security intervention, and emphasis on the positive aspect of services provided to fellow employees. It is a popular trend, for example, to eliminate the traditional security uniform and replace it with a navy and grey business attire. The security manager sets the tone for the organization and should maintain positive interaction with all levels of employees, never using authority as a means of exerting influence or pressure on the employee.

From the guest's point of view, image and reputation will be determined primarily by the nature of the guest's contact, if any, with security while staying in the hotel. Most guests do not have any direct contact with security and those who do will have such contact under adverse conditions, such as a theft of property or some other negative situation. Therefore, the reputation of security will depend on: first, the outward appearance as discussed above and second, the way in which the security officer helps the guest in the adverse situation. Any victim, regardless of the seriousness of the loss or inconvenience, will behave a in fairly predictable manner, frequently blaming the security officer for not doing her or his job. It is essential that security personnel recognize that the guest response is a normal reaction to stress. Furthermore, most victims know that it is unlikely that the harm they suffered cannot be undone and therefore, really only look to the officer for some compassion and information. Most victims simply want someone to listen to their plight, which suggests that security officers must develop good listening skills.

To the outside world, the public at large, the reputation of a hotel's security department will depend on all the factors cited above and the relationship that the department and its members, especially the manager, have established in the community. That community includes other hotels, the local police, the courts, and the criminals themselves, who have been known to have their own information system. Prostitutes, for example, know which hotels are easy to work in and which ones are not. The security manager should make every effort to become involved in community activities related to crime and the hotel's business. As mentioned, relations with the local police are extremely important, especially in times of limited staffing in police departments. Problems of crime will occasionally be so serious that police assistance will become necessary. A poor relationship with the local law enforcement authorities will impede solving such problems. The security function must be considered thorough and competent by the general public or the *rent-a-cop* image will attach itself to the reputation of the hotel and security department.

Innkeeper's Liability

There are so many different circumstances in which a hotel or motel may become liable that entire texts have been dedicated to discussion of such vulnerabilities. The reason is due to the myriad of services and general nature of the business that the hospitality industry represents. In effect, hotel operators are telling their prospective customers to stay with them overnight, bring their family, their car, eat, drink, and use all the hotel's facilities while they are guests. Guests are encouraged to entrust their valuables and personal safety with the innkeeper.

The issues explored in this chapter will be limited to the most common types of liability problems that arise and the function of security in these areas.

In the traditional role of hotel security, concern for or involvement in a liability problem was more a result of coincidence. Management was often reluctant to involve security because they had questionable confidence in the staff. With the phenomenal growth of liability throughout the country and in the hospitality industry in particular, security has taken on major responsibilities to investigate, report, and develop loss prevention programs for many of these liability problems. It makes considerable sense that the security manager in a hotel would have the responsibility for overseeing and managing most liability problems, including the loss of guest property, claims of guest injury, damage to or loss of guest vehicles, and employee accidents. Particularly in view of the fact that security is usually the first department to which losses and injuries are reported, the chances are better that information gathered at the initial stages will be more reliable and the methods of data collection can be controlled with the greatest consistency.

Guest Property

The most common claim against a hotel made by a guest is that property has been stolen from the room. Security personnel must be cautioned not to assume that all reports of lost property involve stolen property. It just may be missing. It is common for victims of a loss to blame someone else; therefore, adequate information must be gathered to determine whether the item was in fact stolen or simply misplaced by the guest. The reason is fairly obvious —lost property will require one investigative approach, while stolen property another.

To determine whether a claim of stolen guest property is valid, a number of prerequistites must be met before the security investigator can recommend which action is to be taken by the hotel (i.e., to pay damages or deny liability for the loss). Guest status must first be established by showing that the individual was in fact a guest of the hotel/motel at the time the loss occurred. Such status is important, for case law has been clear that without the requisite status, the standard of care due an individual not a guest will not be as great. Liability will therefore be less likely to follow. Such status, however, is not as obvious as one may think. In the 1957 case of *Fidelman-Danziger, Inc. v Statler Management*, 136 A2nd 119, the court noted that "to establish the relationship of innkeeper and guest, the parties must intend to do so; the person accomodated must be received as a guest and must procure accomodations in that capacity, although it is not essential that he register."

The second concern is whether the guest property alleged to be stolen was in custody of the innkeeper at the time of its disappearance. The legal doctrine *infra hospitium*, meaning within the inn, states that the property of a guest must either come within the walls of the inn or come under the innkeeper's care, thus is a rule of care and custody (*Davidson v. Madison Corp.*, 247 NYS 789, 795, 1931). Care and custody more commonly arises under the concept of bailment, which is defined by Blacks' Law Dictionary as:

> A delivery of goods or personal property, by one person to another, in trust for the execution of a special object upon or in relation to such goods, beneficial either to the bailor or bailee or both, and upon a contract, express or implied, to perform the trust and carry out such object, and thereupon to redeliver the goods to the bailor or otherwise dispose of the same in conformity with the purpose of the trust.

The nature of a bailment is the entrustment of property from one person to another.

Traditionally, the courts have held innkeepers absolutely liable for any losses sustained by guests while at the inn with the intent of protecting the guest/traveler from the dishonesty or negligence of the innkeeper or servants. In response to what amounts to strict liability for guest property, innkeepers have been aided by most state's legislatures enacting legislation limiting an innkeeper's liability for guest property to a fixed amount. This amount varies from state to state (e.g., $300 in Massachusetts, $750 in Nevada, etc.), and determination of fault also varies, with some states requiring that the guest/claimant prove negligence on the innkeeper's part, while other states simply requiring that a guest prove to have been a guest (see *guest status* above) at the time of the loss and that an item or items were missing.

An innkeeper and security manager should be familiar with the governing statute of the state, for in addition to amounts and issues of fault generally there are requirements that a copy of the statute be posted for each guest to view and therefore be put on notice or made aware of such limitations if the inn is going to enjoy the benefit of the limited liability.

The legislature's reasoning for passage of such laws relates to the difficulty of innkeepers to prevent crimes involving personal property and the likelihood of fraudulent claims (*Morris v. Hotel Riviera, Inc.*, 704 F.2nd 1113, 1983). Although the validity of these statutes has been challenged on numerous occasions, including the preceding case cited, the courts have consistently upheld the statutory limits, even where the claims for losses were for thousands of dollars. The implications for the security investigator are to be sure that the statute was in fact posted at the time of the loss, determine if the claimant was a guest (per the above discussion) and if the property alleged to be stolen was indeed within the inn.

Guest Injuries

Guest injuries are common and result from a great variety of hazards including wet floors, worn carpets, electrical wires improperly laid along a floor, and so on. The type of guest injury of greatest concern to security professionals is allegations of unsafe conditions based on the argument that the hotel had an inadequate level of security protection. With the publicity received by the so-called *Connie Francis* case *Garzilli v. Howard Johnson's Motor Lodges, Inc.*, 419 F. Supp. 1210 (1976), there has been a string of similiar cases against hotels and motels for injuries received by guests as a result of attacks by third party criminals. It is fair to say that this is a popular theory of liability now that the public is aware of the necessity for security protection in a hotel and the lack of public/jury tolerance to perceived inadequate levels of protection.

An innkeeper has the duty to exercise reasonable care for guests. If a hotel is experiencing an increase in crime or if the surrounding neighborhood is having problems with crime in a way that could affect the hotel's guests, and the innkeeper fails to respond to that threat by improving or increasing security measures, the innkeeper may be liable for any injuries suffered by guests at the hands of a third party criminal. Even if the offender is not caught, providing that the injury occurred in or immediately adjacent to the hotel's property, the hotel may pay.

Liability in these cases is predicated on the theory of negligence, in that it was reasonably foreseeable that the injury would occur if corrective action were not taken. If the innkeeper knew or should have known of the threat, the *reasonable care* rule will apply, establishing that the innkeeper had a duty to protect the guests, failed to do so and as a result a guest was harmed.

Implications for Security

The implications for security and management focus on prevention. Once the injury has occurred, a routine investigation will be conducted and the offender may or may not be caught. The objective must be placed, however, on preventing the loss in the first place. In view of the above and the standards by which hotels are measured, it is incumbent upon the innkeeper through the security department to keep informed of recent trends or changes in crime patterns both within the hotel and in the surrounding neighborhood. Maintaining a positive working relationship with the local police is essential for a regular flow of information regarding crime in the neighborhood. The security department's reporting system should include both monthly and semiannual reports which will summarize all criminal activity for the time period covered and will make everyone aware of emerging patterns or changes.

What Is Adequate Security?

There is no acceptable answer to that question due to the lack of established industry standards. Courts make the determination of adequacy on a case-by-case basis. It is helpful to look at other hotels, motels, and inns which are similarly situated and of similiar size to try to assess what is acceptable within a given community. This comparison is referred to as the *community standard* and is also used by the courts to assess the property being sued. Community standards can be misleading if it can be shown that the entire industry was below acceptable levels of protection in view of the crime problems and cost of preventive measures.

to leave the hotel and drive a car. If a security officer is called for assistance in such circumstances, everything possible should be done to discourage the guest from driving. A hotel may even want to consider the extreme measure of refusing to return the guest's keys or calling the local police department. This action may cause other problems regarding the inappropriateness of refusing to give a person back her or his property, but the hotel operator must weigh the risks and decide on the safest option.

Security frequently receives calls from bartenders who may need assistance discontinuing service to a guest who in the bartender's opinion has had enough to drink. The security officer must respond promptly to such calls, for the situation is apt to get out of hand quickly. The security officer should obtain all relevent information from the bartender and without question, back up the decision to discontinue service. Security should never try to overrule the server's decision. The same rule applies to any member of management who is called in for assistance in such circumstances.

On the subsequent report and follow-up investigation, it is essential that all pertinent facts, names, and addresses of witnesses and documentation (e.g., bar tabs) be obtained for future reference.

Alcohol Server Liability

Alcohol server liability is a form of third-party liability which arises when the server of alcoholic beverages is held liable for the actions of an intoxicated person who consumed alcoholic beverages in the liquor serving establishment and, as a result of the intoxication, causes injury to self or another person. It is usually food and beverage department employees such as bartenders, waitresses, and managers who are directly in control and responsible for alcohol consumption. Security does have a function to perform, however, in helping to prevent intoxicated patrons from driving a vehicle or becoming injured while walking around the hotel property. Not all alcohol liability cases are limited to motor vehicle accidents. Guests can become injured by falling down in an otherwise safe area due to intoxication. The innkeeper is not in the position in such cases, unfortunately, of being able to claim the defense that it was the guest's intoxication that caused the injury—when it was the innkeeper in the first place who allowed him to get drunk.

The type of situation which presents the greatest risk of alcohol liability arises when the guest attempts

Theft

Theft, which occurs in a variety of ways, represents the single greatest problem for security in a hotel. Almost all of the items found in a hotel or motel and used by the guest can also be used at home. Guests usually travel with cash, credit cards, jewelry, cameras, and other valuables which appeal to a thief. Further, the hospitality industry is a labor intensive industry. Large urban hotels employ several hundred employees, creating ample opportunity for a thief to steal the employee's personal property as well. Although every type of theft situation—hotel, guest, or employee property, will require a different response or prevention measure by security, all are forms of theft and as such are a major concern to innkeepers.

Hotel Property

Contrary to common belief, hotels do not generally build a certain dollar amount into their budgets for anticipated losses. Losses are not acceptable, nor are they part of the operating costs of a business to be offset by increased room rates. Certain types of losses are tolerated, however, due to the sensitive nature of the circumstances surrounding the theft. An example of such a sensitive problem is the theft of linens and other items from a guest room by the guest. Innkeepers are often reluctant to question a guest suspected of removing towels or other objects. The risk of liability for slander and/or ill will affecting future business is usually seen as far outweighing the benefit of recovering items believed stolen.

The security manager's task is to determine which of the hotel's property is vulnerable to theft, and then devise a program which will reduce the risk. The usual approach is to conduct a risk analysis of the property, identifying the likelihood of loss and determining what measures may be employed to reduce risk. In addition to the actual physical survey, the security manager should review all reports of losses for the previous year to assess where and possibly how items were stolen. This approach will help to identify the most vulnerable areas of the hotel.

The key to the protection of all property is to obtain cooperation and assistance from all hotel employees. The security manager who tries to do the job alone is making a dreadful mistake. Security is everyone's job.

Successful loss prevention depends on a well designed program which involves input from all concerned. The security manager's objective is to get cooperation from all the employees. This is done by working with them in a variety of ways:

1. Educate employees about the seriousness of theft and how it affects them.
2. Attempt to develop their understanding of the theft problem in terms that they as members of the organization can relate to and appreciate. For example, if it is made clear that losses cut into the profitability of the property and reduced profits may result in lower wages or even the layoff of employees, they may realize that they too have a financial stake in the prevention of losses.
3. Assistance from employees may be elicited by encouraging them to call and report suspicious people or circumstances and make an extra effort to secure valuables when leaving a work area (e.g., storage rooms).
4. Maintain a good working relationship with all employees. Avoid morale problems created by poor communications and a lack of understanding. Employees suffering from poor morale are not going to be very concerned about protecting the assets of the employer.
5. Work with all other department supervisors to get their input on a variety of solutions to the problem of losses. Some security personnel may think that the solution to the loss problem is locking up and limiting access to the hotel's assets, but if a measure interferes with the normal operation of a hotel, it will be resisted and is a guaranteed failure.

Program design is only half of the task of crime prevention. The other half, equally important, is follow-up. The security manager should repeat the risk survey every six months to ensure compliance with the program and to discover any changes in the physical structure or the hotel operation that may negate the original plan. The program must always remain flexible to meet the needs of the business. New and better methods should always be sought for improved loss prevention. When losses do occur, as they will, evaluate the loss and adjust the program accordingly.

Guest Property

The greatest reason why hotel operators are concerned about the theft of guest property is the potential impact upon future business. As discussed in the liability section, most states have limited an innkeeper's liability to a fixed amount. Generally that amount is not very high. A history of serious guest losses can, however, drastically affect the hotel's reputation and may indicate other weaknesses in the security system which could lead to more serious crimes. For example, if poor key control was to blame for a number of guest losses, that same lack of key control can increase the chances of an assault on a guest. Such criminal activity is a common area for lawsuits against innkeepers.

There are two aspects of guest room thefts that are appropriate for discussion: investigation and prevention. Investigation of such losses should include an examination of the locking hardware on the room to ensure that it is working properly, since guests frequently may not close their doors tightly when leaving or when running down the hall for ice or soda. Next, determine which employees have access to the room under investigation and note same for further follow-up. If good records are kept, after a short while the same employee's name will begin to appear. During my first month as a new security director in a large urban hotel, similar recordkeeping revealed the same name in over twenty thefts. Once that information was discovered, the investigation could be focused more on that individual. Eventually, the employee was caught and prosecuted. Third, find out when the locks were last changed or rekeyed. There may be an overall problem with key control. And fourth, check for missing room or master keys which anyone could use to gain entry to guest rooms.

Prevention of guest room thefts depends on cooperation of employee and guest. Employee cooperation and assistance was discussed above and applies to guest losses as well. Guest cooperation is more difficult to obtain due to guests' transient nature but can be achieved to some reasonable extent. The type of cooperation sought is generally limited to two areas: use of existing risk-reducing facilities such as safe deposit boxes and in-room safes, and making sure that guests close and properly lock room doors when they leave or go to sleep. Some properties will put small signs or fliers in the guest rooms advertising the availability of safe deposit boxes and the like. Other hotels have used the cable system on the room television to promote safety and security measures.

Employee Property

Employee personal property, uniforms, and equipment are subject to theft in locker rooms and work areas. Although the theft of such property is not generally seen as a liability problem, it is of concern from the standpoint of employee morale. Poor employee morale will reduce productivity and will fail to promote the kind of cooperation that is needed by the security department to prevent other losses.

The theft of any employee property should be treated with the same level of concern and priority as any other loss. The two most common areas for losses to occur are locker rooms and office areas. Lockers must be equipped with adequate locks and a sound key control system. Security can check locker rooms frequently during their rounds, both night and day. If local law and/or the employment agreement allows, employees' lockers should be inspected. Such inspections reveal health problems (e.g., pests) and problems such as stockpiling of uniforms or other hotel equipment. Consistent with this approach is the need for a company policy on what may be kept in employees' lockers and the company's right to inspect them periodically. In office areas, the most frequent loss occurs when female employees leave their handbags unattended in one of the lower, unlocked drawers. The security manager needs to be sure that the desks are capable of being locked and to encourage those employees to utilize those locks to protect their personal property.

Prostitution

Prostitution has been around as long as have inns. Some say prostitution is a victimless crime and society should not care what consenting adults do in privacy. During my tenure as a hotel security director, several people actually asked me if it was true that large hotels often provided such services for their guests. Well, if any do I am not aware of them and none should, for as far as hotels are concerned, prostitution is not a victimless crime. It is a problem that can ruin a hotel's reputation. Prostitution cannot and will not ever be eliminated. The security objective, however, should be to reduce such activity to an absolute minimum. A hotel's reputation is, without doubt, its single most important asset. No matter how professional the services may be, no one respectable will stay at your hotel if it is identified with prostitution.

Theft of the guest's property, assaults on the guest, and drugs are but a sample of the problems that can arise from prostitution. It is in the best interests of the hotel operator and guests alike that prostitution be controlled.

There are a number of simple methods of control that a hotel, motel, or inn can employ. First, guest registration procedures must be established and followed closely. All guests must be required to show valid photo identification such as a driver's license and preferably a credit card, even if they are paying cash in advance for their stay. The reason for the tight registration process is that most prostitutes will not carry identification. The more sophisticated call girl type of prostitute, however, will have identification.

The second method of control can be utilized in those properties which have computerized front desks and can print out a complete list of all guests by room as well as list the number of occupants registered in that particular room. Once armed with that list, the security officer, particularly at night when guests are returning to their rooms with a visitor they may have just met in the local bar, can screen all incoming guests at the elevators (hotel policy regarding visitor access after a certain hour should be implemented prior to taking this action). This method has two benefits: the innkeeper is entitled as a matter of law to the additional revenue that a second person staying in a room represents; and if in fact the visitor is a prostitute, she or he may not have identification or the guest may not want to risk having a second person listed on the registration folio. The final two methods of prevention require employee cooperation. The housekeeping department will know if a certain guest room has heavy traffic going to and from it and can alert security. And hotel managers must maintain a *no tolerance* attitude toward prostitution while supporting security in its effort to deal with the problem.

Bars and Restaurants

There are three general types of problems associated with bars and restaurants in a hotel. They are intoxicated patrons, theft of food and beverage, and credit card theft. Incidents involving intoxicated patrons are probably the most common for security to handle, especially at night, and are not limited to the lounge areas. It seems as though alcohol consumption is an element in a majority of circumstances requiring security intervention or assistance in a hotel.

Removing an intoxicated patron from a lounge is a difficult and dangerous situation. The potential for injuries to other patrons, the security officer, employees, and the person intoxicated is ever-present. It is essential that security personnel receive the proper training for the handling of such persons. Force should be avoided and patience exercised to the limit. The objective is avoiding confrontation and then dealing with the individual in such a manner as to avoid the type of liability previously discussed in the alcohol liability section.

Walk-outs are nonpaying customers who have enjoyed a meal or drinks in a hotel and then proceed to leave without offering to pay for the items consumed. Unfortunately, this type of crime is only a misdimeanor in most states, and the innkeeper is very limited in what to do after the fact. Frequently, if the patron is caught leaving, they can be confronted and if they refuse to pay, be held long enough for security or the police to arrive so that the thief can be identified for future reference or prosecution. The amount of the loss usually is not worth the time required to prosecute, so it is recommended that the name and description of the individual be recorded and distributed to all outlets in the hotel. If the person tries to obtain food or drink again, they can be refused until they pay for the previous bill.

Credit cards are used very often due to their convenience and the generation of tax records for business customers. This convenience, however, has caused some problems for the hotel operator. Problems ranging from use of stolen cards, cards over their allowable limit, inflating gratuities, or even writing in gratuities by dishonest employees, are just some of the concerns facing hotel management. The best means of avoiding these and other scams is to develop detailed policies and procedures for authorization of use and training employees to follow the procedures. Management must then follow-up to ensure compliance with the rules; any violators should be dealt with appropriately. If security is called in to investigate alleged wrongdoing by employees, swift and severe action should be planned.

Drugs

The sales and use of illicit drugs are a major issue in any industry but appear more complicated in the hotel industry. The innkeeper must be concerned with both employees and guests who become involved with the use and selling of drugs.

The use of drugs by employees either on or off the job presents some serious consequences affecting both performance and safety. With the cost of worker's compensation insurance rising to high and representing one of the largest liability expenses, employers must address the increased use of drugs by employees. Only recently have employers begun to consider the possibility that some employee accidents are directly related to drug use on or off the job. Even employees who do not use drugs but who are aware of others who do are affected. Morale problems are likely to arise among those employees who want to do their jobs well, but see others not performing because of drug use. Furthermore, a work environment that allows employees to sell drugs to other employees will create the impression that management does not care.

Although still controversial at this stage, employers may wish to consider the possibility of drug testing on employees whose performance levels decline, or who have exhibited unusual behavior (e.g., excitability, moodiness, changes in personality, mood swings). Legal counsel should be consulted before this action is considered.

Drug use or sale by guests is not an unusual problem, but is one that hotel operators should be prepared to resolve. It is very unlikely that a hotel manager will be able to prevent a guest from using drugs in the privacy of a room. If, however, management becomes aware of such use, action should be taken. Limited use (e.g., smoking a marijuana cigarette) may best be dealt with by warning the guest that repeated activity will result in removal from the hotel. Evidence of more serious use (e.g., large parties with cocaine left out in the open) may require assistance from the local police. It is common for drug dealers to use large urban hotels to conduct their drug deals, for the largeness of the hotel helps ensure their anonymity. Certainly, no hotel operator wants to see their hotel aired on the evening news as a result of a major drug bust.

Medical Emergencies

Hotels are not required to provide medical services for guests. In busy hotels, however, a large percentage of the calls for security services will be medically related. These calls will range from simple abrasions or burns to serious trauma. Even if your hotel is in the middle of a large city with several hospitals and ambulance services, significant delays can be expected before expert assistance will arrive. Although many of the calls that security will receive in a given day require some medical assistance, they may not justify the cost of a full time physician who would need to be available twenty-four hours a day, seven days a week.

The answer is training. Security personnel and/or any hotel employee who are likely to respond to calls for such services must be properly trained. The key is to train for the level of services which you are prepared to provide and no more. Once medical assistance has commenced, you are expected to carry it out properly. The legal liabilities that arise against any person in such situations generally relate to how they performed the services rendered. You are expected to render only that level of aid for which you are duly qualified. In short, a security officer must not attempt to perform CPR on a victim unless certified to perform CPR.

The level of training recommended for most hotels is basic first aid and CPR, both of which are generally available through the local chapter of the American Red Cross and are of nominal expense. In a very large property, some security employees can be trained as emergency medical technicians (EMTs). The risk, however, of such advanced training is that guests may come to expect medical care all the time.

Protection of Public Figures

Depending upon the size and location of a hotel, certain types of public figures are likely to stay as guests on occasion. In larger, convention class hotels, political figures are often in attendance for some social event. Public figures often draw crowds. These crowds may be admirers or demonstrators protesting the appearance of the individual. In either instance, security is faced with the difficult task of controlling the crowd while protecting the public figure. One should not assume that the local police will always take care of these problems. Instead, the security manager should be prepared to work with the authorities in planning for the person's arrival, stay and departure.

Planning is the key. Occasionally, hotel security will not receive any advance notice of the person's arrival. In that case, some general planning such as previously established policies and procedures activity should be in place. Generally, there will be varying amounts of lead time available before the upcoming event. With adequate lead time, detailed planning will avert risk with minimal difficulty.

The following are issues for consideration during the planning stage:

- Communications systems—be sure all equipment is operating or rent additional radios if needed
- Transportation—transportation of the celebrity to and from the hotel/airport, with alternate routes worked out and alternate escape routes from within the hotel
- Emergency equipment—medical equipment and an ambulance if needed standing by
- Access control—limit access to the areas where this person will be located, by both bystanders and other hotel employees; work out provisions for crowd control—roping off areas, using mounted police officers, etc.
- Intelligence—if this public figure has been at other hotels or events recently, contact the security and/or police in that area for any information that might be helpful in planning for the upcoming event (e.g., were there demonstrators, who were they, any violence, etc.)

Basic Hotel and Motel Premises Protection

WALTER J. BUZBY, II

One of the most important considerations in a security program for any hotel is access control—access not only to the general overall premises, which may include golf courses, beaches, and other outdoor areas, but also to parking areas and the actual buildings that constitute the hotel complex. The innkeeper is in a business and therefore must have patrons and guests. The innkeeper not only sells rooms, but also requires customers for restaurants, shops, bars and other facilities. The innkeeper invites the public onto the premises and, indeed, under Federal Civil Rights Laws, the innkeeper cannot refuse to accept a customer except where accommodations or facilities are not available or where the customer fails to meet the restrictive and definite exceptions as set forth by the law. Intoxicated persons or known criminals, persons deemed to be filthy, profane, behaving indecently or improperly, or one suffering from contagious or communicable diseases may be refused, as well as those who have no legitimate business being on the property.

In addition to the access afforded the public, there is also the access required for employees and staff of the hotel, and this must also be controlled. Purveyors, tradesmen and outside contractors of various types will enter the premises and must be controlled. Just how can access control be provided?

The answer to that question can come only from a survey of the property and the conditions that exist at that particular property. Some of the factors that will be taken into consideration during the survey will be:

1. The geographic location of the hotel—city and neighborhood
2. The size of the property and the actual physical layout of the building(s) to be protected
3. The type of clientele frequenting the hotel
4. The number of employees available or normally on duty at any given time
5. The availability of local law enforcement, fire fighting, and other emergency services and their normal response time
6. Local, state, or federal regulations that may have a bearing on what can or cannot be done by the innkeeper or what may be required by law
7. The local crime index
8. The availability of a labor force and the type of people that may be expected to comprise that labor force
9. The history of the particular hotel, as it relates to security problems and incidents of crime

The answer to each of these survey factors, as well as many others, will help to identify the circumstances that exist and will aid in establishing what problems presently exist, what must be changed, and what must be added in order to provide basic premises protection. To merely add hardware or labor power without knowing what purpose will be served and what benefits will accrue is money and time wasted and probably will not result in the desired results.

Bearing in mind that the object of premises protection is to protect the physical plant, as well as all who may be in or upon the property, after the survey is completed a decision can be made as to how we will accomplish this task and what tools are required. Consideration will have to be given to the possible need of fences, lighting, gates, alarm systems, communication systems, locks, safes, closed circuit television, tamper-proof screws, antitheft bolting devices, various monitoring devices, wheel locks, timers, and many other mechanical or physical items considered security hardware. In addition will be considerations of training programs for employees, preplanned emergency programs, policy decisions pertaining to operations in relation to security, internal audit and accounting controls, rules and regulations for employees, and rules and regulations that will apply to guests and patrons. Finally, decisions will have to be made relative to the amount of labor power that will be expended to properly operate the security program.

One very important factor that must never be forgotten is that once decisions are made and a program is instituted, the job is not completed. The program is not static and must constantly be reviewed, inspected, tested, maintained and updated when necessary. The court records are filled with examples of defendants who had good systems on paper and had good hardware available, but failed to activate and practice those systems or failed to maintain that equipment. If a system or a piece of equipment or a human being fails to provide the security for which it was intended, either through improper maintenance, improper use, or lack of training or of supervision, negligence will be charged. The unknowing hotel guest is relying on the visible signs of security. When locking the door at night upon retiring, a guest expects that lock to prevent unauthorized entry to the room, short of someone actually breaking the door down to gain entry.

In summary, when premises protection is being considered, we must take into consideration what problems are inherent to this property, what can those problems lead to (foreseeability), and what can we do to correct the situation by the application of hardware, planning, training or labor power. Once we have made these decisions, we must then implement, install, or constantly reevaluate all factors within the program.

Motel Holdup Vulnerability

In a motel, a holdup is most likely to occur during evening and nighttime hours when there are few witnesses available. Several precautions can be taken by the motel operator, including the following:

1. Remove any excess cash to a night depository or safe location, and maintain a very limited night bank.
2. Lock the office or lobby door after a given hour so that it can be opened only by the release of an electric lock by the night clerk after a caller has revealed identity.
3. Install silent alarm systems connected to municipal police departments or central stations. The alarm can be triggered easily by the clerk in attendance. Such systems are valuable to large or small operators at any time of the day or night and are worth the expense.
4. Advertise a policy of accepting as little cash as possible, indicating to the potential thief that the rewards of a holdup would not outweigh the risks involved.
5. Do not rely solely on a sticker attached to the office window, advertising a security service, to ensure security. This practice may deter a potential thief, but it is not complete protection.
6. Keeping a large dog, such as a German shepherd, on the premises may increase protection from attack.

Special Security Situations

There well may be times when an everyday security program will not be capable of handling certain situations. However, part of the overall security program established in any hotel should contain plans which have been developed, tested, and formalized in advance so that they can instantly be placed into operation without taking time to make decisions at the last moment. Decisions that are made under stress of within a time limitation are often poor ones that could result in dangerous situations. The problem should be approached just as the basic premises protection plan is approached—with an identification of the problem, the possible consequences that could develop, and then a decision as to how the situation might best be handled. What type of situations are we talking about?

1. An unusual increase in activities on the property. This may be due to a convention, a gathering of people for a sporting event at the establishment or in the area, a holiday or a special local celebration, the visit to the area or to the establishment of a VIP, or merely seasonal changes.
2. Civil disturbances, labor problems (within the hotel or in the general area)—again, any event or incident that will increase the numbers of people in the area or in any way change the type, attitude or emotions of people from what is normally expected on the premises or in the immediate area.
3. Incidents of disasters or other emergency situations in the area or within the specific establishment.
4. During construction or renovation, when the normal security of an establishment may be compromised by additional means of access, increased hazards are created by the construction. Materials and supplies may be stored on the site and must be protected from theft.

These are but a few examples of the "unusual" situations that may develop.

Where conventions with private meetings or exhibit facilities are involved, often arrangements can be made with the convention officials to share or assume the added costs of providing security. Outside contract guard services can be employed directly by the convention officials, or the hotel can provide the added security for an agreed fee. The hotel still has its own security responsibilities above and beyond any special considerations or requirements created by the convention or the meeting.

Depending on the VIP involved, added security may be provided by local, state or federal agencies; however, there must be close cooperation with the hotel security forces. The outside agencies are concerned with the protection of the VIP and not as concerned with the hotel property or other guests within the establishment. That is the responsibility of the innkeeper. Preplanning and cooperation cannot be overemphasized.

Fire Protection and Evacuation Procedures

At the top of any list of priorities for an innkeeper should be the need for fire protection and an emergency plan to automatically commence operations when such an incident occurs. Many of the available systems developed to prevent, detect, report, control, and extinguish fires are now required by law to be present in hotels. Some of these requirements pertain only to buildings constructed after a certain date or to be constructed in the future. Many older hotels are exempt from complying with current regulations or portions thereof. Each hotel will require individual consideration as to what is required due to size, physical layout, age, type of construction, as well as legal requirements.

Bear in mind one very important factor. The occupants of a hotel are, for the most part, strangers to the establishment. They enter through the front door, go to their room via an elevator, walk down a hallway to the room assigned to them, and are unaware of any other means of reaching or leaving their room. When awakened in the middle of the night by the smell of smoke or the sound of an alarm bell, they have no idea how to escape, or what to do for their own protection. The problems to be encountered in a two- or three-story building are tremendously increased in a high-rise hotel of 50, 60 and more stories, yet, just as many fires occur and just as many people die in the two-story building as in the high-rise. Were it not for heroic actions on the part of many fire and rescue workers, the numbers of those who have died in hotel fires across the nation would be much higher.

Records show a number of causes for hotel fires with careless smoking being one of the leaders. Mechanical malfunctions involving electricity or other utilities may cause ignition, or it may be an act of arson that brings the fire into existence. In most cases, however, the initial fire is small and confined and could be extinguished easily except for conditions that prevent such action.

Causes of Hotel Fires and Injuries

Even a very small fire, however, can be extremely dangerous. Most people do not die or are not injured by the flames, but rather by the by-products of the fire—the smoke, heat, toxic gases produced by the burning material, and by panic. If we understand how and why these conditions develop, perhaps we will be able to understand what to do to prevent such conditions from developing.

Some of the factors involved in hotel fire deaths and injuries would be:

1. *Delayed discovery.* Most serious fires occur at night when guests are sleeping and the number of employees on duty is at a minimum. In the absence of heat/smoke detectors or other sensing devices, the fire goes undetected until discovered by someone smelling smoke, a casual passerby or someone who just happens to stumble on the scene.
2. *Failure to notify the fire department.* Even though pull boxes may be available and utilized by the person discovering the fire, many such boxes are only proprietary alarms, sounding alarms or bells in the immediate area, a central location, or even throughout the entire complex, but they are not connected to the local fire department or to a central station. Consequently, no one calls the fire department.
3. *Failure to notify the guests.* Often this occurs because there is no alarm system, or the system does not operate properly, or is improperly designed. Many instances have occurred wherein the fire alarm bells operated, but the ringing bell could not be heard by the guests asleep in their rooms. In some cases, the system itself failed to work because it was not maintained properly and had not been tested to ensure its operation prior to use.
4. *Use of highly combustible materials in decoration or design.* The presence of highly combustible materials accelerates the fire. Often in the name of beauty or distinctive and attractive visual effects, safety is compromised and dangers are created.
5. *Improper or lack of training of employees.* Often employees are placed in charge but no effort has been made to instruct them in their duties as they apply to emergency situations. This training should be provided for every employee.
6. *Failure to comply with existing codes.* The hotel has failed to comply with legal requirements through either oversight or pure neglect, and such violations have not been cited by inspecting authorities or, for some reason, the requirements have not been enforced. Such requirements might include the location and number of emergency exits, directional signs, emergency lighting, watch patrol service, unprotected openings, presence of fire doors, extinguishers and other first aid fire fighting equipment, posting of instructions for each guest room, control of elevators, and many others.
7. *Failure to have sufficient personnel on duty to provide services.* It is one thing to have a program of emergency response and all the tools and hardware to ensure an adequate response to the situation, but people are still required.

Normally, it is not just one factor that results in a serious fire or emergency situation, but rather a combination of conditions and events. A fire is discovered at some point in time. No one calls the fire department immediately. The fire spreads rapidly, feeding on combustible materials and through openings where fire doors had been propped open or where wedges have been used. The fusible link is rendered inoperative. Guests are not awakened, as they cannot hear the fire bells or the telephone operator has not gotten around to calling their rooms to warn them. If and when they do become aware of the fire and attempt to escape from their bedrooms, they are confronted by dense, black smoke which blinds them and prevents them from finding the nearest exit. How would you like to escape from such a situation, make your way to a fire tower, and stumble down the stairs—only to be confronted by an emergency exit locked with a chain and padlock? It has happened and many excuses have been offered as to why the condition existed, but no excuse will ever be accepted for the loss of a human life under such circumstances.

Evacuation of a burning building will depend to a great extent on the size of the building and the amount of fire involved. The decision to evacuate in the event of a fire will usually rest with the senior fire officer present. In a smaller establishment, and certainly in any size building in the area of the fire, evacuation must take place, and the hotel must know in advance exactly how this will be accomplished. Escape routes must be preplanned, assistance must be provided, and plans must be made as to where the evacuees will be placed. Decisions such as this cannot be left for the moment of the incident but must be made in advance. The hotel should be aware of any potential special problems, such as the presence of handicapped persons who may require assistance. There must be a method developed to provide that special aid.

The decision to evacuate due to a bomb threat is the responsibility of management. It is strongly recommended that the advice of law enforcement personnel be solicited, but the final decision cannot be delegated. If evacuation is necessary, all escape routes should be inspected prior to the commencement of evacuation so that the evacuees are not being routed directly into the most dangerous areas. In addition, evacuees must be removed far enough from the scene to prevent danger from any explosion that might take place. Window glass and walls being blown out of a 15-story building will scatter debris over a large area, and those standing in the street will certainly be in danger.

Probably the most effective tool in any emergency situation is some form of voice communication capable of reaching certain zones or the entire premises simultaneously. Information can be given as to the type of emergency existing and what action the occupants of the building should take for their own protection. Simple, predesigned messages should be prepared either on cassettes or for live reading. The calm, clear voice issuing information and instructions can go a long way in controlling any panic that might otherwise develop.

Internal Theft Control

As in any business or building, theft control will be accomplished in two basic ways—by the use of internal audit controls and by locks or some form of locking device. Thefts are quite often perpetrated by people who are not really criminals at heart, but who see an opportunity and feel that no one really cares anyway. If people are held accountable for goods and materials entrusted to them, and if they know that someone is watching them, the temptation will be greatly reduced. By separating responsibility and accountability, the opportunity is also decreased. For example, if one trusted employee is responsible for purchasing, receiving, storing, issuing, and approving payments for goods purchased, the opportunity for stealing and fraud is present. Good accounting practices involve more than the way the books of the company are kept. They also provide internal controls—a system of checks and counterchecks—that will create accountability.

Internal Audit Controls

The term *trusted employee* is one that deserves some thought. Just because an employee has been with the hotel for years does not ensure complete honesty. Having been with the hotel, this person knows the system, the procedures, and is in a far better position to commit a crime or a fraud and get away with it than the new employee, who is unfamiliar with the system. It is not necessary that we create a feeling of mistrust of employees, but by applying good, sound business practices and having a policy that creates accountability, the employees will know exactly where they stand and what is expected of them.

Certain employees must enter guest rooms for the purpose of cleaning, repairs, etc. It is hoped that these employees will all be honest and trustworthy. Unfortunately, such is not always the case. No one has a better opportunity to steal from a guest room than those with a legitimate reason to be in the room. When a group of employees band together in a conspiracy, the problem is intensified. For example, suppose a maid steals some valuables from a guest's room while in the process of cleaning. He or she hides the stolen merchandise in a pocket, on the maid's cart or in the supply closet. A housekeeper delivering clean linen to the maid picks up the stolen merchandise and hides it in her or his locker, or perhaps passes it on to a gardener. The gardener, having access to the outside of the building, passes it along to a confederate on the street. The theft is completed, and it is going to be a difficult case to solve. As mentioned before, complete reports should be kept of such cases to see if a pattern develops.

Supply closets and maid's carts are the property of the hotel and can be inspected by supervisory personnel at any time. Inspectors and security officers should make routine checks regularly. At best, these inspections may be a deterrent; at least the employees will know that someone is watching them. Some employees may object, stating that they are under suspicion and being treated as common criminals. It can be pointed out to these employees that the inspection is for their own good, as a third party may have hidden stolen goods in their equipment or storage rooms. If and when a loss is reported by a guest, an immediate search should be made of all possible hiding places in the vicinity.

It should be the policy of the hotel that every guest room must be inspected at least once every twenty-four hours regardless of whether the room is occupied. There are a number of reasons for the policy:

1. Mechanical breakdowns may occur that should be discovered as soon as possible, such as a lavatory that will not stop running or a leak in a pipe.
2. Lights or television sets may be left on inadvertently.
3. The room may have been disturbed in some way and be in no condition to receive a guest.
4. Where a *Do Not Disturb* sign remains on a room door over twenty-four hours and no activity has been seen or heard in the room, management should be notified and the matter investigated. Deaths often occur under such circumstances, and it is most desirable that such incidents be discovered as soon as possible.
5. *Skippers* or thefts of hotel property from rooms can be discovered.

Normally, a room count is made by the maid as soon as she or he comes on duty. Where there is some indication that a guest is in the room—the *Do Not Disturb* sign is on the door or the indicator button on the lock shows the room is occupied—the maid merely reports the room occupied. Where there is no such indication, the maid enters the room. If it is vacant, this is indicated on the report. These reports are forwarded to the office, where they are checked against the hotel register. Any discrepancies should be investigated at once. This investigation is usually handled by the housekeeping department and constitutes merely a recheck of conditions. If a real discrepancy does occur, the security officer can immediately look into the matter.

In the process of a maid's work in a guest room, a great deal of information can be gathered which might be of considerable help to the security department. Anything of an unusual nature should be reported by the maid to the supervisor or to the security department. Such observation might include:

1. The presence of luggage without the normal contents, such as clothing. Many persons intent upon stealing property from a hotel room will fill their luggage with bricks, old telephone books or other useless material, dispose of these items and pack their bags with the hotel's linens, blankets, television parts, etc. If a maid spots such empty suitcases, a report of the incident can put the security department on the alert.
2. Firearms or weapons of any type seen in the guest room. They may be legitimate; however, it is better to have the matter investigated by the security department.
3. Evidence of any illegal activity. Gambling equipment, drug paraphernalia, etc., should be reported.
4. Any evidence of damage to the room, such as broken furniture or equipment, a fire, or evidence that someone has suffered physical harm, such as blood spattered on the walls.
5. Evidence that hotel property has been removed and that the guest is no longer an occupant of the room.
6. Any condition in the room that might be harmful to the guest.

Locking Devices

Another second method of theft control is use of locks or locking devices. This can extend from the simple padlock or mortice lock on a storeroom door to electronic alarm systems, CCTV, electronic locking devices, camera surveillance, or a human guard at a post. The decision as to the type of hardware to be used will depend on the value of the property or material to be protected. There is a correlation between the value of an item and the chances of it being stolen.

Locks used on guest room doors should be of sufficient quality to preclude the use of a celluloid strip or simple skeleton key to open the door. The most expensive lock is not necessarily the best, but the least expensive will most likely not provide the security desired. A competent locksmith can be of great assistance in choosing a lock.

An additional matter must be considered when selecting hardware for doors. We not only want to prevent illegal entry into a guest room in the absence of the guest, but we also want to give protection and privacy while guests are in their rooms. Every bedroom door should lock automatically when closed. Whee it is required to put a key in the lock in order to activate the locked position, guests going out quite often forget and leave their doors unlocked. Locking the door from the inside with the use of a key is acceptable, and most hotel bedroom locks operate in this manner. When this is done, a signal device on the outside indicates the room is occupied and prevents maids from attempting to enter. In addition, it is well to provide the guest with a chain or a dead latch.

Any number of devices can be installed to ensure complete security and prevent anyone from entering the room short of breaking down the entire door. Unfortunately, these extremes are impractical, as the innkeeper must be able to enter the room in the event of an emergency. People may become ill, commit suicide, die of natural casues, or become incapacitated and unable to open the door to admit help. Therefore, the locking devices used should give a degree of protection to the guest but, at the same time, be accessible to management.

With the increase in the number of forcible guest room robberies committed in the presence of the guest, it would be well to consider the use of one-way peepholes in bedroom doors so that the guest can see who is knocking at the door without having to open it.

Where sliding glass doors are installed, they should have locking devices on the inside only. An additional security device can be provided by way of a metal bar which lies in the track of the door and prevents it from being opened.

Where French doors are used on outside balconies, the glass in them should be something other than normal window glass which can be easily broken, permitting the thief to reach in and unlock any device on the interior. Most areas have fire codes or life safety codes outlawing the use of transoms; however, some older buildings may still have transoms. These present a security hazard since a small, thin person can sometimes slip through them. In addition, they are very dangerous from a fire protection standpoint. A louvered door is also vulnerable to intrusion. Normally the louver frame can be knocked out easily, permitting admittance.

Entry to a room can be made through a window as well as a door. If a room happens to be on the nineteenth floor and there is no outside ledge, the opportunity of gaining admittance in this manner is greatly reduced. In many cases where air conditioning has been installed, the windows are sealed and admittance cannot be gained short of breaking the window. Even where air conditioning does exist, quite often one window is left unlocked or unsealed so that the occupant can obtain fresh air. In such cases, some type of device must be provided so the window can be locked. A catch installed in the center of a double-hung window provides very little protection, as only a small break in the glass is necessary to allow the thief to reach in, unfasten the catch and lift the window.

There are any number of devices that offer considerable window protection. It is not necessary to make the device one that management can circumvent, as is the case with the door. A window device can and should be absolutely secure short of breaking the window itself to gain entry.

Windows present a particular hazard where air conditioning does not exist and the window is the only form of ventilation. This is true quite often in some older buildings. To make matters worse, these windows may open onto a common balcony accessible to anyone. There is no real answer to this problem. Heavy metal screens, windows that will lock in a given position not large enough to allow entry, or bars on the windows, are all possibilities but are objectionable to many people. The only answer from management's standpoint is to provide an acceptable lock for the window; if the guest decides not to keep the window locked, that becomes her or his risk.

In a motel, especially if units are arranged at ground level, illegal breaking and entering present a special problem. Attractive, heavy window screens can be employed, or windows can be alarmed to prevent illegal entry through breaking. Proper lighting along exterior walls and the absence of shrubs and other cover material will reduce the natural cover for the window thief. Occasional patrols around the property by the proprietor at unannounced times may indicate attempted entries or problem areas, and may deter the potential thief from an attempt.

Key Control

Whatever system of protection is used, the primary concern is that the integrity of the device or system is protected. Key control is vital, protection of combinations, constant monitoring of CCTV, fail-safe alarm systems along with immediate action in the event there is any indication that there has been any compromise of a system.

Probably the greatest problem confronting hotels is key control, not only keys to storerooms and other parts of the hotel, but more important, keys to guest rooms. The individual room keys are lost or stolen and new replacements are made by the thousands. Submasters and grand master keys are lost, stolen, or sold on the street. Think for a moment of the potential for a common thief who can get a grand master key that will open any guest room door in a hotel of 1,000 rooms. Is it any wonder that a thief would be willing to pay $1,000 on the street for such a key? The truly amazing fact is that many hotels, in spite of this loss of individual room keys, do not change the room locks *ever*! Some hotels are even more helpful to the potential thief by attaching to the key a tag with the hotel's name and room number printed on it. Talk about opportunity—what greater invitation could there be to commit a theft?

In the past few years, a number of systems have appeared on the market to combat this serious problem. An electronic system has been developed in which a coded card is used in place of a key with a computer that changes the combination whenever management wishes, certainly whenever the occupancy of a guest room changes or a master key card is lost. A new type cylinder lock has been designed wherein the keying of the lock can be changed without removing the cylinder in a matter of seconds, thereby providing each guest with a new lock as soon as he or she takes possession of the room. New cylinders have been developed in which the entire cylinder can be removed and a new one installed in a matter of seconds. Such advances are vital to the industry; no innkeeper can afford to ignore these developments. Any innkeeper, regardless of the age or size of property, can afford to continue operating with guest room door locks that can be opened with a credit card or a screwdriver. Nor can one continue operating using the same locks that have been in place for the last ten years.

In spite of these advances which protect the assets of the hotel, as well as the person and property of the guest or patron, someone in the organization still has access to and control over the system and key control or system control is still vital. No employee should have any more access to the system than is absolutely necessary, and accountability must be of the very highest degree. The hardware itself is no better than the people who maintain, control and supervise its operation.

Safekeeping for Guests' Property

Provisions must be made for the safekeeping of moneys and other valuables belonging to the guests. Normally, such protection is in the form of safe deposit boxes like those used in banks. The hotel has a master key and each box has an individual key; both keys are required to open the box. Once items have been placed in the box, the guest retains the individual key. If extra keys are maintained for the individual boxes, they should be located in a bank deposit or at some location where they are under maximum security and available only to management.

Where such deposit boxes are not available, deposit envelopes are often provided with a claim tag attached. These envelopes may be stored together in one safe, with each depositor holding a separate claim ticket. Upon presentation of this claim ticket and a signature, the corresponding envelope is returned to the ticket holder.

Because the law states that furs and other valuables must be protected by the hotel if requested by the guest, provisions must be made for the storage of these larger items which would not fit into an envelope or safe deposit box. Therefore a theft-resistant, fireproof vault, or cabinet of some type must be provided.

These safes, boxes, and vaults must never be left unlocked and/or unattended. They deserve the utmost degree of security, and any steps deemed necessary to establish that security must be undertaken if economically feasible. While the hotel has a limited liability for items stolen from a safe deposit box, it would be far more satisfactory if the theft could be prevented in the first place.

A series of thefts took place in New York City a few years ago wherein a group of professional thieves entered a hotel, overpowered the employees, and proceeded to open a number of safe deposit boxes containing the valuables of the guests. While their work took more than one hour and required drilling and other methods of breaking open the boxes, they went undetected and were able to escape without incident. Although this seems almost impossible, it happened not once but several times.

Extra Protection for Valuables. The following precautions should be considered for protecting valuables:

- The use of commercial, silent alarm systems such as those used by banks, which can be triggered by employees without alerting the thief.
- The use of alarm systems designed to detect any action such as moving, hammering, drilling, or in any manner tampering with the safe.
- The use of CCTV in connection with the opening of any safe or vault. This device can be installed to

record on tape the presence of any person coming within the perimeter of the protected area. This would at least provide a filmed record of who opened the safe or deposit box.
- The positioning of deposit boxes or vaults so that they may be observed easily by passing patrols.

Fraud Committed Against the Hotel

When the subject of fraud is mentioned, usually the first thought that comes to mind is the use of stolen credit cards. A few very simple steps on the part of the cashier or desk clerk can in many cases eliminate problems in this area.

1. Check the card to ensure that it is valid, that the card has not expired.
2. Check the signature that appears on the card. Is there any indication of alteration? Does the name agree with the person presenting the card? If necessary, ask for additional identification, such as a driver's license.
3. Most credit card companies have a toll-free number where a card can be checked for validity. Do not hesitate to use this service if there is any doubt.
4. At the time of check-in, determine how the account will be paid and if a credit card is to be used, take the card information at that time. This will give you additional time to make inquiries, if necessary.

Travelers' checks also present a problem on occasion, but again, with a few simple precautions, many frauds can be averted. Ensure that the checks are countersigned in the presence of the cashier. A forger may have practiced a signature for some time and can do a good job when he takes his time, but will fail if required to write that signature in a hurry or under pressure. Watch the countersigning and compare the signatures. Demand additional identification, if required.

Another type of fraud is the skipper, the guest who merely walks out without paying his account. There is nothing wrong, if a guest is staying overnight, to demand payment in advance for the one night stay. If the guest is going to remain for an extended period of time, the charges that appear on that account may be an indication of a possible skipper. One intending to defraud the hotel usually will go all the way. There will be charges from the restaurant, bar, barber shop (or beauty parlor), drug store, health club—every possible facility within the hotel. Cashiers or posting clerks often get a gut feeling about certain accounts when such charges appear. All should be trained to be on the lookout for such activity and the matter should be brought to the attention of the credit manager or security for possible further investigation. Most skippers and others who attempt to defraud the hotel as guests are con artists and they have a line of patter and an extroverted attitude. Their very conduct and behavior often is a clue to their intended actions. Alertness and training on the part of all employees in such matters is vital.

Internal Fraud

Not all fraud taking place at the office cashier's location is perpetrated by the guests. The greatest bulk of the hotel revenue passes through these cashiers, and the possibilities for juggling accounts, falsifying accounts, and absconding with funds are numerous. Without question, these employees should be bonded. The accounting department auditors should perform their duties in a careful and diligent manner.

A case recently occurred in a hotel that experienced a large amount of late-night, one-night business. The night clerk, who acted as night auditor and cashier, would register a guest late at night, collect in advance, and pocket the money, never making up a guest folio. Unfortunately, the serial numbers of the accounts were not checked, so even if a bill was made up and the hotel copy destroyed, the hotel was unaware of what had taken place. The clerk, however, neglected to destroy the registration card. A question developed concerning one guest and, while that registration card was located, the copy of the folio was not. A long and detailed investigation uncovered many similar incidents. While the investigators were never able to be certain they had uncovered all such incidents, an amount in excess of $15,000 was found to have been pocketed by this clerk in a period of two years. When confronted with the evidence, the clerk admitted his theft and the hotel was repaid by the bonding company.

As with other cashiers, the banks of the front office cashiers should be checked regularly. Normally, these banks are the largest used in the hotel and may be a temptation to a dishonest person. If a cashier is absent for an unexplained reason for more than two days, the bank should be checked immediately.

Fraud is also committed by others against the hotel, being perpetrated by purveyors, contractors, and others providing goods and services to the hotel.

These frauds may be accomplished in many instances in cooperation with employees of the hotel. Goods and services ordered or contracted for are delivered or provided, but not in the quantity, quality, or condition requested. Again, it becomes a matter of an opportunity presenting itself. For example, a hotel buyer orders ten prime, oven ready ribs of beef. The purveyor delivers ten ribs and an invoice indicating a total weight of 315 pounds at the price of prime meat. The receiving clerk, through either ignorance, inattentiveness, or perhaps knowledge of the kickback he or she will receive, does not examine the meat and, therefore, does not see that it is not of prime quality. He does not weigh the meat, so he does not know that actually only 275 pounds are being delivered. All he does is sign the delivery slip for 10 ribs which he received. With the price of meat today, 40 pounds short and of poorer quality means many dollars out of the innkeeper's pocket. If the purveyor knows that no one is going to check his deliveries or that he can buy off an employee, he certainly has a golden opportunity.

Obviously, the only answer to such problems is knowing with whom you are dealing, strict regulations relative to receiving goods which will require inspection, weighing, counting, and training in applying standards of quality formulated by the innkeeper. Where services are involved, careful supervision and recordkeeping are required to ensure that services being billed were actually performed and in the quality expected.

Food and Beverage Controls and Procedures

Entire volumes have been written on this subject. Anyone in the food and beverage service business, whether in a hotel, a restaurant, an office building or a roadside stand, knows the many and various problems that can develop. Theft of goods, waste, overproduction, manipulation of records, guest checks, and outright stealing of money from the patrons or the cash drawer are just some of the problems. Employees, as well as patrons, may be involved. It may be an individual effort or it may be a conspiracy between any number of people, employees or patrons. It may involve one drink at a time or one steak, or it may involve case lots.

The answer can only come through a combination of efforts involving systems as well as people. Internal audits and controls (paperwork) is a must, as well as a definite set policy that is administered without compromise. People must be held accountable and employees must know exactly what they may or may not do in the performance of their jobs. If they fail to comply with the rules and regulations of management, their employment must be terminated without question (union regulations permitting) so that they become an example to others. As indicated in the discussion of internal theft control, accountability must be spread out so that no one person has complete control over the entire operation. Certainly, the person responsible for the storage and issue of alcoholic beverages should not be the person who takes the physical inventory of these goods. Waiters and waitresses must be held accountable for meal checks issued to them. Bartenders must be required to ring up every drink served and the size of the drink must be predetermined and enforced. Controls must be established, understood by those involved, and enforced. Without enforcement, opportunity will emerge and the system will fail. Another word for enforcement is *supervision*, and without supervisors who will also be held accountable, again, the system will fail.

The hotel security officer may not be trained in food and beverage operations and may not have any training in accounting, but he or she should be an investigator and able to cooperate with and assist those who find their systems fail to obtain the desired results as far as providing security for the assets of the hotel, the guest, or the patron.

Help

Many innkeepers, faced with the problems of security in their establishments today, seek help. It is not entirely a matter of money or how many employees are present or available, but rather what you do with what you have and what can you do to help yourself. Too often this feeling of frustration has resulted in no action, no concern, or no attempt to inquire as to possible solutions, systems, or items of hardwood that could be utilized at little or no cost. For those who feel lost, we would suggest consideration of the following possible steps:

1. Use of local security and safety agencies. Many police departments now have crime prevention units which are funded and operated under special grants, but which are concerned only with assisting and advising business people, as well as private homeowners in improving security. These units have expertise in such matters and have contacts with other agencies who can also provide assistance on any subject dealing with security. In addition, there isn't a fire department in the

country that is not willing to inspect and advise on fire prevention matters.
2. Insurance company safety engineers are available through an insurance carrier as a service to the policyholders. Many valuable suggestions and much good advice can be gained from consulting with these experts. You pay an insurance premium—why not get your money's worth by taking advantage of such a service?
3. Maintain membership in associations that represent your industry and subscribe to their publications. Almost monthly, articles can be found in the various trade magazines that deal with phases of security, building maintenance, etc., that can be of great value.
4. Outside security agencies are available (naturally at a fee) for consultation, surveys and advice. Check with the local police department for those that are reliable and consult with them if need be.
5. Attend seminars, lectures, even school programs, dealing with security that may be available in the area. The more you know about the subject, the easier it will be to spot problems or potential problems and provide a solution.

In any litigation involving a claim against a hotel where negligence or lack of security is charged, questions will be asked of the defendant on the very subject we have just discussed. "Have you ever had a security survey conducted of your property?" "Do you subscribe to any professional trade magazines?" "Have you or has anyone in your organization had any training in security or attended any programs dealing with security?" Negative answers to these questions certainly do not indicate any concern on the part of the innkeeper toward the subject of security or reasonable care toward the person of guests or patrons. All innkeepers can learn a lesson from a recent court case where the plaintiff was awarded $6,000,000 for damages and injuries. The details of the crime are not what is important, but rather the attitude of the defendant innkeeper and statements made are what, in the eyes of those attending the trial, resulted in this large award.

The defendant innkeeper had a contract security service provide him with guards and, as a part of the initial proposal for this service, a survey had been conducted outlining a total security program. The defendant indicated the number of guards he desired and rejected other recommendations offered. His reasoning was that the contractor was in business to make money and to sell as much of a service as possible. The fact that the purchase of additional equipment and hardware, suggested as a result of the survey, would not have added one penny to the profit of the contractor, was not considered. Within a very short period of time, the contractor advised the defendant that he could not operate under the conditions set forth by the defendant and still do a good job. This was a very reputable firm that felt an obligation to uphold their reputation. Another firm was hired, and the recommendations that they made relative to security were almost exactly the same as what the first company had recommended. Again, the innkeeper refused to take the advice and recommendations offered. His reasoning, stated loud and clear at time of trial, "I know more about security than those people. I've been in the hotel business for 40 years, and I know what is needed. All they [the security firms] know is how to stand around banks or other buildings." The result: Had the recommendation made by both companies been accepted, the crime would never have occurred. The cost to have complied with the recommendation: less than $5,000. Would you spend $5,000 to save $6,000,000? Of course, most of that award was paid by the insurance company, but every innkeeper pays a portion of that loss. Premiums go up each and every time an insurance company must pay out a loss of such magnitude and that is only one of many such losses. The attitude of "I don't care" or "I know more about security than anyone" or "my insurance will cover my loss" or worst of all, "I never had a problem, so why should I worry?" will be the undoing in the days ahead.

Summary

Hotels are, by their very nature of operations, prime targets for many illegal acts and potential serious losses of life and property. The innkeeper is not the insurer of the guest but does have an obligation clearly set forth in the law of the land to provide reasonable care. Hotels have long hidden behind the defense that certain incidents were unforeseeable and, therefore, could not be protected against. Those days are gone forever, as the public becomes more demanding and our lifestyles more sophisticated. Security is in part common sense but, more, it is the application of tried and true principles that will provide that degree of reasonable care that must be provided by innkeepers.

Chapter 7
Campus Security and Crime Prevention

LAWRENCE J. FENNELLY

An effective college or university police department is a multifaceted, service oriented, professional police and security agency. The department is comprised of police officers, security guards, and student security personnel who service the entire university on a seven-day week, twenty-four-hour-a-day basis. The police and guard force are responsible for the preservation of peace, prevention and detection of crime, enforcement of university regulations and state laws, protection of life and property, and the performance of nonregulatory police services. Calls for service are monitored by trained communications specialists who dispatch police and/or security units via cruiser or two-way portable radios. This direct communication enhances response time to scenes of emergencies and other distressful situations.

Instituting the Program

For crime prevention to succeed in any community, two things are needed: first, total support and commitment from the chief of police; and second, total support and commitment from the community.

Crimes occur because the criminal element has the opportunity and the desire to commit an illegal act. Since we cannot crawl into the head of the criminal who has the desire to commit a crime, we must work at reducing and eliminating the opportunity to commit a crime. This cannot be achieved without a total commitment from members of the community and law enforcement agency. Listed below are duties and responsibilities of a crime prevention unit.

1. Crime analysis—the analysis of crime reports to determine crime trends and patterns. This responsibility includes recommending programs or projects to counteract crime problem areas.
2. Conduct formal and informal security surveys for university facilities. The need for such security surveys may be generated through crime analysis, a request by a university administrator, upon recommendation by uniform officers, after a major theft, or after a crisis situation.
3. The design and supervision of installation of electronic alarms, CCTV, access control systems, and other related electronic security systems.
4. Develop architectural and environmental security recommendations for new or renovated university facilities.
5. Advise the community (faculty, staff, students, and employees) on effective crime prevention and security measures and enlist their cooperation and involvement in crime prevention programs.
6. Develop crime prevention programs and establish procedures to monitor and evaluate the effectiveness of these programs.
7. Gather and disseminate information on new security systems and new crime prevention techniques.
8. Maintain and supervise the operation of installation of service, of all intrusion alarms which your department is expected to respond to.
9. Cooperate and provide staff assistance to watch commanders in the area of crime prevention.
10. Program and implement university-wide crime prevention programs (including the design of publications, displays, posters, etc.).
11. Ensure continued liaison with security service and product manufacturers, the state office of crime prevention and the National Crime Prevention Institute.

12. Maintain a crime prevention library to include product information files and reference data concerning physical security.
13. Coordinate with the training section continued crime prevention training within the department.
14. Perform any other related duties which may be assigned from time to time which will further contribute to the reduction of crime.

The police need to apply the most effective methods of patrol and deployment to optimize crime prevention efforts. The methods to be selected depend on the type of crimes occurring and the modus operandi of the criminal.

A crime prevention officer must analyze the community's crime patterns before implementing any action. Identifying the main problem is an important tool. College crime prevention is different from that in a city or town in that every four years, 75 percent of the community has been replaced. Therefore, the educational program for crime prevention must begin with the freshman class.

Freshman Program

First impressions are always lasting, and in dealing with a freshman class, this is emphasized. In September, when they arrive, campus police officers are first observed by their parents who are obviously worried about their child's safety. They may inquire about crime on campus, so the officers should be prepared.

Next, a meeting should be planned with the freshmen during orientation week. This is generally the first time the chief of police meets these students. The speech should be brief and to the point, generally describing the department and instructing them in ways to avoid being a victim of a crime. One should conclude by stating that the department's Crime Prevention Unit (CPU) will be in touch with them in the future. Then, the crime prevention officer takes their orientation program within the dormitories.

Upper Classmen and Graduate Student Programs

A representative from the CPU should make at least two visits to each dormitory every year to promote the various crime prevention programs. The officer should contact the dormitory secretary to learn the night of the house committee meeting.

A house committee meeting will probably have an audience of 33 to 75 students, depending on the size of the dormitory. A computer printout shows the crime rate problem in each house. After the crime prevention talk has been delivered, the next phase is the question-and-answer period on any topic which a student may want to discuss.

If a topic arises where added information is needed, it is extremely important to get that answer back promptly to the chairperson of that specific house.

Summer School Program

Crime prevention on the college campus does not cease during the summer months. It should begin with the orientation of summer school proctors. Then a presentation is given by the chief of police, who invites all the proctors to an open house. This open house program will provide a chance to introduce members of the department, other police services, and offer a better understanding of overall police operations.

Secondly, have flyers printed in multilanguage to accommodate foreign students. Foreign students should be advised of all aspects of the law and how they can avoid being the victim of a crime.

Whistle Program

The Whistle Program is a safety program to signal trouble. Whistles are blown when trouble occurs or threatens to occur. Anyone hearing a whistle responds by calling the police emergency number.

A person blows the whistle to signal trouble. Someone who hears a whistle knows someone else is in trouble. At the sound of the whistle, the police department should be called and given the details, location, problem, and any other possible information. Always advise the recipient when to use the whistle.

Rape Awareness Program

An essential part of any crime prevention unit is a rape awareness program. All students, not just the women, should be educated in workshops relating the various aspects of this violent crime (see Figure 7–1). The key to having a rape awareness program is to have speakers who are specialists in each topic of discussion.

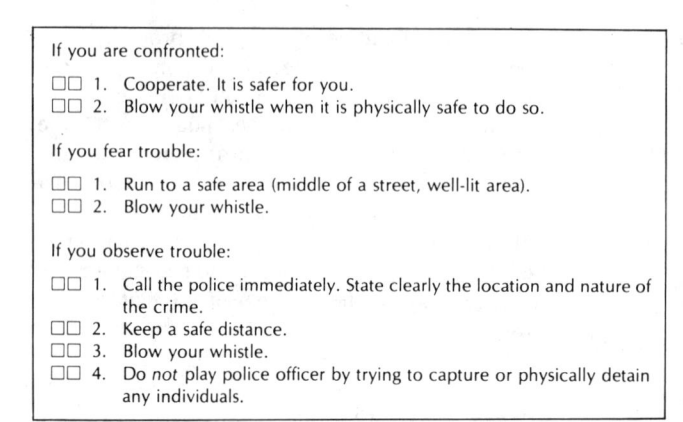

If you are confronted:
- ☐☐ 1. Cooperate. It is safer for you.
- ☐☐ 2. Blow your whistle when it is physically safe to do so.

If you fear trouble:
- ☐☐ 1. Run to a safe area (middle of a street, well-lit area).
- ☐☐ 2. Blow your whistle.

If you observe trouble:
- ☐☐ 1. Call the police immediately. State clearly the location and nature of the crime.
- ☐☐ 2. Keep a safe distance.
- ☐☐ 3. Blow your whistle.
- ☐☐ 4. Do *not* play police officer by trying to capture or physically detain any individuals.

Figure 7–1. Tips for dealing with an assailant.

A rape prevention seminar can be divided into four sessions:

- Crime prevention; specific information concerning the city or town, trauma of rape; medical, psychological, and social aspects
- Psychology of the assailant/rapist
- Legal aspects of assault and rape crimes, reporting, investigation, prosecution, and victim services
- How to avoid being assaulted; self-defense techniques and methods of escape; participation, practice and role-playing session

Neighborhood Watch on the College Campus

A neighborhood watch is, generally speaking, not a program geared for the college community. However, the principles of the program can be applied.
The objectives are:

- To request individual cooperation and to look out for one another's property
- To report suspicious persons
- To report crime regardless of the amount stolen
- To increase the community's awareness through educational programs on how to reduce crime

Phones on Campus

Outside phones on campus seem to be a trend left over from the seventies. Phones strategically located on main footpaths throughout the campus provide an instant means of communication between the community and the police. This is part of the neighborhood watch program. The community should always be urged to report all crime and suspicious persons they may observe.

Thefts in the Library

A recent survey among several colleges throughout the country showed that the biggest problem among officials was the theft of wallets while studying at the college library. This is attributed to negligence on the part of the victim. Frequently the victim reports that he or she left a coat, backpack, or purse containing money on an unoccupied table while going to locate a book. Upon returning, belongings are found in disarray and the wallet or money is missing. One answer to this problem is to give out a flyer to everyone who enters the library advising of the crime problem.

Bicycle Theft Prevention

A recent crime analysis study revealed that bikes with poor quality locks are stolen most frequently. To combat this crime problem, two flyers can be distributed.

The first flyer alerts the owner to the problem and urges the owner of the bike to obtain a better quality lock.

It also highlights the following crime prevention tips on securing bicycles:

1. Locks are worthless if not used properly.
2. When not in use, lock the bike to an immovable object such as a bike rack.
3. Lock the bike through both wheels and around the frame, then secure it to an immovable object.
4. Bikes locked outside should be left in a well-lighted and frequently traveled area.
5. Bikes brought inside should be locked in the dormitory room or in areas of high activity.

The second flyer urges students to register their

bicycles. If bikes are registered, there is a good chance they will be returned if stolen.

Administration Office Security Programs

According to a survey conducted on five college campuses, the most common items stolen are: wallets, cash and credit cards, calculators, IBM typewriters, petty cash, and computers.

Wallets are generally taken because of carelessness, being left in an unsecured and unoccupied area. At times, the theft of a wallet goes beyond that of just cash when credit cards and checks which were inside the wallet are used. Defensive measures for this type of larceny are to conduct meetings informing students and staff of the problem, and seeking their support, in addition to conducting a security survey.

The theft of office equipment generally can be prevented by bolting down typewriters, calculators, etc., and utilizing the operation identification program which is geared to deterring such thefts.

Petty cash thefts for the past year should be analyzed as to why and how such thefts were occurring. Members of the crime prevention unit should meet with members of the controller's office where the crime prevention security tips can be passed on to those handling petty cash. These tips should include the following:

1. Do not maintain a petty cash fund in an amount larger than necessary. Most departments have found that a fund in the $25–$50 range is adequate.
2. Keep the fund in a properly secured place. Avoid desk drawers or cabinets which have locks that are easily opened by a common key.
3. As a matter of policy, the person designated on the controller's records as the petty cash fund custodian should be the only person with access to the fund.
4. In no instance should the responsibility for a petty cash fund be given to a parttime employee or anyone hired on a contractual basis.

Thefts occur because thieves are given an opportunity to steal. If preventive measures are implemented, losses can be prevented.

Operation Identification

Operation Identification started in 1963 in Monterey Park, California, as a burglary prevention program. Probably it is the most frequently discussed crime prevention program in the country. A unique identifier is placed on items to be used to link recovered property to a specific crime or criminal. The deterrent effects of operation identification rely upon the burglar's assumed belief that operation identification items marked properly increase the risk of apprehension and lower the economic gain.

Intrusion Alarms

Intrusion alarms are necessary equipment for proper security. Unfortunately, the installation of such a system is accompanied by numerous false alarms.

Basically, there are five causes of false alarms:

1. Faulty installation of equipment
2. Equipment failure
3. Telephone line malfunction
4. Subscriber's error (human problem)
5. Such natural acts as rain, high winds, or electrical storms

On a national scale, more than 90 percent of all false alarms are caused by the above factors. Again on a national scale, more than 50 percent of all false alarms are caused by the subscriber's error. Such simple matters as neglecting to properly secure doors or windows which are alarmed with magnetic contacts, or by permitting personnel to enter alarmed areas by an alarm key, are a couple of common errors on the subscriber's part.

Again, let's look closer at the five causes of false alarms:

1. Faulty installation of equipment
   a. Incapable installer or installer's helpers
   b. Installation does not meet the requirements of Underwriters' Laboratory standards
   c. Wrong equipment designed for intended job.
2. Equipment failure
   a. Dead batteries
   b. Breakdown in control panel, wires, relays, or shunt-key cylinder
   c. Poorly designed equipment which doesn't have a UL listing
3. Telephone line malfunction
   a. Telephone company causing an open circuit in the lines
   b. Telephone company repairmen working in the general area of alarmed site
4. Subscriber's error
   a. Complex not properly secured and alarm turned on

b. Operation of alarm never properly stated to subscriber by alarm company
5. Natural acts
   a. Electrical storms and blackouts
   b. Heavy rain, winds and floods could affect alarmed area or telephone wires

What can be done to reduce false alarms? The existing alarm system can be modified. All control panels installed on campus should be of the finest quality, as listed by Underwriters' Laboratory.

Then, look carefully at other features: a locked panel to prevent unauthorized entry and possible tampering with the alarm, an entrance/exit delay feature which would allow the subscriber thirty seconds to enter or leave before the panel registers an intrusion, and the hookup of an audible horn. This feature allows the subscriber, if leaving the property and failing to observe the meter on the control panel, thirty seconds before the audible horn goes off, and an additional thirty seconds for the subscriber to shut off the alarm before the signal is sent to the police station.

On all construction or renovation projects, standards should be developed with the planning office regarding the installation of alarms.

The intrusion alarm, then, will be installed by an Underwriters' Laboratory listed alarm vendor; however, if the general contractor has the responsibility for said installation, then options are to use a union or nonunion UL listed alarm vendor, or to use one's own electricians.

If it is decided to use the latter, the system must be inspected by a UL listed alarm vendor before being tied into the police board. In this way, it is certain the products and installation are done correctly.

Additionally, it is helpful to educate users of the alarm system in the proper method of handling it in order to reduce the rate of false alarms.

Safety in Numbers Program

Because the victim of a crime is usually one who is alone, it is advisable to draft an educational program for the community which advises them that there is *Safety in Numbers*. Such a program would suggest that they should also travel in well-lit areas and be familiar with the locations of emergency and outside centrex phones.

Some college and university police departments provide a motor vehicle escort service. This service is for both students and employees who would prefer not to walk alone from one section of the campus to another.

Lock Your Door Program

As in residential thefts, losses occur in the college dormitory at times because a door or window is not secured. Most thefts occur when property is left unattended, for example, in unlocked dormitory rooms, coatrooms, dining halls, or libraries. The following five crime prevention tips should always be given as part of a presentation to students.

1. Do not leave wallets, purses, wristwatches, money, calculators, and other items of value exposed on desks or dressers.
2. Require identification and authorization from people who desire access to your room to do work or to remove equipment for servicing.
3. Never lend your key to strangers.
4. When leaving your room or office unoccupied, be sure that:
   a. All doors are closed and locked—even if you will only be gone for a few moments. Those few moments are when most thefts occur.
   b. All easily removable valuables are stored out of sight.
   c. All windows are closed and locked.
5. When leaving your dorm room to take a shower, lock your door behind you, but do not leave your room key in a pocket of your clothing or robe. Keep your key with you in the shower, to prevent someone from removing it from your pocket to gain access to your room. This has happened to numerous students in the past.

Follow-up to Thefts

Members of the crime prevention unit, as a means of further reducing crime, should conduct follow-up investigations of larcenies, making recommendations to either harden the target or create awareness to lack of good security procedures to reduce losses.

These follow-up reports are conducted for thefts occurring in both administrative and dormitory buildings. Granted, this program is responding after the fact; nevertheless, it will be found to be worthwhile.

Flyers and Posters

Flyers have been mentioned several times in this chapter as a means to promote a certain program. Posters and flyers are a means of selling your program and your department. It is important to realize this from the beginning.

Keep in mind these helpful hints for putting together posters and flyers:

- Size of poster. It should be large enough to be visible from a distance.
- Color and thickness of stock. Is it eye-catching?
- The department's logo and telephone number are included.
- The message is brief and to the point.
- Proofread before printing.
- Proofed copy is approved by the chief of police.

Conclusion

Crime prevention, in order to be effective, must be viewed as everyone's responsibility. The police cannot assume the burden alone; neither can the citizen. The general impact of a crime prevention program is most effectively seen as a long-term change in the nature of both police and community roles, as they relate to crime.

Citizen involvement in preventing crime is not a new concept. Over a thousand years ago the peace was kept for the most part by the entire community, not by just a few.

There is a new recognition of the need for increased citizen participation in dealing with crime today. This awareness has surfaced as a result of the realization that government alone cannot resolve all of society's problems. All available resources, both public and private, must be utilized to effectively deter crime. Every citizen must devote time, energy and imagination to this effort.

The goal of crime prevention is to reduce crime through public awareness and educational programs. There is no overall secret to preventing crimes from ever occurring, but losses can be reduced and kept to a minimum. The programs described in this chapter have been proven to be effective and I hope they can work for you.

Appendix 7a
Directed Patrol*

Most colleges and universities throughout the United States have recognized the necessity of developing effective campus law enforcement and security agencies. The need to protect the academic community from the criminal element of society has become obvious; meaningful research and education cannot coexist in an environment overshadowed by the fear of crime.

The historical approach for providing protection to the college campus usually centers around the employment of a service or support unit within the institution responsible for the security and/or law enforcement function. Such departments operate under a variety of titles: campus security, university police, and public safety, to name a few. In many cases, this department also has the responsibility for parking management.

Most campus law enforcement agencies are organized in accordance with rather traditional law enforcement concepts. They may be staffed by at least one administrator, a number of supervisors, one or more investigators, possibly a specialist in the area of training or records, and the remainder of the force will be in patrol operations. Such departments are often structured with a paramilitary form of organization.

A relatively new addition to this more or less accepted model for campus law enforcement agencies has been the crime prevention specialist. This specialist, in most cases a law enforcement officer, is usually charged with the responsibility for developing printed crime prevention material (brochures, posters, decals), giving crime prevention speeches or lectures to various interest groups on campus, coordinating crime prevention programs, such as Operation Identification, conducting security surveys of campus buildings, statistical analysis of crime and, in general, serves as the department's crime prevention expert.

The primary concerns of most college and univer-

*Reprinted with permission obtained from Chief Daniel P. Keller, Director of Public Safety, University of Louisville (Kentucky).

sity administrators are education and research and, most importantly, the absence of crime on campus. The majority of college and university administrators would agree that they prefer their campus protection agency spend most of its time and resources in the prevention of crime and the reduction of criminal opportunities.

Herein lies the point of conflict: While most campus law enforcement agencies are organized and function in a manner not unlike the traditional reactive police model, the needs of an academic community and its environment are more closely allied with a form of protection strategy stressing positive proactive measures. What appears to be in order then is a marriage of the concepts of law enforcement, security, and the newest addition, crime prevention.

At the University of Louisville, we have developed a new concept of operations for the agency responsible for campus protection, the department of public safety. We call this concept simply *directed patrol*. Although labeled as new, this approach is really a combination or consolidation of the disciplines of law enforcement, crime prevention, security, and investigation.

The principle modification of operations for the department of public safety with regard to the concept of directed patrol is the use of personnel resources in the employment of more preventive measures directed at eliminating or reducing criminal opportunities at the university. No longer is crime prevention a specialized function practiced only by a few select officers. Instead, every law enforcement officer has become thoroughly involved and active in the department's comprehensive crime prevention program. The successful adoption of contemporary crime prevention practices and projects is the number one priority of the department.

Criminal investigation, as a function of the department, is modified within the parameters of directed patrol. Uniformed police officers now have a wider responsibility for follow-up and scientific investigation. They no longer serve merely as report-takers and initial, on-the-scene investigators. They are, instead, held more accountable for follow-up investigation and contact for the criminal incident reports they initiate.

The redirection of operational emphasis by directed patrol does not preempt or diminish the need for traditional preventive patrol tactics. Instead, the concept of Directed Patrol consists of a union of accepted preventive patrol techniques and contemporary crime prevention measures. Directed patrol also integrates the principles of preventive patrol, and modern crime prevention measures with a broader investigative responsibility for the uniformed police officer.

Through this directed approach, we are attempting to marshal the skills and talents of every officer in the department toward the objective of eliminating or reducing criminal opportunities. By virtue of our proprietary relationship with the university, we have a unique opportunity to affect a comprehensive crime prevention program for the institution. To do so, however, requires the combined coordinated efforts of every member of the department.

The Role of Police Officers in Directed Patrol

Uniformed Officers: Crime Prevention Practitioners

Each uniformed officer has become a knowledgeable crime prevention *practitioner*. Through training and education, uniformed officers have been exposed to the most contemporary crime prevention concepts. Furnished with necessary support materials, they are, in turn, expected to aggressively employ these concepts and techniques to reduce criminal opportunities on campus. Since uniformed police officers naturally experience the greatest degree of public contact, they have the best opportunity to interface their crime prevention skills with the needs of the university community.

Among the areas of responsibility for uniformed police officers within the concept of Directed Patrol are the following:

a. *Criminal Incidents*. In addition to taking and recording the necessary information and coordinating the appropriate investigative techniques, uniformed officers have additional crime prevention responsibilities at the scene of a reported criminal incident:
1. Officers should determine why, from a crime prevention standpoint, the criminal incident took place. Once determined, the information should be entered either on the original incident report, or, if not appropriate, on a follow-up document.
2. Officers should advise the victim of possible crime prevention measures or precautions to be taken.
3. In some cases, officers may make recommendations to the department of public safety concerning crime prevention measures or projects to eliminate or reduce the probability of other such incidents in the future.

b. *Public Speaking.* Uniformed police officers are requested to give crime prevention presentations to various campus groups or organizations throughout the university.
c. *Operation Identification.* The principle responsibility for carrying out Operation Identification rests with uniformed police officers. They are specifically involved in the promotion of Operation Identification in university dormitories and office buildings.
d. *Crime Specific.* Uniformed police officers are provided with an analysis of discernible crime trends on campus by the Crime Prevention Unit. The officers, in turn, take measures to alert members of the university community in the areas where trends are noted and seek active cooperation and assistance to remedy the problem. Entitled *crime specific*, measures may include the distribution and posting of flyers, meetings with staff employees, plainclothes patrol, etc.
e. *Crime Prevention Reporting.* Uniformed police officers are vitally instrumental in the reporting of crime prevention problems or hazards. They report inoperative exterior security lighting, file security advisory reports indicating physical security deficiencies, and leave *Rip Off* cards in offices or vehicles found unsecured. Thus, an important aspect of their role as crime prevention practitioners is to remain alert for criminal opportunities that can be corrected before the crime occurs.

Crime Prevention Specialist

The department of public safety currently has one police officer working exclusively as a crime prevention *specialist*. This officer has a variety of crime prevention responsibilities, among which are the following:

a. Crime analysis—the analysis of crime reports to determine crime trends and patterns. This responsibility includes recommending programs or projects to counteract crime problem areas.
b. Conducting formal and informal security surveys for university facilities—the need for security surveys may be generated through crime analysis, a request by a university administrator, or upon recommendation by uniform officers.
c. The design and supervision of installation of electronic alarms, CCTV, access control systems, and other related electronic security systems.
d. Developing architectural and environmental security recommendations for new or renovated university facilities.
e. Programming and implementing university-wide crime prevention programs (including the design of publications, displays, posters, etc.).
f. Ensuring continued liaison with security service and product manufacturers, the state office of crime prevention and the national crime prevention institute.
g. Maintaining a crime prevention library to include product information files and reference data concerning physical security.
h. Coordinating with the training section continued crime prevention training within the department.
i. The preparation of a bimonthly crime prevention management report for the director.

Investigator

The department of public safety utilizes one sworn police officer as a fulltime investigator. In addition to serving as a crime prevention practitioner, this officer is responsible for the following areas relating to criminal investigation:

a. Follow-up investigation of criminal incidents requiring immediate attention.
b. Coordination of initial and/or follow-up investigation of major and/or sensitive criminal incidents.
c. Assist and counsel uniformed officers pursuing follow-up investigation of criminal incident reports.
d. Liaison with the various elements of the local criminal justice system—courts, prosecutors, police agencies, etc.
e. Maintenance of court and prosecution records.
f. Maintenance of department evidence and property files.
g. Coordination with the training section for criminal investigation training within the department.

The Role of Supervisors in Directed Patrol

Department supervisors play a pivotal role in the success of directed patrol. Without their active support, the degree of success of the program would be minimal. The sergeant-shift commanders are graduates of the National Crime Prevention Institute and bear increased responsibilities to ensure that crime prevention activities are carried out by the uniformed officers under their supervision.

Management By Objectives

Each year a number of clearly defined crime control objectives are established for the department of public safety. These strategic objectives are designed to complement the department's long range goals and its comprehensive crime prevention program. Officers of the department are actively involved in the development of these objectives.

The role of directed patrol is to serve as a vehicle or means of accomplishing the strategic crime control objectives of the department of public safety. Rather than merely reacting to incidents of crime, through directed patrol, we take the initiative by establishing strategic crime control objectives and then carrying them out to a successful end. Examples of such objectives may include:

- To reduce thefts from dormitories by 20 percent
- To have 80 percent of the dormitory residents participating in Operation Identification
- To reduce thefts from motor vehicles by 10 percent
- To reduce thefts from offices by 30 percent
- To have at least 90 percent of the bicycles on campus registered

Supervisors and Management by Objectives

The four sergeant-shift commanders currently rotate three times a year (on a semester basis). As previously stated, these uniformed supervisors play a vital role in the success of directed patrol. Prior to their rotation at the end of each trimester, shift commanders confer with the director and crime prevention specialist and develop directed patrol objectives for the following four months. These directed patrol objectives consequently serve as guides for the shift commanders and their officers. Supervisors are then evaluated on the accomplishment of the objectives established for their shifts.

After directed patrol objectives have been established by the sergeant shift-commanders, they meet with the patrolmen assigned to their shift and, working together, develop a performance plan for the accomplishment of their objectives. This performance plan will serve as a strategy for the meeting of their directed patrol objectives. The plan may entail special assignments, speaking engagements, modification of schedules, etc. The important consideration is that the sergeant-shift commander works together with patrol officers to design a plan of action to accomplish their directed patrol objectives.

Shift commanders are held accountable for the successful completion of their directed patrol objectives. They each submit a monthly report to the director delineating the directed patrol activities and accomplishments of the previous month.

Evaluation and Feedback

The success of directed patrol is, to a large degree, dependent upon evaluation and feedback. It is anticipated that not every crime prevention project we develop will prove to be either beneficial or cost-effective. Of those programs developed by the department, some will continue, some will be modified and others may be discontinued altogether. New projects will be added from time to time. To achieve maximum utility from departmental resources, however, it is essential that evaluation and feedback be designed into the overall crime prevention program. Methodology for the generation of such information is indicated as follows:

1. *Incident Reports.* As previously indicated, police officers are required to include additional supplementary crime prevention information on original criminal incident reports.
2. *Statistical Analysis.* Through the employment of uniform crime reports, pin maps, the crime line, and crime statistics generated for internal use, statistical analyses of crime trends or rates serve to indicate the success of directed patrol.
3. *Supervisor's Activity Reports.* Shift commanders include crime prevention activities in their daily supervisor's activity reports.
4. *Public Service Questionnaire.* The department mails public service questionnaires to victims of criminal incidents. Included as a segment of this questionnaire is a section requesting the victim to assess the reporting officer's crime prevention efforts.
5. *Supervisor's Monthly Reports.* Shift commanders file a monthly report citing progress toward the fulfillment of their established directed patrol objectives. Since some of these objectives may be difficult to evaluate empirically, this report serves as an important evaluation and feedback medium.

Training and Materials

Each sworn police officer employed by the department of public safety receives a minimum of 40 hours of formal crime prevention training. This training is provided by either the Kentucky Bureau of Training, the National Crime Prevention Institute, or by our own training section. In addition to formal classroom training, each officer is exposed to numerous hours of in-service crime prevention training. The department's audiovisual, multimedia training program often includes crime prevention topics and programs. Finally, departmental training bulletins covering a variety of crime prevention subjects are issued on a periodic basis.

A briefcase containing crime prevention brochures, pamphlets, decals, engravers for Operation Identification, posters, etc., is carried in each patrol vehicle. The purpose of these briefcases is to make crime prevention support material available to the uniformed officers at the time most needed—while on patrol.

Criminal Investigations

One of the primary objectives of directed patrol is that of improved criminal investigation. Traditionally, uniformed police officers completed criminal reports at the scene of a criminal incident and conducted to varying degrees initial investigation of the incident. If the criminal offense was not solved through initial investigation and there were leads or information requiring follow-up or if the incident was serious enough to warrant such follow-up, a specialized plainclothes investigator was assigned to the case. In most instances, the uniformed officer who responded to the incident and completed the initial crime reports lost contact with the victim and the incident. Through directed patrol, however, we have endeavored to expand the investigative efforts and involvement of uniformed police officers. Subsequently, the following guidelines have been developed concerning criminal investigations.

1. Uniformed officers are permitted and encouraged, whenever possible, to conduct follow-up investigations of those criminal incidents to which they initially respond. Such investigations may be conducted independently or in conjunction with the department's investigator. Follow-up investigations may necessitate a changing of shifts, a plainclothes assignment, a stakeout, or other changes in routine responsibilities; thus, coordinating follow-up activities with the patrol lieutenant and the appropriate shift commander is necessary.

2. Whenever a member of the university community is the victim of a criminal incident, a special effort is made by the initial investigating officer to maintain or reestablish contact with the party involved. The primary purpose of this procedure is to reassure the victim of the officer's interest and concern about the incident. A secondary benefit of such contact is that new information concerning the incident often comes to light.

The results and benefits of these new procedures are numerous. Uniformed patrol officers have a greater degree of self-actualization through their involvement in a criminal incident from its beginning to some type of conclusion. Since the cases investigated are their own, patrol officers are expected to do a more thorough job in the initial phases of the criminal investigation and the preparation of preliminary reports. Although our current statistics are inconclusive, we hope that such procedures will result in higher clearance rates for criminal offenses. Finally, we feel that a more positive rapport with the university community has been achieved through officers' follow-up with the victims of criminal incidents.

Summary

Directed patrol is designed as a program to utilize to the maximum degree possible the resources of the department of public safety in its crime control activities. Rather than relying solely or even primarily upon the traditional concepts or principles of preventive patrol, Directed patrol gives us the opportunity to take the initiative and employ a proactive, comprehensive attack against crime.

Directed patrol allows us the opportunity to more fully exploit the education, training, and skills of our officers. If the specific strategies and approaches developed for directed patrol are adopted and adhered to, the following results should be achieved:

1. A reduction in the number of preventable crimes on campus
2. An improved relationship and level of communication between the university community and the department of public safety
3. An increase in the reported crime rate due to improved public confidence in the department
4. A greater degree of job satisfaction on the part of the officers of the department of public safety
5. A higher rate of interception of crimes in progress
6. Higher ratios of recovered stolen property
7. Increased community support for the department through a better understanding of campus crime problems
8. Perhaps most importantly, an increased recognition of the responsibilities of each member of the university community to endeavor to reduce crime hazards

Appendix 7b
Campus Security Manual*

Protection of Person and Property

Following the precautions outlined in this section can substantially decrease your chances of becoming a crime statistic.

General

1. BE AWARE—Recognize your vulnerability.
2. Report all suspicious persons, vehicles and activities to Campus Security IMMEDIATELY, via the Campus Emergency Line.
3. Use the "buddy system," and watch out for your neighbor. If you see someone being victimized, get involved and notify Campus Security.

Walking

1. Avoid traveling alone after dark.
2. Confine walking to well-lighted, regularly traveled walks and pathways. Avoid short-cuts and keep away from shrubbery, bushes, alleyways, or any other areas where an assailant might be lurking.

---

*Originally titled "Security at Amherst is Everyone's Business," 1979, by William Dion, Chief of Police. Reprinted with permission.

3. Don't hitchhike or accept rides from casual acquaintances.
4. Report lights that are out and any other hazardous conditions to the Physical Plant Department's Service Desk.
5. When walking to your vehicle or residence, have your keys ready in hand.
6. When being dropped off at your residence by taxi or private vehicle, ask the driver to wait until you get inside.
7. If threatened by an approaching vehicle, run in the opposite direction. The vehicle will have to turn around before it can follow.
8. When getting out of a car or off a public conveyance, take a look around to make sure that you are not being followed.
9. If you think you are being followed, cross the street and, if necessary, keep crossing back and forth. If you are pursued, scream and run to a lighted business or residence; enlist the aid of a passerby; flag down a passing motorist; or, as a last resort, break a window in a residence or pull a fire alarm. Do anything that might attract attention or summon assistance.
10. If you find yourself confronted by an assailant, you must remember that while screaming and struggling may in some instances frighten off the assailant, in other instances such actions may further antagonize the assailant and bring forth a more violent reaction. Above all you must keep your head and assess the situation before choosing your course of action. Whether or not the assailant is armed or has made threats against your life will, obviously, be a determining factor in your decision. The key word in this type of situation is **survival.**

Driving

1. Don't pick up hitchhikers.
2. Whenever possible, limit traveling to well-lighted, heavily traveled roads.
3. Keep your windows closed and doors locked.
4. When stopped at traffic lights or stop signs, keep your vehicle in gear. If threatened, keep sounding your horn and drive away as soon as possible.
5. Avoid stopping in poorly lighted, out-of-the-way places.
6. If your vehicle breaks down, signal for assistance by raising your hood and by tying a white handkerchief to the radio antenna or door handle. Stay inside your vehicle with the windows closed and the doors locked. If a roadside Samaritan stops, roll down your window just enough to talk and ask that he just call the police. If the person appears to be a threat, keep sounding your horn and flashing your lights until he leaves.
7. If you think you are being followed, keep out of desolate areas. Look for a place where there are people, then stop and let the vehicle behind pass you.
8. If the vehicle continues to follow, drive to the nearest location where you can get assistance, i.e., gas stations, shopping centers, police or fire stations, etc.
9. If you are followed into your driveway or parking lot, stay locked inside your vehicle until you can identify the occupants of the vehicle. If threatened, keep sounding your horn until you attract attention or the vehicle leaves.
10. When parking at night, choose well-lighted areas. Before getting out of your vehicle, check for people loitering.
11. Always remove your ignition keys. Lock the vehicle whenever it is unattended.
12. Before entering your vehicle, always check the interior, paying particular attention to the floor and rear seat.

Dormitories and Private Residences

1. Keep the door(s) to your residence and windows accessible from the outside locked at all times.
2. Never sleep in an unlocked room or house.
3. Don't put your name and address on key rings.
4. Don't keep your residence and vehicle keys on the same ring.
5. If you lose the keys to your residence, have the lock(s) changed. On-campus residents should notify Campus Security

immediately—for a fee, arrangements can be made to have your lock changed.

6. Women living alone should not use the prefixes "Ms.," "Miss," or "Mrs." on their doors or mailboxes. Instead, use the first initial and last name. This is also advisable for telephone directory listings.
7. Don't study in poorly-lighted, secluded areas.
8. Require callers to identify themselves before opening your door. Off-campus residents should require official identification from all repair or service personnel.
9. Don't let strangers in to use your telephone. Offer to make the call for them or direct them to a public telephone.
10. If you receive obscene or harassing telephone calls, or several calls with no one on the other end, immediately notify Campus Security and the College Operator (college housing), or the town police (private residences).
11. If you find that your room has been entered, **don't go inside.** Go to a neighbor and call Campus Security (college housing), or the town police (private residences). If you are already inside, **don't touch anything.** In so doing, you may disturb evidence that is important to police investigation.
12. If you are awakened by an intruder inside your room, don't try to apprehend him. He may be armed or may easily arm himself with something inside the room. If he poses an immediate threat, get out of the room; if he does not, common sense may dictate pretending you're still asleep.
13. If you see a suspicious person or vehicle on campus or in your neighborhood, immediately contact Campus Security or the police. Try to get the license plate number.
14. Those in private residences should consider installing "peep holes" and "intruder chains" on outside doors. Also, if returning after dark, leave a light on at the entrance to your residence.
15. Students in campus housing particularly are advised against blocking open the entrance to dormitories or other College buildings. Defective locks on windows and doors should be reported to the Physical Plant Department's Service Desk.

Upwards of 60% of the crimes committed on the Amherst College campus are "rip-offs." Larcenies are crimes of opportunity and occur primarily when property is left in unlocked and unattended areas. In an attempt to alleviate this problem, Campus Security makes the following additional suggestions:

16. Avoid bringing large amounts of cash or other valuables to campus or your residence.
17. Keep items of value out of sight.
18. Never lend the key to your residence.
19. Don't hide keys under mats, above doors, in mailboxes, or anywhere else where they can be easily found.
20. If you live in a dormitory, take your room key into the shower with you. Don't leave it in your robe or other clothing, where someone going through your pockets can find it.
21. When leaving your vehicle at a service station or parking garage, leave only the ignition key.
22. When having keys made, have them made in your presence.
23. Participate in Operation Identification.
24. When leaving for vacations or Interterm, store valuables such as stereo equipment, televisions, etc., out of sight. During the summer recess, **do not** leave valuables in student storage rooms. These areas are not secure, and the College is not responsible for the loss of property.
25. Check with your family insurance agent to determine if your property is covered under your parents' homeowner's insurance. If not, you should consider purchasing insurance. Information on personal property insurance for students is available at the Office of the Dean of Students.

The following suggestions apply especially to those in private residences:

26. When advertising something for sale in a newspaper, don't include your name and address.
27. Invest in an electric timer, which can be pre-set to turn lights and other appliances on and off. This device can be used when you go out for the evening or when your residence is unoccupied for extended periods.
28. Don't give strangers information about your neighbors, particularly how long they will be away.
29. When leaving your residence unattended, give it the "occupied" look:
    - Don't leave notes advertising your absence.
    - When gone after dark, leave a light and a radio or television on.
    - Take the time to check that all windows and doors are secure.
    - Leaves shades and blinds partly open.
    - Close the garage door.
30. Take these additional precautions during *extended* absences:
    - Stop all deliveries (mail, newspapers, milk, etc.).
    - Inform your neighbors that you are leaving and ask them to keep an eye on your residence.
    - Place all valuables in a bank safe deposit box.
    - Don't publicize your plans.
    - Notify the police of your absence. Many police agencies, including Campus Security, will upon request make periodic checks of the exterior of your residence. *Important:* notify the police agency immediately upon your return.
    - In the summer make arrangements for someone to cut your lawn, and in the winter to shovel your walks and driveway.

Offices and Laboratories

1. Keep all offices and laboratories locked when not in use.
2. Make sure that all locking devices are in proper working order.
3. Utilize a key control system. All department keys should be signed out and collected when not in use. Keys should be issued only when absolutely necessary.
4. Don't label keys with their use. Use a code system instead.
5. Keep desks, cabinets, etc., locked when not in use.

6. Typewriters and other portable office machines should be bolted or locked down. Smaller items, such as calculators, tape recorders, etc. should be kept locked up when not in use.
7. Avoid bringing valuable personal property with you.
8. Petty cash should be kept to a minimum.
9. Women should keep their pocketbooks locked up, and men should not hang up coats or jackets with wallets or other valuables in the pockets.
10. Before leaving, check to make sure that no one is hiding and that your area is properly secure.
11. Watch out for your neighbor. If someone forgets to secure their area or property, keep an eye on it for them.
12. Report all suspicious persons or improperly secured areas to Campus Security immediately.

Athletic Facilities

Of special importance to women . . .

1. Avoid using the athletic facilities alone, especially after dark or during off-hours.
2. Use the "buddy system." Work out with a friend, and make arrangements to go to and from the gym together.
3. Confine your running and jogging to the daylight hours and to open, well-traveled areas.
4. Make sure the entrances to the women's locker room are locked upon entering and leaving.

5. Become familiar with the location and operation of the "Intruder Alarms" and the emergency telephones located in the women's locker room and other areas of the athletic complex.
6. Avoid showering if you are alone in the locker room. Shower back at your residence.
7. If there is another woman in the locker room, ask her to wait for you. If you are with a male companion, ask him to wait for you just outside the locker room.
8. If you encounter a male intruder inside the women's locker room:
   - Scream for help. Your call for assistance should carry to the men's side of the locker room, to the trainer's room, and possibly upstairs.
   Activate the "Intruder Alarm."
   - Keep out of the intruder's way, and do not attempt to prevent him from leaving.
   - Formulate a description of the intruder in your mind.
   - Notify Campus Security on the emergency telephone.
9. Report all incidents of voyeurism to Campus Security *immediately.*

For protecting your property. . . .
10. Avoid bringing cash, wallets, watches, or other valuables to the athletic facilities.
11. Avoid storing valuable sports equipment in your locker.
12. Keep your locker locked whenever unattended. This includes those times when you leave briefly to shower, visit the trainer's or the equipment room, etc. Most of the thefts at the athletic facilities are from unlocked lockers.
13. Report suspicious persons and incidents of theft to Campus Security immediately.

Bicycles

1. Invest in a good bicycle lock or in a strong padlock and chain. Chains should be case-hardened steel with links at least 5/8" in diameter.
2. Always lock your bicycle. Bicycles should be locked around the frame and through both wheels to an object such as a telephone pole, sign or lamp post, bicycle rack, etc.
3. Whenever possible, keep your bicycle inside.
4. If you must leave your bicycle outside, choose a well-lighted, heavily traveled location.

5. Find out if your bicycle is covered under your parents' insurance policy. If not, it would be advisable to insure it.
6. Campus Security provides a secured facility for the summer and winter storage of bicycles, available to all members of the College community.

> 7. Participate in the Bicycle Registration Program! The program, an off-shoot of Operation Identification, has greatly deterred bicycle theft on the Amherst College campus.
>
>   To participate in this program, bring your bicycle to the Campus Security Office. An officer will engrave your social security number on the frame of the bicycle, record bicycle identification information, and affix an identification decal to the bicycle.
>
>   This free service, like Operation Identification, is available to all members of the Amherst College community.

Elevators

1. If, while waiting for an elevator, you find yourself alone with a stranger, let him take the elevator and wait for its return.
2. If you are on an elevator with someone who makes you feel uneasy, get off at the next floor.
3. Always stand near the control panel, where you have access to the alarm and floor buttons.

Motor Vehicles

Protection
1. Report all suspicious persons or vehicles seen around parking areas to Campus Security immediately.
2. Keep your vehicle locked and the windows rolled up tight.
3. Never leave your vehicle running when unattended.
4. When parking, choose a well-lighted, heavily trafficked area.
5. Packages, luggage, and other valuables should be locked in the trunk.
6. Stereo tape players and C.B. radios should be mounted either out of sight or with slide-out brackets. These brackets will permit the removal of the unit for taking with you or for

securing in the trunk when the vehicle is unattended. C.B. radio antennas should be magnetic or detachable, and should also be stored in the trunk or taken with you.
7. Stereo tape players, C.B. radios, and other auto accessories should be marked with your social security number and registered under Operation Identification.
8. Consider having your vehicle's ignition, doors, and trunk keyed differently.
9. Keep spare keys in your wallet or purse, not inside the vehicle where the professional thief can easily find them.
10. Consider the installation of anti-theft devices, such as alarm systems; hidden ignition or fuel "kill" switches; steering column ignition switch protectors; steering wheel to brake pedal bar locks; tapered door lock buttons; and locks on hoods and accessory items, such as gas caps, mag wheels, spare tires, etc.
11. Unless required by law, don't keep registration, insurance or title certificates in the glove compartment.
12. Keep a record of your vehicle identification number (VIN), registration plate number, and title certificate number.
13. When leaving your vehicle parked for extended periods, such as over the Interterm recess, the following precautions can be taken to immobilize it:
    a. Disconnect the battery and lock it in the trunk.
    b. Remove the high-tension distributor/coil wire.
       Caution—To avoid shock this must only be done with the engine **off**.
14. Campus Security advises against bringing any motorcycle-type vehicle to school with you because of their extreme vulnerability to theft: such vehicles are easily loaded on to a pickup truck or inside a van. If you do bring a motorcycle-type vehicle, choose a well-lighted, heavily trafficked parking area and strongly secure it to an immovable object.

College Regulations

The Amherst College motor vehicle regulations are in effect throughout the year, including during the summer, spring, Interterm, and holiday recesses. Failure to comply with the regulations can result in ticketing, towing at the owner's expense, and revocation of campus parking/operating privileges. It is incumbent upon you to inform your guests of these regulations.

1. All student motor vehicles must be registered within 48 hours of their arrival.

2. All visitor vehicles wishing to park on Amherst College property must obtain a Visitor's Parking Permit at the Campus Security Office. When the Campus Security Office is closed, arrangements for obtaining a Visitor's Parking Permit can be made by telephoning (542)-2291, the Campus Security Business Line.
3. There are a number of "NO WARNING TOW ZONES" on campus. Take heed of any sign marked "Cars Parked Here Will Be Towed" or "Unauthorized Vehicles Will Be Towed." These signs mean what they say.
4. Violations of the 15 mph Campus Speed Limit, wrong-way operation, driving or parking on sidewalks or lawns, and the reckless use of a motor vehicle will not be tolerated.
5. By vote of the faculty, no motorcycle, motorscooter, or motorized bicycle may be operated on campus or stored on campus or in any college building, including fraternities. Such vehicles are restricted to parking in the Alumni, O'Connell, Hills or the fraternity lots.
6. During the "Snow Season," November 30 to April 1, all motor vehicles are prohibited from parking overnight on campus, including weekends.

Additional information regarding parking and motor vehicle regulations can be obtained at the Campus Security Office.

Sensitive Crimes

If you are the victim of a sensitive crime, i.e., rape or sexual assault, it is important that you notify Campus Security, regardless of whether you wish to pursue the matter further, for the following reasons:

1. If the assailant is allowed to remain at large, he is a potential danger not only to you but also to the other members of the community.
2. Campus Security can assist you in obtaining medical, counseling, and other available supportive services.

It should be emphasized that all information that you provide will be kept in the strictest confidence, and it will be your choice of whether or not to pursue the matter further. You should remember that Campus Security will do everything possible to assist you.

Although acts of sexual perversion such as voyeurism and exhibitionism are not classified as sensitive crimes, they should be immediately reported to Campus Security. Verbal threats or encounters should be similarly reported.

Further information on this subject is available at the Campus Security Office and the Student Health Center.

What You Can Do

Your involvement is essential to the prevention of crime on campus. Disinterest and complacency are the prime contributors to the success of crime. The burden of crime prevention rests not only with Campus Security but also with each member of the Amherst College community. President Ward summed this up when he said, "Security is everybody's business."

Campus Security is not omnipresent, and therefore is dependent upon you to recognize and report incidents of suspicious and criminal activity. The extent of your cooperation will greatly influence Campus Security's effectiveness in combating crime. Doing your part means:
1. BEING AWARE of your vulnerability and following the suggestions outlined in this pamphlet to protect yourself and your property.
2. BEING ALERT for suspicious or criminal activity and conditions that may represent a hazard to the community.
3. GETTING INVOLVED by becoming more security conscious and by reporting all incidents of criminal or suspicious activity, no matter how insignificant, to Campus Security *immediately*.

Remember that unreported crimes cannot be solved, and that by not reporting crimes you allow the perpetrators an opportunity to commit additional and perhaps more serious crimes.

Many times crime solving depends upon how accurately and promptly the incident is reported. Therefore, when reporting an incident, it is important that you be able to provide as much of the following information as possible:
1. Nature of the incident.
2. When the incident occurred.

3. Where the incident occurred.
4. Persons involved (names, sex, race, age, height, hair style/color, complexion, distinctive characteristcs, i.e., facial features, scars, physical defects, glasses, clothing, etc.).
5. Direction and method of travel.
6. Vehicles involved (color, type, make, model, license plate number and state, distinctive characteristics, i.e., decals, bumper stickers, damage, number of occupants, etc.).
7. Description of stolen property (item, manufacturer, model number, serial number, value, color, dimensions, etc.).
8. Any other applicable information.

Chapter 8
Hospital Security*

RUSSELL COLLING, CPP

The terms *security* and *protection for health care facilities* seem vague and elusive to many. It is in fact a relatively ill-defined concept that has taken on different connotations in many different settings. In our context, however, protection of health care facilities may be simply defined as *a system of safeguards designed to protect the physical property of the facility and to achieve relative safety for all persons interacting within the organization and its environment.*

One of the problems that immediately arises is to define safety. What is safety today may not be safety tomorrow. And it is often very difficult to evaluate the environment of a particular facility to determine if safety has in fact been achieved. Protection or security is intended to reduce the *probability* of detrimental incidents, not to eliminate all risks. Security, then, is not static and can be viewed as a state or condition that fluctuates within a continuum. As environmental factors and personal conditions change, so does the status of protection. It is this phenomenon that requires constant evaluation and reevaluation of any system of protection.

Many persons in the field attempt to define security too strictly. The organization being served is the entity which provides the ultimate definition of the protection system. The organization provides the funding. This is not to say that the protection program in place, and the philosophy and objectives of the principal security administrator, will not have a strong influence on molding the organizational definition.

A common error is for hospital organizations to view security as being too closely restricted to the law enforcement function. While there may be some common ground between the security and law enforcement function, at least 90 percent of their activities are completely different. Security must be viewed as an internal protection concept, while law enforcement is viewed as an external concept attempting to uphold the law for the whole of a given society.

Rationale for Hospital Security

Among several basic reasons for providing a protection system, the first is a moral responsibility. Any organization, and especially one serving the public, has an obligation to manage is environment in such a way that it minimizes the possibility of injury or death to the patient, the employee, the physician, and other persons visiting the premise for various purposes. It is, of course, also a moral responsibility to take reasonable steps to preclude the destruction, misuse, or theft of property, so that the physical facility remains intact to carry on its business of rendering patient care.

A second basic justification for secure premises is the area of legal responsibility. A hospital corporation operates through its governing board. Its members, as in any other corporation, have the duty to exercise care and skill in the day-to-day management of corporate affairs. A specific example of this general obligation is the duty to preserve the property by keeping well-regulated premises in terms of fire hazards, safety hazards, and the protection of all persons from the actions of malefactors.

The hospital's obligation to its patients is contractual, in that it assumes certain responsibilities toward them. The duty of protection becomes even greater when the patient is unable to take care of herself or himself, as in the case of the critically ill, the elderly, infants, and children.

The liability factor in the management of patient

---

*From *Hospital Security*, 2nd ed, by Russell Colling (Stoneham, MA: Butterworths, 1982).

care facilities is becoming more and more acute. There was a time when hospitals were relatively immune from legal actions. As charitable institutions, devoted to the public good, hospitals were afforded a special protection by the courts. The reasoning was that hospital funds were in trust to be utilized for the public good and should not be subject to pay the claims of an individual. Court decisions, however, have eliminated the so-called doctrine of charitable immunity as a protection against liability judgments in most circumstances.

Negligence for which a hospital may be liable is negligence of the individual employee under the doctrine of *respondent superior* or under the corporate negligence theory.

In terms of employee negligence, two factors are requisite to imposing liability on the corporation. There must exist an employer-employee relationship, and the employee's act or failure to act must occur within the scope of that person's employment. It should be noted that some persons performing duties or services in a hospital are not employees per se, and can be classified as independent contractors. Such a person differs from the employee in that the hospital does not have the right to control his or her actions. Although relevant statutes vary from state to state, the attending physician and the private or special duty nurse are examples of independent contractors performing services within the hospital. In general, the resident and intern have been considered to fall under the doctrine of *respondent superior*, although the question of control by the hospital is open to some debate.

Corporate negligence occurs when the hospital maintains its building and grounds in a negligent fashion, furnishes defective supplies or equipment, hires incompetent employees, or in some other manner fails to meet accepted standards and such failure results in harm or injury to a person to whom the hospital owes a duty.

A third important reason for maintaining a safe and secure environment is the practical responsibility to comply with requirements imposed by the Joint Commission on Accreditation of Hospitals (JCAH), the Occupational Safety and Health Act of 1970 (OSHA), and other federal, state, and local codes.

While the JCAH has not truly come to grips with the important aspect of protecting the hospital from a security viewpoint it has made some feeble beginnings in this regard. The 1981 Manual on Accreditation states:

> Security measures will be taken to provide security for patients, personnel, and the public, consistent with the conditions and risks inherent in the hospital's location. When used, these measures shall be uniformly applied. Based on administrative decision, these measures may include, but are not necessarily limited to, the following:

- Effective screening and observation of new employees
- Identification badges for all hospital personnel
- Exit/entry control, including good lighting
- Internal traffic control, including the use of visitor passes
- A written plan for managing bomb threats or civil disturbances. This plan should be coordinated with, and may be a part of, the hospital's internal disaster and evacuation plan.
- Use of security guards
- Package control, to deter theft and to prevent introduction of unauthorized items
- Well-lighted walkways and employee and visitor parking areas
- Use of surveillance equipment such as visual monitors (mirrors and closed circuit television) and alarm systems
- Management of prisoner-patients as required

It is difficult to understand why this important element of providing health care has been neglected in the standards; however, it is assumed that this deficiency will be addressed in the near future. On the other hand, the JCAH has given considerable attention to the areas of fire safety, work safety, and patient safety, in terms of standards and in their field inspection activity.

A fourth rationale for a sound protection system is the enhancement of the profitability of the organizations engaged in health care delivery on an economic basis. In this connection, it is pertinent to recall that health care has been faced with mounting criticism, especially in regard to the rapidly escalating costs of delivering quality medical care. Critics often cite the lack of cost containment measures, which play a major role in a sound security program in preventing theft and waste of supplies and equipment. Depending on which authority is acknowledged, there are estimates that between 3 and 20 percent of hospital expenditures could be saved if proper security controls were implemented. In most cases, the protection budget for health care facilities is below the 3 percent level. As costs continue to rise, it is reasonable to expect that greater public and governmental pressures will come to bear on the whole issue of cost containment.

It would seem that the economic stimulus, that of increasing profits, would be especially important to proprietary hospitals, yet they as a group are no further ahead of other hospitals in providing adequate protection systems.

Last, a safe and secure environment is required to maintain good public and employee relations. While this reason does not appear as important as each of the others, it probably has produced more dollars for the security budget than the other four justifications combined. The hospital administrator faced with bad media coverage concerning a security problem or the restless employees threatening to walk out over a security incident somehow finds the money to make necessary adjustments in the protection plan. Unfortunately, these quick fixes are generally not cost-effective and may be counterproductive. This crisis management phenomenon, with its resulting funding, can be productive provided the security administrator has previously cited a deficiency with a planned solution.

Unique Aspects of Hospitals

A hospital operates twenty-four hours a day. Service —or, in the language of the industrial complex, production—cannot be shut down at 5:00 P.M. and resumed again at 7:00 A.M. the next work day. The facility must remain relatively open to admit the sick and injured at any hour, to permit patients their visitors, and to carry on the normal business of the gift shops, the cashiers, the pharmacy, the physicians' offices, and the like. With people entering and exiting through numerous entrances at all hours, it is extremely difficult to determine who belongs and who doesn't belong.

The hospital patient and the visitor present unique problems not found in most other social settings. It has been said that the patient is an involuntary consumer, since one generally has no desire to be in the hospital or undergoing major treatment. Basically the patient resents being sick. Often it is the physician who selects the facility, although many patients requiring hospitalization select their physician on the basis of the facility they desire. The patient is paying the bill, and her or his wants must be considered within the organizational framework of the institution. This is no simple matter, since the patient is presumed to be not completely himself—if he were, he probably would not be in the hospital undergoing treatment.

Similar considerations can be applied to the hospital visitor. When a member of one's family or a close friend is ill, the visitor's actions and reactions to management practices may not always be completely rational. Thus tolerance and allowances for peculiarities and abnormal human behavior under stress must be major considerations in any hospital protection program. Think of the nonambulatory patient as one who cannot move from place to place without the assistance of another person or persons. The patient in restraints and the patient in orthopedic traction are two common examples.

The newborn child is certainly in a helpless state from every aspect. And the pediatric patient can also be classed as helpless in many situations, either from being nonambulatory or from being unable to define clearly a course of action in a situation affecting welfare and safety.

Health Care Security Expanding

Although we have focused on hospital security, other areas of health care security are either growing or have recently emerged.

The extended care facility, of which the largest segment is often referred to as the nursing home, is constantly upgrading protection efforts and is faced with the same general problems confronting the hospital industry. The magnitude of the problem is decreased, however, as the size and scope of activity are considerably less than the general hospital.

The Medicare and Medicaid programs have produced significant security concerns. In July 1980 Oliver Revell, Assistant Director of the FBI's Criminal Investigative Division, testified before the Senate Finance Committee's health subcommittee that agents had identified a large number of doctors, hospitals, and clinics who would directly solicit kickbacks as a precondition for any business arrangements. Further, a survey conducted late in 1978 by the House Select Committee on Aging revealed widespread involvement of organized crime in Medicare/Medicaid Programs. This involvement, according to the survey, was most prevalent in nursing homes and pharmacies but included prepaid health plans, clinical laboratories, and hospitals as well.

Along with the growth of organizations providing direct health care, allied professional associations have also been increasing in size and sophistication with resulting protection problems. An example is the American Medical Association, which met their security problems by establishing the position of director of corporate security in late 1980.

According to Keith R. MacKellar, the first AMA Security Director, "Many of the current security programs are based on solutions to problems developed over the years in hospital security." The corporate security function covers the planning and development of security programs for all physical property and establishes policies and procedures covering physical security, internal security, loss prevention, and emergency preparedness to protect employees and corporate assets.

Through the use of closed circuit television, alarms, and computer card access systems, control measures were established for increased protection of people, information, and other resources. Investigative activities include all losses, cases of security violations, conflicts of interest, and extensive background investigations. In addition to maintaining liaison with local, state, and federal law enforcement agencies, the corporate security program also covers personal protection of the AMA executives and their homes.

The AMA program is but one of the forerunners of new types of security program opportunities which will emerge in the health care field in the 1990s.

Hospital Security Vulnerabilities

All hospitals, regardless of size, are faced with the same basic security vulnerability. It is only the magnitude of the vulnerability, which will vary to a considerable degree, that determines the threat to the organization. The following vulnerabilities need to be assessed by each and every hospital organization.

Assaults

There are several facets to this extremely acute problem facing all facilities. This vulnerability exists both inside hospital buildings and on the grounds. It involves not only the employee but the visitor and patient as well. It is this security vulnerability, when it occurs, that gets immediate administrative attention and accounts for the majority of lawsuits against hospitals involving security. The patient is particularly vulnerable to attack due to his physical condition. Recorded incidents range from simple assault to murder. A person would never check into a hotel and go to sleep at night with the corridor door open, yet in a hospital the patient does just that, with the added problem of being sick or injured and thus less able to protect himself.

A recent case of such an incident involved a young female who was a patient in a Montana hospital. The nineteen-year-old patient was admitted for minor ear surgery. In the early morning hours at approximately 4:00 A.M. an unknown male entered the patient's room. He explained that he was a doctor and at this point sexually assaulted the patient. Although somewhat groggy from sleep and medication, the patient was able to activate the nurse call system. The perpetrator left the room and the building undetected.

This type of security situation can be devastating to the hospital from several viewpoints, including community relations, employee morale, and litigation. While this type of event has occurred in numerous hospitals, the same Montana hospital was again the victim of a similar situation less than two years later when five more female patients were sexually assaulted.

It is not always the patient who is the subject of an assault inside the hospital walls. While reports of assaults are numerous in all sections of the country, examples include a nurse who was pulled from a stairwell into a construction area and raped. In another case, a female employee was sexually assaulted in a washroom just off the main lobby at about 5:00 P.M. in the afternoon.

Although sexual assaults command the most attention, other assaults occur between patients and staff, between staff and visitors, and various other combinations. Within the hospital, there are remote areas where females often work alone at night. On some nursing units, there may be one nurse and one nursing assistant. When either is off the unit on an errand or at work, the lone female is quite unprotected. Security patrols are necessary to provide a reasonable degree of safety.

The assault problem is also evident outside the facility itself. The streets surrounding many of our health care facilities are becoming more and more unsafe, especially in large cities. Some organizations are experiencing difficulty in recruiting for the evening and night shifts because of the personal safety problem. More and more organizations are providing patrols on streets in the hospital vicinity, and many are offering escort services to bus stops and nearby housing facilities for employees and visitors alike. This escort service takes the form of walking persons to their destination or providing a vehicle shuttle service. The vehicle shuttle service is especially prevalent at larger hospitals where parking lots may be several blocks from the hospital.

Theft

The pilferage or theft of supplies, equipment, and personal property is a major problem for all health care organizations. There are as many estimates of the extent of theft in hospitals as there are estimators. At the upper end of the scale, one estimate attributes as much as 20 percent of all hospital costs to losses from pilferage, theft, and waste. Norman Jaspan, head of a firm which has specialized in uncovering theft for over forty-five years, believes that, "if anything, stealing by employees is worse in hospitals than in business." He cites a $700,000 shortage in drugs at one hospital as a "typical" example.

It is virtually impossible to calculate the specific losses for any single hospital. Like an iceberg, only a small part of the problem actually surfaces, and it is often difficult to allocate the loss to theft, waste, or loss of accountability. In the opinion of many knowledgeable administrators, however, that loss is a substantial item in the cost of operating any medical care facility, regardless of size or location.

William O. Cleverly is an associate professor in the Hospital and Health Services Administration, Ohio State University. As Editor of Topics in *Health Care Financing* (Vol. 5, No. 2, Winter 1978) he states:

> Whatever the actual dollar loss in the health care industry may be, it is in all probability quite large. Certainly health care executives should be vitally concerned with this cost. It may possibly represent the single best opportunity we have for reducing costs in our industry.

The general feeling expressed by Cleverly is basically the same for most hospital administrators, security administrators, researchers, law enforcement officials, and insurance companies. Yet, how much is really being done to correct the problem? A case in point is that no school of hospital administration, to this writer's knowledge, offers a course in security or organization arrest protection. The majority of hospital administrators are reluctant to implement even the most basic elements of theft control.

What Is Being Taken?

Since there are over 3,000 items purchased by hospitals which are usable in the home the list of items being taken from hospitals is indeed lengthy. Items which top the list are drugs, linens, food supplies, and maintenance parts and materials.

A study in St. Louis, Missouri, indicated an average hospital loses 10 percent of its purchased inventory each year. When you are talking about purchases of $2 or $3 million per year the figures represent a lot of loss. Another study at Chicago's 1,418-bed Cook County Hospital said that "pilferage of hospital property appears to be a way of life." An example presented in this study was that of bath towels. Of over 6,000 bath towels issued weekly to patients, over 2,000 were reported lost or unaccounted for. At $2.00 each, the loss in towels alone would exceed $200,000 annually. It is not always the obvious that is being stolen from hospitals. In Cologne, West Germany, physicians, nurses, and administrators have been implicated in a scheme to remove pacemakers from the dead, according to the public prosecutor's office. The devices were being sold to a marketing subsidiary of a Dutch pacemaker manufacturer for $1,400. After resterilization, the pacemakers were being resold as new for $6,500.

In another case, a pathologist and three other persons were charged with the illegal sale of human organs, tissues, and fluids which were in the custody of a Veterans Administration Hospital in California. These human parts of the body, removed during autopsies, were sold to biomedical companies. The biomedical firms utilize human organs and fluids to do research and develop diagnostic tests.

The items of theft getting a lot of media attention in the early 1980s have been scrub suits, silver from development sludge, and old X-ray film. The number of scrub suits being stolen (blues and greens) is of epidemic proportions. Several large hospitals that have done inventory loss studies report losses in the $70,000 to $80,000 a year range. Denver General Hospital, a modest 200-bed facility, reports spending approximately $30,000 per year to replace scrub suits.

Employee Theft

There is common agreement among security practitioners that the majority of hospital losses from theft can be attributed to the employee. The employee thief works day in and day out, and this creates a constant drain on resources—as opposed to persons from outside the organization who generally do not have the everyday opportunity. There have been numerous ingenious employee theft schemes revealed; however, the biggest loss comes from the employee who simply dips into the supplies without the slightest chance of being caught.

An in-depth research into employee theft in hospitals was been conducted by the Sociology Department of the University of Minnesota under a Law Enforcement Assistance Administration grant. The study surveyed nine retail operations, ten electronic manufacturing firms, and sixteen hospitals—all in the general Minneapolis area.

In hospitals some 37 percent of the employees responding to the survey indicated they had been involved in taking hospital supplies. Of this 37 percent, 11 percent admitted engaging in the theft of supplies on a monthly or more frequent basis.

In addition to the taking of general supplies several other areas of interest to the hospital security administrator were revealed:

Take tools or equipment	8 percent
Excess expense reimbursement	2 percent
Paid more for time than worked	9 percent
Take or use medication	10 percent

The study findings concerning the amount of misuse was taken from questionnaires sent to a cross-section of employees. In this respect, the employees returned the written questionnaire (over 50 percent return) to the research organization. Although the number of employees revealing misdeeds is high, the percentage of actual malefactors may be higher than those admitted. A certain number of the respondents would not be sure the document could not be traced back to them, even though the study process was completely anonymous. Also, most people know right from wrong and would tend to deny wrongdoing as a matter of course. These two factors alone would certainly account for some understating of the real magnitude of the problem. On the other hand, it is doubtful that too many respondents would admit to these negative acts if they in fact were not involved. While this research indicates the number of persons taking advantage of the system, it does not provide any information concerning the amount of theft.

Factors in Employee Theft

The Minnesota study also had an objective of determining the circumstances under which employee theft was most likely to occur. Although the study did not tend to focus on counterproductive behavior, its incidence appears to have a direct correlation to theft. The employees who reported above-average theft levels were also quite prone to above-average counterproductive behavior such as taking extra long lunch and coffee breaks, slow or poor work, use of drugs or alcohol at work, and abusing sick leave.

The study conclusion indicates the younger and never-married employee has the highest levels of theft involvement. An explanation for this is that they are less vulnerable to management sanctions including dismissal, since they have no dependents and little if any seniority built up.

Job dissatisfaction also leads to higher theft involvement, especially in the younger work force. The most consistent source of employee dissatisfaction was found to be in the employee-supervisor relationship. In terms of relationships, the study further concluded that employees who frequently got together with coworkers after work hours were also in a higher theft level category.

The most consistent predictor of theft involvement found by the study was the employee's perceived chance of being caught. The source of potential sanction, when caught, appears to play a role in theft levels. When there was opposition to theft, on the part of management and coworkers, the amount of theft decreased. It is of interest that the inferred sanctions imposed by coworkers appear to have a much stronger influence in shaping behavior than does the more formal action of management.

Effect of Controls on Theft

Another important aspect of the Minnesota study was to determine how effective certain actions or controls imposed by the organization were on reducing the amount of employee theft. The controls reviewed were the security department, management policy, inventory, financial procedures, and preemployment screening.

The study indicates that the level of sophistication of a security department has little effect on theft reduction. The major thrust of the majority of security operations reviewed addressed the problem of external theft control, grounds safety, employee safety, and other nonemployee theft-related activities. I believe that further study will not support this conclusion; rather, the higher the level of security operations the greater chance that detection will result. The study itself indicates that the probability of being caught is a strong influence on theft reduction. The reader must keep in mind that the study initially involved only organizations in one locality and there was no reporting on what the level of security operations might be as viewed on a national scale.

As one would expect, clearly defined employee theft policy, inventory programs geared to detecting losses, and effective preemployment screening of job applicants all tend to lower the level of theft in an organization. Although these controls seem simple and straightforward, the health care industry is extremely weak in all these categories.

Dale Systems Survey

A survey conducted by Dale Systems, Inc., a business-security research organization, established the fact that clever employees, with larceny in their hearts, have devised 415 ways to steal money, goods, and material from their employer. It is also a fact that new twists on existing methods are being discovered with greater frequency. Their conclusion is that theft exists solely because management does not make a serious effort to stop it. Employee theft is much more serious than hospital administrators might dare imagine.

Reporting Hospital Property Losses

It is extremely difficult to develop a program of theft control unless one can measure the extent of the problem. One method of such measurement is through a good reporting system. This requires all employees to fulfill their responsibility in the reporting of missing hospital property. It is in this area, however, that hospitals have really failed. It is estimated that security departments receive reports of hospital property losses in only 1 to 2 percent of losses. There are many reasons for this poor showing, chief of which is that the supply lines are so open that stocks are quickly replenished. Employees, too, do not really want to get involved, especially when the hospital does not seem to care. Property losses can also be viewed as supervisory failures and thus there is a tendency to ignore the problem when possible. As stated, the extent of the hospital property theft problem will vary from facility to facility. Sanford Beck of Royal-Schutt International, an investigations firm, states that his firm feels national estimates of $1,200 to $1,300 per bed per year are grossly *understated*. Beck believes the figure should be closer to $2,300 per bed per year.

Patient Property Losses

Loss of patient property is a continual problem for all hospitals. Although the value of most patient property lost is minimal when compared to the losses of hospital property, most hospitals place more emphasis and more effort on protecting the patients' property and investigating their losses than on their own asset protection. When a patient loses property, it seriously affects public relations, and the patient can interpret this problem as an indication that the medical care may also be inferior. There are many instances of a patient's becoming so upset over a theft that a medical condition was adversely affected.

The health care employee is not always to blame for patient property losses. There is one case in which a patient in a nursing home could not locate his false teeth. Investigation solved the loss when the teeth were discovered in the mouth of another patient. Although it happened a number of years ago another case of missing false teeth reveals the interaction of the visitor in the area of lost property. A patient's false teeth could not be located despite the search efforts of security and the medical care staff. After five days a dentist was called in to take impressions for a new set of dentures. At this juncture the second patient in the room was discharged, with the parting comment that "I have the missing dentures. As soon as my bottle of scotch is returned, I'll return the teeth." It seemed that, accidentally, visitors to the patient who lost the teeth had taken the gift bottle of scotch. A ten-dollar bill produced the missing teeth.

One of the serious questions facing the organization when a loss is reported is whether the matter is one of a loss of accountability, or a bonafide loss. Many patients' belongings are reported stolen when in actuality the property has been misplaced or accidentally discarded. This loss of accountability appears to increase in relation to the number of transfers occurring within the hospital. Often, too, reportedly stolen items have actually been taken to the patient's home by a friend or relative. In such cases the claim of theft is usually made in good faith, and most patients will advise the hospital if the item is found. This is not always true, however. Some patients rationalize that, since the hospital thinks something was stolen, it might reimburse them for the supposed loss. It is not uncommon for such patients to feel that they are paying an exorbitant room rate, and that anything the hospital pays to them, they actually have coming. Another type of patient may be embarrassed to have caused trouble over nothing and thus hesitate to confess that the lost item has been found.

The question of responsibility or liability is usually raised in the investigation of a loss that is not successfully recovered. Laws vary, of course, but generally the hospital cannot be held responsible for all property brought into the facility by a patient. The hospital must take every reasonable precaution to see that the patient's property is safeguarded so that losses do not occur, but it is generally considered that any loss is the responsibility of the patient and that the hospital should not automatically reimburse patients for losses. An exception exists where the hospital has a valuables protection system. In such a system, where the patient has actually been given a receipt for the property, and the property is subsequently controlled by the hospital, the organization would, of course, be liable for any loss.

A complicating factor concerning patient property is that the property in the custody of the patient is not constant. Rarely does a patient go home with the same property as when she or he checked into the hospital. Visitors to patients often bring gifts, and the patient also requests additional items to be brought into the hospital by relatives or friends. In the reverse these same visitors are taking things home for the patient. Thus the property belonging to the patient is in a constant state of change.

Employee Property Losses

Loss of employee property is another problem that confronts virtually all hospitals, clinics, and long-term-care facilities. This vulnerability can generally be correlated to the size of the organization. In the large facility, the anonymity factor and the erosion of interpersonal relationships combine to create an environment that tends to increase theft of employee property. This does not imply that the larger organization necessarily has more theft. The effective application of safeguards can manage this vulnerability, resulting in a good record of theft prevention.

The most common loss among employees is the purse left on a shelf at the nursing station or under the typing return of the office clerk. Other commonly stolen items include shoes and articles of clothing. Although theft can occur at any time, experience shows that it can sometimes be predictable. A prime example is the theft of winter coats and car batteries in geographical areas that have fairly severe cold weather. There is generally a rash of these thefts at the onset of the winter season, tapering off as the thieves get what they need to see them through the winter. The holiday seasons also provide more opportunity for the thief.

As with patient property, the dollar volume of loss of employees' property is quite low compared to organizational losses. Regardless of dollar amounts, however, employee property loss is often a key factor in employee relations. No one wants to work in a facility where losses, distrust, and suspicions are an ongoing problem. In contrast to the extremely low reporting level of hospital property losses it is estimated that security receives reports on 98 percent of all employee property losses.

Armed Robbery

The armed robbery both within the hospital and on the grounds is a growing vulnerability. This increase is apparently a direct result of the increased use and thus demand for drugs as the drug-abuse problem continues to grow. The main threat to the hospital involves the pharmacy, which routinely stocks large quantities of the desired narcotics and dangerous drugs. While most armed robberies have involved the pharmacy there are also reports of roberies occurring in the emergency room as well as at nursing stations themselves. The nursing station was the target of such a robbery in a hospital in Greenville, South Carolina. A patient and two friends took drugs from a medicine tray after binding up four nurses and a secretary. The patient was able to advise the friends of the routine, number of personnel on duty, and even obtained the hospital tape used to bind the employees. The drug loss amounted to about $1,000 in street value.

Another target of the armed robbery is cash. While the main cashier is a likely target there have been robberies of cafeterias, gift shops, outpatient cashiers, and other areas where cash is handled or stored. The robbers' timing, however, is not always right. In one case two armed persons held up a cafeteria cashier at 3:30 in the afternoon and escaped with less than twenty dollars. Just minutes before, the breakfast and lunch cashier had cashed out the register drawer for the afternoon cashier to take over. Cash in the drawer had been in excess of $1,200 as opposed to the twenty-five-dollar beginning bank.

Robbery is also of major concern on the hospital grounds. Victims are often visitors to patients, and this sets up a scenario of negative public relations. The patient becomes upset, especially if the victim is a spouse. Other patients quickly learn of the incident and word goes out to countless relatives and friends. Thus the hospital organization itself becomes a secondary victim.

The survey conducted of some 200 hospitals revealed that there were fifty-nine armed robberies within twelve months. Of this number, which included internal as well as external robberies, 28 percent were directed to drugs while the remaining incidents involved money.

Burglary

Loss of assets is not always through theft and robbery. The burglary incident is all too common in our hospitals. Although one might suspect the main pharmacy would be a primary target, the advent of many twenty-four-hour pharmacies and the alarm systems of closed pharmacies have reduced the number of such burglaries to a very few. There are numerous targets for burglary in the hospital; however, the operating-room narcotics supply tops the target list. The operating room is generally fairly well isolated and shut down for the most part at night. As with the armed robbery any place where cash is stored, whether it be large or small amounts, is the frequent scene of the hospital burglary.

Kidnappings

The threat of abductions is a security problem that must be seriously considered in analyzing the organization's vulnerabilities. This risk is generally associated with the newborn and pediatric patient; however, there have been incidents of adult patient and employee abductions.

A hospital in Fort Lauderdale, Florida, experienced litigation from the disappearance of a three-month-premature baby who died sixteen hours after his birth. The infant was tagged and left in the hospital morgue to be picked up by a funeral home. The baby vanished from the morgue before the pickup could be effected.

An incident which reveals the ease of abducting a baby occurred at Grady Memorial Hospital in Atlanta, Georgia, in August 1981. An unknown woman appeared on the maternity ward of the hospital. She took a baby, at which time the mother snatched back the baby from the intruder. A few minutes later an infant who was only twelve hours old disappeared from a crib beside her mother's bed. A hospital spokesman stated, "There is no special precaution on the maternity unit because there never really has been any need to."

Although there have been numerous incidents of infant abductions, the Fronczak baby kidnapping in Chicago has received the most publicity in recent times. The two-day-old baby was taken from the maternity ward by a female dressed in a white uniform. Litigation against the hospital, charging poor security, asked for $1 million. The suit was settled out of court for an undisclosed sum.

Homicides

This is a vulnerability that occurs more frequently than most persons think. Each year numerous homicides occur within our nation's hospitals.

At LaGuardia Hospital in New York City a female patient was stabbed to death as she lay sleeping in her hospital bed at 3:00 A.M. Police later arrested a male neighbor who had visited the patient during regular visiting hours.

Hahnemann Hospital in Philadelphia was the scene of a murder in the glass-enclosed telephone switchboard room located on the top floor of the nineteen-story structure. A telephone operator was stabbed to death shortly after beginning her night shift.

At 161-bed Coffee General Hospital in Douglas, Georgia, a maintenance man was found strangled to death in the hospital maintenance shop. Death occurred sometime between 5:00 P.M. and 9:00 P.M. with no indication of forced entry or signs of a struggle.

Perhaps the most tragic case of hospital-related homicide took place in Chicago in 1966. Eight student nurses were killed by Richard F. Speck in the townhouse living quarters provided by the hospital. As a result of this tragedy, a lawsuit was filed charging the hospital with failure to provide adequate security. The suit asked for just over $2 million and specifically stated that the hospital failed to:

- Exercise reasonable care and diligence for the personal safety of the nurses and their guests
- Adequately inspect, supervise, and guard the townhouse
- Provide adequate locks on the doors and windows of the townhouse to prevent persons from breaking into it
- Provide an emergency alarm system in the townhouse

The suit also said that "in the immediate vicinity of the townhouses; rowdy persons habitually congregated who did frequently breach the peace and were a menace to the safety of persons in the homes in said vicinity." It is, of course, impossible to guard against all possible contingencies; however, it can be argued, as this suit did, that assessment of vulnerabilities and reasonable safeguards might have prevented this horrendous incident.

External Disasters

Security vulnerabilities during an external disaster are centered around visitor control, patient valuables control, external traffic control of both pedestrians and vehicles, providing extra security support to the treatment areas, and designating an area for the dissemination of information concerning injured persons.

As disasters strike, new and often unthought-of factors present themselves to challenge even the most comprehensive and well developed hospital disaster plans. A prime example is the Three Mile

Island, Pennsylvania, nuclear plant accident which occurred in March 1979. Most disasters cover a rather small affected area; however, the affected area is what caused many problems for hospitals in the area of Three Mile Island. The first reports were that persons within a radius of twenty miles were in danger, and then a radius of 100 miles was declared to be in danger. Another unique aspect of this emergency is that the hazard could not be concretely identified. People heard many, and often conflicting, reports of the "uncontrolled release of radioactive gas" from the nuclear plant. An estimated 150,000 persons left the area. Hospitals were faced with the decision of total evacuation, while some almost had to close their doors as employees left town as well.

The closest hospital was the Milton S. Hershey Medical Center, which is located less than ten miles from the reactor site. Working around the clock for two days, the hospital was able to cut the patient census by two-thirds and implemented plans to evacuate the remaining 101 patients. As William DeMuth, Jr., a Hershey surgeon, stated, "many of the patients to be discharged didn't require much prodding."

In July 1977 Lee Hospital in Johnstown, Pennsylvania, was inundated during a major flood which filled the entire basement. The emergency department became unusable, telephones ceased to function, and the flooding cut the hospital off from outside help for over twelve hours. Once the water did subside townspeople started appearing seeking treatment, shelter, and food.

The Hyatt Regency Hotel in Kansas City, Missouri, was the scene of a tragedy the night of July 17, 1981. The collapse of an overhead walkway resulted in the death of 111 persons and 188 others injured. Two hospitals were located within two blocks of the hotel. In all, twenty-one hospitals were pressed into emergency service.

Strikes

One of the longest hospital strikes on record occurred in 1970 at Chicago Wesley Memorial Hospital (now named Northwestern Memorial Hospital), where a picket line was in effect for approximately six months. The hospital had to maintain its own delivery and trash removal systems amid constant confrontation and disruptive activities that included smashed windshields, fire-bombed trucks, deliberate accidents, injuries, and intimidations despite the assignment of a sizable police detail. One of the main supply problems was that of liquid oxygen, which had to be brought in small containers in a quantity to serve the 650-bed hospital. There was also a major problem in terms of trash removal, which had to be hand loaded and unloaded.

In another strike at Community General Hospital of Sullivan County in Harris, New York, the union decided to boycott and picket some of the commercial ventures of several members of the hospital's board of trustees. The businesses were two resort hotels and a department store located in the Catskill Mountain area of New York.

The protection problems encountered during a strike are somewhat similiar to those to be managed during the civil disturbance. These problems include:

1. The disruption of services from within
2. The disruption of external services such as deliveries and trash removal
3. Malicious destruction of facility-owned property as well as employee personal property such as parked vehicles
4. Intimidation and assault of known promanagement employees
5. Harassment in general, such as illegal secondary boycotts
6. Altercations between prounion and antiunion employees

Bomb Threats/Bombings

The bomb threat has practically become a way of life for the health care industry. The motives for such threats range from extortion and revenge to just the excitement of creating a problem.

A caller to one large western hospital stated that he had placed a bomb on the fourth floor to go off at 4:00 A.M. Reason: "Your people killed my wife!" Despite two succeeding calls, no bomb was found. In another case, a call was received by the administrator of the Sacred Heart General Hospital in Eugene, Oregon. The caller demanded $250,000 to reveal the location of a bomb he had placed in the hospital. This incident was concluded with the arrest of two persons who were charged with violation of the Hobbs Act, which deals with the use of interstate commerce in attempted extortion. No bomb was found.

Fires

Fires in medical care facilities are an everyday threat which contributes to high loss of life and property.

Prior to 1961 it was a widely accepted premise that modern fire-resistive construction would preclude large loss of life in fires in hospitals which were so constructed. That premise was proved quite wrong in the landmark Hartford Hospital fire in December 1961, where it was demonstrated that the heat and smoke produced by the burning of interior finishes are extremely lethal.

Another widely accepted premise was proven wrong in a 1970 Marietta, Ohio, nursing home fire which claimed thirty-two lives. This fire occurred in a modern one-story structure with all patient rooms on a ground-level floor with windows to the outside. A study of this fire leaves no doubt that having a certain building construction layout and a fire-detection system is not sufficient for good fire safety. There must also be trained personnel on duty who know how to react to a fire situation, however small the threat may be.

The major cause of fires in health care facilities is the careless use of matches and smoking materials. The second major cause is electrical.

Fires, of course, can start anywhere; however, the most common location of fires in hospitals and nursing homes is the patient room.

A fire in 1980 at the Gorden State Community Hospital in Marlton, New Jersey, forced the evacuation of all 180 patients plus fourteen newborn babies. This fire did not result in any deaths but indicates that it can happen that all patients need to be evacuated. The fire started in an elevator shaft of the four-story facility.

The Touro Infirmary located in New Orleans, Louisiana, was the scene of a five-alarm fire in July 1979. Allegedly the work of an arsonist, the fire started in a tenth-floor waiting area of an outpatient clinic which was closed for the holiday. Approximately 125 patients were evacuated to other areas of the hospital, including relocating all intensive-care patients. Although no lives were lost, the fire caused close to a quarter of a million dollars in damage.

Fire safety in hospitals and nursing homes is attracting considerable attention from various governmental agencies. In one recent case a hospital was inspected by a State Health Department acting as an agent for the Social Security Administration's Bureau of Health Insurance. This inspection was part of a highly questionable nationwide validation of the JCAH inspection program. The inspection revealed twenty-six separate alleged violations, resulting in a letter from the Social Security Administration that the hospital no longer met Medicare standards by virtue of its accreditation by the Joint Commission for Accreditation of Hospitals. In this instance, the inspection was typically bureaucratic and by law was inspected according to the outdated 1967 Life Safety Code.

In late 1974 the United States Department of Justice filed suit against the Commonwealth of Pennsylvania, charging that the state had failed to enforce fire safety standards in nursing homes. The suit filed in the United States District Court charged state officials with violating federal and state statutes and regulations that require skilled nursing homes approved for Medicare and Medicaid reimbursement to meet the Life Safety Code of the National Fire Protection Association. The Justice Department states that at least 147 nursing homes in Pennsylvania do not meet the standards.

In addition to accidental fires in hospitals, the problem of arson is very real. A portion of the Robert Lee survey dealt with such fires. The 160 hospitals surveyed reported a total of eighty arson fires, fifty-six of which were attributed to the psychiatric patient. Of the remaining twenty-four fires, over half were attributed to employees or former employees. The second leading cause was said to be union tactics. In one case, records indicate that the arrest of two employees on strike and a union organizer were involved. Areas requiring close attention in arson prevention are storerooms, employee locker rooms, lounge areas, machine rooms, linen storage areas, loading docks, and both employee and public washrooms.

Drug Abuse

An alarming security vulnerability which has surfaced in the last decade is that of drug abuse among employees. Although drug abuse is not unique to employees of the health care field, here it is a matter of increased concern because the patient is placed in a dangerous position. It is a well-established fact that a great number of the employees employed in health care facilities have purposely selected the facility with the intent of diverting drugs for their own use or for resale. Today's health care facilities are a major source of the nation's illicit drug traffic, especially in the area of barbiturates and amphetamines. These drugs are obtained in a variety of ways, including falsified drug requisitions and just plain helping oneself to the uncontrolled supply.

A routine locker check at one facility discovered sixty to seventy *red devils* (secobarbital sodium) in an employee's locker. A check of the employee's status revealed that he worked in the housekeeping department. It turned out that one of his duties was to sweep the pharmacy, which he did with utmost diligence. Every day he was able to reach into a large bottle of the *red devils*. Even though this one individual had taken several thousand capsules over a period of time, the pharmacy never reported missing any.

Another case suggests that the employee far removed from any drug handling can still be involved in procurement of drugs from a facility. This case involved three food service workers. The three would take a tour of the nursing units during the peak employee cafeteria time in the evening when a maximum number of unit nursing personnel were off the unit for their meal break. When remaining unit personnel were caring for patients, leaving the nursing station unattended, the thieves would check to see if the drug cabinet was unlocked. Two would act as lookouts while the third helped himself. Many nights they returned to their work assignment empty-handed, but on just as many occasions they successfully obtained limited quantities of drugs.

Drug abuse and narcotic addiction are not necessarily one and the same thing. Certain of the drugs listed in the dangerous drug category are not necessarily addicting. Narcotic drugs are generally derived from opium; perhaps the most well known are heroin, codeine, and morphine. Science has synthesized drugs called opiates which have properties similar to these three narcotics. One of the most widely used opiates is meperidine, otherwise known as Demerol. (Federal law also classifies the cocoa leaf and a chemical derived from it, cocaine, as a narcotic. Medical science, however, classifies cocaine as a stimulant and does not consider it to be a narcotic.)

Narcotics addiction among doctors and nurses is a much more serious problem than is generally recognized. The ease with which these persons can obtain narcotic drugs contributes significantly to the problem. In 1965, Doctor Solomon Garb, Missouri School of Medicine, Columbia, Missouri, analyzed data from thirteen states in which various agencies keep records in Demerol addiction. His admittedly understated projections at that time indicated that there were close to 400 known addict nurses, which is greater than the number of nurses graduated per year from thirteen average size nursing schools. Given the increase in the number of nurses since that time, the growing drug abuse problem, and less stringently enforced controls, the number of actual, not necessarily just known, addicts is staggering.

There are numerous methods by which a nurse can divert narcotics for personal use or for resale. Among these methods are charting a narcotic drug as administered and either administering a partial amount or none at all. If a patient is lucid enough to know if he received an injection, a substitute, even water, can be injected. It is extremely difficult to catch the perpetrator in the act.

A recently reported case reveals how one nurse was caught. It was suspected that the nurse was not administering the drugs she was charting as given. A nursing supervisor was able to identify a discarded syringe just after the suspected nurse had apparently given an injection of a narcotic drug. The syringe was immediately taken to the laboratory for analysis. The results were that the syringe contained vistaril.

There have even been cases where a narcotic was charted and there was no physician's order for the drug. Another very common method is not to destroy excess drugs. This occurs when a smaller amount of a narcotic is ordered than the normal packaged amount stocked on the unit. The ampule which is dropped and broken or accidentally contaminated is still another ploy used to obtain drugs fraudulently.

The outright shortage is also becoming more prevalent as facility management apparently becomes less concerned about strict accountability. It was not too long ago when there was only one key to the unit narcotic supply, controlled by the charge nurse. Now it is not uncommon for two or three keys from one unit to be in the possession of as many as six or eight persons during a given shift, greatly reducing the degree of control and accountability. In short, the physician, the pharmacist, and the nurse have become too complacent concerning the control of drugs.

Destruction of Property (Vandalism)

The malicious destruction of property is a constant threat for all health care facilities. In general, the facility is relatively open and it is extremely difficult to guard against the destructive act, which can take place at any time, quickly and unexpectedly. Reported incidents of malicious destruction range from writing on walls to the complete shutdown of a computer operation.

One case of continuing destruction suggests the bizarre turns that vandalism can take. This case involved the continuing removal of vinyl wall covering from the cab of an elevator. The perpetrator would cut a two- to three-inch strip of the vinyl covering from the interior of the cab of an elevator. This event occurred every day over a period of time and the guilty person was never apprehended.

Another malicious act which has occurred in

numerous facilities is exploding firehoses within standpipe fire hose cabinets. The perpetrator opens the water control valve so just a trickle of water is forced into the folded hoses. The length of time before the hose pressure becomes great enough actually to explode varies considerably, but it can take several hours.

The disgruntled employee and the ex-employee who has left under less than favorable conditions are prime sources of the malicious act.

Internal Disturbances

A continuing security problem for all medical care facilities is the internal disturbance. This problem can occur virtually anywhere within the facility and can involve patients, visitors, or employees. Many incidents of disturbance simply involve verbal arguments of varying degrees; however, all too frequently these verbal arguments lead to assaults and the destruction of property. The number of disturbances taking place can generally be correlated to size and type of facility.

A common area of disturbance incidents is in the emergency outpatient department. The large urban hospital with a busy emergency room will experience incidents on almost a daily basis. Many such facilities have found it necessary to station a security officer in the emergency area on a twenty-four-hour basis. The combative patient in the emergency room can usually be classified as intoxicated, drug overdosed, or in police custody.

The patient is not the only source of the disturbance problem; the person or persons accompanying the patient are often involved. Many such incidents are precipitated by long waits for medical treatment or the alleged lack of proper treatment. It is my experience that many incidents which have ended up with fists flying and property destroyed could have been averted if the medical personnel had taken a little extra effort to inform the patient or accompanying persons of the reason for delayed treatment or the type of treatment being rendered.

As stated, disturbances occur everywhere, including parking lots, employee locker rooms, lobbies, work areas, and even patient rooms.

A particularly dangerous source of trouble is the husband or wife who comes to the facility looking for a spouse to continue a feud. These incidents often lead to serious injuries, and the security officer can easily become injured in attempting to deal with the problem.

Another fairly common occurrence involves the intoxicated visitor or the visitor under the influence of drugs who demands to visit a patient at three o'clock in the morning. This type of person can usually be escorted from the property before he or she wakes up half the hospital with threats to sue the facility and everyone connected with it. This type of incident often requires a show of force to quell the disturbance quickly.

Information Loss

The loss of confidential or privileged information is a security vulnerability that is often overlooked in medical clinics as well as hospitals. Of primary importance are the medical records of the patient. These records often contain valuable information concerning lawsuits and other legal action. Other types of proprietary data that must be properly safeguarded are unusual incident reports, bid specifications, and certain financial records.

In one case the hospital was getting continuous complaints from maternity patients that a diaper service was calling them to offer their services, sometimes within hours after a delivery. An investigation revealed that a department records clerk was phoning the patient data to the diaper service in exchange for a fee. Although information concerning newborns is generally available publicly, the speed of getting the information certainly gave the diaper service an edge over competitors. This same type of information is often obtained through clinic sources, enabling the supplier to contact the prospective customer well in advance of the newborn delivery date.

In late 1976 an investigation into a large-scale medical record theft operation led to a number of grand jury indictments. This investigation, spearheaded by Denver District Attorney Dale Tooley, revealed that for a quarter of a century at least one firm, Factual Services Bureau, Inc., engaged in a nationwide, multi-million-dollar-a-year business of obtaining private citizen's medical records without authorization and by false pretenses. The firm made up to $3 million annually and operated out of regional offices in fifteen large cities throughout the nation. Its customers were more than 100 of the largest and most prominent insurance companies in our nation and often the legal counsels representing these companies. David Linowes, Chairman of the United States Privacy Protection Study Committee, described this widespread practice as "the insurance industry's Watergate."

The general mode of operation by investigators

of the Factual Services Bureau was to pose as doctors to obtain the information. The firm even had a training book on how to be an effective impostor in obtaining medical record information. These false-pretense methods were successful 99 percent of the time. In order to build business, the firm sent out hundreds of form letters to insurance companies soliciting investigation business. The following direct quotation from one such letter leaves little doubt as to their practices: "In essence our investigation reports cover all medical aspects of a claim whether you have authorization or not."

As a result of this investigation Colorado is one of the few states which has passed a law which renders the securing of medical records without proper authorization felony theft. The act further places a duty on every corporation or person promptly to report suspected crime to law enforcement authorities. This law was signed in June 1979.

Kickbacks/Fraud/Embezzlement

There are numerous forms of kickbacks, frauds, and embezzlement occurring in almost all medical care organizations. One of the problems in dealing with the kickback situation is the definition of what is a kickback and what is not. Most organizations do not regard the free lunch or dinner as a kickback, but just where is the line to be drawn? What about the TV sets, microwaves, and video recorders that were offered to chief X-ray technicians if they would sell the hospital's old X-ray film or silver flake? While the offer was basically an incentive rather than a kickback it leaves the door open to abuse.

One of the most vulnerable kickback areas in the hospital setting is the food service operation. Consultant Norman Jaspan cites a case where the food buyer in one hospital pocketed $170,000 in kickbacks from suppliers in just three years. In a New York City hospital a food service director was indicted on seventy-eight counts of soliciting and accepting bribes from thirteen food suppliers whose sales to the hospital accounted for 60 to 75 percent of all food purchased. The district attorney stated that initially the suppliers paid the director of food service 2 percent of their gross sales, and this demand was increased as business increased to a point where the director was taking up to 10 percent from some suppliers.

Therefore, of course, hundreds of schemes are used to defraud organizations. One of the more common methods is to sign for, and cause the organization to be subsequently invoiced for, more goods than are actually received. Another common method is to accept a lower-quality product than that shown on the invoice.

A relatively new scam involves the sending of phony invoices where no goods were ever sent. There are numerous reports of phony company names. In a recent case postal inspectors intercepted over 650 bogus bills being sent to hospitals by a nonexistent firm. The invoices from General Diagnostics Industries, Inc., checked out to originate from a phony mailbox at a vacant apartment. It is believed many hospitals paid the invoices, and the perpetrators got out ahead of the law.

Embezzlements do not always happen at the cashier's station or in complicated bookkeeping manipulations. Two parking garage employees at St. Paul Hospital in Dallas were charged with embezzling between $400 and $500 per day. Their arrest involves twelve counts of grand larceny.

All points where money is handled present a vulnerability that requires close scrutiny by the security administrator.

Gambling

Gambling activity is a problem generally associated with the larger facility. In respect to this vulnerability the organized ring often preys on workers in the support areas such as laundry, housekeeping, and food service. Although organized gambling has been a problem in some medical care facilities, it has not been of the magnitude found in many manufacturing organizations. One hospital case, however, did involve a security director who was *running numbers*.

Safety

All facilities must be concerned with employee, visitor, and patient safety. Safety in this regard refers principally to preventing injury caused by unsafe physical conditions or the inadvertent act of the victim or another person.

It is somewhat paradoxical that the safety record of some health care facilities is inferior to that of the major industries that send accident victims to them for care. In treating a wide variety of accident cases, it would seem that the hospital organization would have a special awareness of the need to maintain an adequate safety program. Many of the accidents involving the healthcare employee are similar in nature to those occurring in industry, with falls and improper handling of materials and patients being high on the most-common list.

The courts have consistently ruled that all medical care facilities are required to maintain a safe premise. The courts, and the Joint Commission of Accreditation of Hospitals, say that costs of medical care, loss of services, cost of malpractice and public liability insurance, loss of materials and equipment, human suffering, and loss of public good will are just some of the reasons why medical care facilities must constantly guard against accidents.

Impostors

The impostor represents still another security vulnerability that is occurring with increased frequency. The nurse's aid or Licensed Practical Nurse who has fraudulent identification indicating the status of a Registered Nurse appears in all segments of the health care industry. There have been numerous accounts of persons who, claiming to be physicians, have visited patients and given care, interacting in their assumed role for considerable lengths of time before their false identity was discovered.

Vladimir Owens was arrested in New York City in late 1974 on charges of fraudulent use of a doctor's title and criminal possession of a physician's license-renewal certificate. He was held on $500,000 bail. Records indicate that Owens had made at least 164 house calls for a physician referral service.

Parking

Ask a hospital security director which area of responsibility gives her or him the most trouble and chances are you will be told, "It's parking." Parking accommodations connected with health care facilities present unique control and protection problems with a never-ending situation of never having enough parking. The shortage of spaces is most acute in mid-afternoon of the weekday. At this time the day-shift employee is still at work and the arriving second shift must find parking spaces. Once the day shift leaves, there is generally an abundance of parking space. Since security generally has responsibility for enforcing parking regulations, as well as security patrols of the lots, it is security that receives the brunt of employee, visitor, and patient dissatisfactions.

In addition to enforcing parking regulations to effect orderly use the parking lot security effort is being increased due to a growing number of assaults, thefts, and vandalism. There are numerous major lawsuits against hospitals at any given time alleging insufficient hospital parking lot security.

Parking on the side streets surrounding the hospital presents an interesting problem for the security program. This parking is on public property and in a sense exists outside the realm of organizational protection responsibility. In a practical sense, however, the nurse who gets assaulted or the car that is broken into does affect the hospital and its operation almost to the same degree as if it happened on hospital property. Thus it is not uncommon for security to escort an employee or visitor off property. It is recommended that any such escorts be restricted to a maximum of one block—and this restriction should be strongly upheld by hospital administration. Once a security officer is a block from the property he or she is not providing visibility or surveillance activity for the hospital complex. The escort service may be helping one person to the detriment of many others.

Street parking can cause hospitals a community relations problem in addition to the protection problem. Homeowners in the area generally resent cars parked up and down their streets all day, restricting homeowner use as well as bringing increased pedestrian traffic and littering to their streets. Homeowners have been successful in eliminating this street parking by the passage of local parking ordinances. One such case is in the County of Arlington, Virginia. The ordinance provides homeowner relief in the form of issuing parking permits to residents of the neighborhood. Vehicles without permits which are parked in the area are ticketed. Commuters who were used to parking in these neighborhood areas brought suit against the county. While the Virginia Supreme Court held for the commuters the United States Supreme Court reversed the lower court and upheld the county ordinance. While there may not be many ordinances of this type in effect, the wrath of residents around a hospital can develop to the point where a petition for such an ordinance could be the result. In order for the restricted parking system to work the area should include seven to ten blocks, which has been found to be the distance required to discourage on-the-street parking.

In our discussion of employee theft we discussed the natural drive of the employee for the acquisition of property. The hospital organization also exhibits a propensity for acquisition of property. Expansion has become the norm for hospitals and expansion almost always means additional land surrounding the hospital complex. The first use of this land is for additional parking. With the addition of more parking capacity comes a general reorganization of the overall parking plan. If it isn't the acquisition of more property that causes reorganization it is the elimination of a parking area for construction needs that will also cause a review of parking. Between acquisition and elimination of space for parking there is a periodic, if not constant, reevaluation of the parking control plan.

Categories of Parking

Those who utilize facility-provided parking can be segregated into several distinct categories. These include the day-shift employee, the afternoon-shift employee, the night-shift employee, the outpatient, the inpatient, the physician, the visitor, and the vendor or outside service representative. All have specific and individual needs to be considered. The most common approach is to segregate these categories, assigning each to a specific parking area.

Close-in parking during the daytime period is generally assigned to outpatients, physicians, service representatives, and visitors to patients, due to the transient nature of their use of parking space. Physicians as a whole demand, and receive, special attention when it comes to parking. A common physician parking lot control mechanism is the automatic gate, which can be activated with a card. This card is also often the means by which the holder is able to activate the lock on the door of the physicians' entrance to the facility.

The automatic gate is considered to be cost effective for parking areas starting with as few as twenty spaces. The use of the automatic gate with a minicomputer offers many control features including being able to set time parameters for card use, invalidate the card at the output unit, and provide data on the attempted use of invalidated cards. Another feature is to render the card invalid after entering the parking area and to remain at that status until the card is used to exit. This prohibits one card from being utilized by several persons.

Summary

The foregoing vulnerabilities are the most common security problems facing organizations in the health care industry. As time marches on, certain areas become more important in terms of protection programming, and others will take their place in the shadows, awaiting society's ever-changing interactions to shape their impact on the organization. An example is the civil disturbance vulnerability. Barely recognized prior to the mid1960s virtually overnight it became the number-one vulnerability for many of our nation's hospitals and then almost overnight it is again of minor concern at the present time. The theft of X-ray film is another example of a vulnerability that quickly appeared and now has faded into the background as the price of silver went from $40 an ounce down to $8 an ounce.

The conclusion to be drawn is that the organization must be constantly reviewed and evaluate its vulnerabilities so that effective safeguards can be implemented in a general protection plan which can be expanded and relaxed according to current conditions and problems.

Chapter 9
Museum, Library, and Archive Security

GERARD SHIRAR, CPP

The Problem

Museums, libraries, and archives represent dissimilar security environments, and thus present a unique set of problems. One might regard these environments as vastly different from those found in business and industry. Viewed purely from a security standpoint, however, there are more similarities than differences; and I would strongly suggest that the person responsible for the security of a museum, library, or archive read as widely as possible on the subject of security, regardless of the environment addressed.

The theft of art, coins, antiques, and other collectables has increased significantly over the past fifteen years, with the bulk taken in thefts from private homes. Museums, art and antique dealers, archives, and libraries, however, have also borne their share of the attack. There are a number of reasons for the overall increase in this type of crime, but two stand out. First, there is the increase in the number of people who are involved in collecting; and second, the demand for collectables cannot be satisfied in the conventional manner, simply by creating more. Therefore, the thief has come to play an important role in the supply and demand process, by stealing from those who have, so that those who want are satisfied. This is an oversimplification, as the process of passage from the thief to the collector is often through a series of middlemen and interim owners, and involves those who are unwitting of their role in the sale and transport of stolen property, as well as those who are culpable.

The problem grows daily on a national and international scale. Art taken from a home in Cape Cod, Massachusetts, can turn up on the other side of the world within hours after it has been stolen, and within the week can be in the possession of new owner and out of circulation. The vast majority of stolen collectables are rare, not unique, and thus the object does not identify its true owner, as would, for instance, a well-catalogued, museum-owned painting by Rembrandt. The speed at which collectables move, combined with their relative anonymity, represents enormous problems for law enforcement and the recovery effort. Today if an object is stolen, the odds are very much against its recovery. With the situation as it is, the most effective way of dealing with the problem is to prevent the theft in the first place.

Foundations of a Security Program

Security and inconvenience are synonymous. What is convenient for you will also be convenient for the thief. The objective of a well-designed and functioning security system should be to maintain a defense against theft, while providing the activity which the system seeks to protect with sufficient freedom so that what must be done can be done. The balance between the two, at best, is a compromise.

Not everyone connected with an enterprise will agree on the importance of its various activities and functions. Opinions will vary, depending on what each job is, and how that job is affected by the security program. The need for security in a museum, library, or archive is generally recognized by the staff, trustees, and administration. Yet, in the course of the day-to-day activity, it is not uncommon for staff, in the name of expediency, to deliberately circumvent security policy and procedures. Administrators, too, may decline to take the steps necessary to enforce the circumvented policy, if in fact there is a policy in the first place.

The foundation of any good security program is not alarm systems, sophisticated locks, closed circuit television, and the like, but rather written policy and a set of implementing procedures for anyone connected with the institution and fully supported by the person with overall authority for the operation of the institution—the director, president, librarian, etc. This authority must adhere to established policies and actively support those who are charged with enforcement of the policy.

Responsibility, Performance, and Controls

In larger institutions, the responsibility for managing the security program is assigned to a director or chief of security and a professional security staff. In smaller institutions, there may be no security staff as such, with the security functions assigned to a member of the staff as an additional duty. Regardless of the size of the staff, responsibility for managing the security program must be clearly defined and assigned in writing to one individual. The authority of that staff member, and the security program itself, must be set forth in writing, signed and vigorously supported by the senior administrator.

As a minimum, the policy document should cover the following: designation of a specific person by name and title as having day-to-day responsibility for security; and establishment of security hours, a period during which the institution is closed to all staff, except necessary security personnel. If security personnel are not on duty 24 hours a day, a period during which the building is under alarm, no one may enter. All too frequently, an institution will adopt a permissive policy allowing staff to have keys to the building and permission to enter at any hour. This type of permissiveness invites problems, such as alarms not being properly set, false alarms, a general acceptance of activity inside the building at all hours, setting the stage for police and neighbors to ignore a criminal entry. The vast majority of thefts from art museums, as revealed by Bonnie Burnham in her book, *Art Theft: Its Scope, Its Impact and Its Control*, are committed by employees. Quite often the internal thief is an employee of long standing with no previous criminal record who succumbs to financial pressures and opportunity—the classic profile of the embezzler.

In many institutions, exception to security policy is frequently granted to long term and other employees who have flatly refused to be bound by the institution's security rules. Refusal to cooperate usually boils down to an exaggerated concern for position and prerogatives and an assumed implication that an insistence on adherence to security policy is an open expression of a lack of confidence and trust. The truth is that exceptions simply create the opportunity for a thief to circumvent security measures, and are blatantly unfair to those who recognize the need for and support a sound security program. Administrators who allow exceptions to their security policy will usually experience difficulty in enforcing policies in other areas.

There must be control over the unauthorized removal of property or anything which might conceal the unauthorized removal of property. There are three areas which this aspect of the policy should address: the activities of visitors, of the staff, and of the mail room and loading/shipping dock staff. In large institutions, responsibility for each area should be assigned to a department level manager to oversee. Normally the security manager will be responsible for the first two (visitors and staff), while the building superintendent or business manager supervises the third. In small institutions, whoever is assigned the security function should have overall responsibility for all three with assistance as needed. All areas should have the following controls in common:

- Documentation of all property of the institution which enters and leaves the institution
- An inspection process to ensure that what is specified in documentation actually entered or left the institution
- Documentation, prepared and authenticated by (in the case of employees) the employee's superior, except for items in the collection
- Documentation related to the removal or entry of all items in the collection, loans, evaluations, restoration, and other activity should be authenticated by the registrar, or other person charged with overall responsibility for the master records pertaining to the collection. The role of the registrar is a critical one in the security of the collection. Unfortunately, in many institutions, the registrar's role is not exercised as aggressively as it should be.

The type of documentation and the actual procedures required to carry out these principles will depend on the type and size of the institution and its organization.

Records Vital to the Security Program

Accurate records must be maintained reflecting exactly what the institution owns and where it is. This is easier said than done, particularly for institutions with very large collections. Such institutions are most vulnerable to employee theft, since items from the collection can be removed over a period of time by a larcenous employee, and not be discovered for an extended period of time. Large collections should make use of machine records which have two advantages. First, information, if programmed properly, can be recovered more rapidly than through a manual system. Second, the date can be better secured than it would be in a manual system, as the printed record will take up less space and requires special equipment and knowledge to alter. The master file should be secured in a safe or vault in the registrar's office or in a remote location, preferably one specifically designed for records storage. The printed copies of the data on the tape make a convenient working copy for both the registrar and those who work with the collection on a day-to-day basis. Procedures must be established to update the records promptly.

As a minimum, the following should be included in the machine master record:

- Description sufficient to identify the object
- Last date physically inventoried and the name of the person conducting the inventory.
- Collection number
- Department with custody
- Location
- Photograph reference number.

A variety of information is needed on objects in the collection, information which it may not be feasible to place on a machine record. The purpose of the central record is twofold; first, it acts as a complete list of the collection for security and emergency contingency purposes, and second, it supports an audit program. Where feasible, photographs of all objects in the collection should be made, and the negatives kept on file in a secure place. This practice pays off in the event of theft or loss through fire and other occurrences covered by insurance.

The most effective method of detecting internal theft (or mismanagement) is the audit. Ideally, the audit should be performed when the custodian of the objects terminates employment with the institution or is reassigned to other duties, and quarterly by an audit team comprised of disinterested staff members. The members of the team should be knowledgeable of the collection, but not under the immediate supervision of the custodian. The objects selected for inventory should be chosen at random from the master list. The ideal audit team consists of two people. The object should be physically accounted for. For example, an item out on loan can be verified through examination of the documentation pertaining to the loan. The audit will discourage internal theft, since it heightens the possibility of detection.

It would be extremely difficult in the case of a very large collection of several hundred thousand items to do a 100 percent inventory quarterly or upon the change of custodian. However, it is possible to develop a statistical model which will permit a 100 percent inventory while requiring a physical inventory of considerably less than 100 percent of the objects in the collection. The majority of libraries, museums, and archives do not have a formal audit program, and many of the world's large museums do not have an accurate central list of objects.

Personnel Security

There is more mobility found in the workforce today, and museums, archives, and libraries are experiencing a great turnover of employees. A number of security problems result from this trend. It is a necessary part of the security program to obtain adequate verification of all employee backgrounds, including criminal history and other pertinent information relating to an individual's honesty and integrity. An individual deep in debt is more apt to steal than one who is not. A person with a felony conviction who has served time in prison is more likely to come under criminal influence. Although federal and state privacy of information laws restrict access to criminal records and place limitations on the use of personal information, the investigation of an individual's background is not precluded, and is a prudent, low-cost investment in security. The investigation and its scope as a minimum should include:

- Employment check, going back ten years; if military service is involved, the DD Form 214 should be requested from the applicant.
- Verification of education if specialized training or a level of formal education is a job prerequisite.
- A check of the individual's credit history. This will frequently surface a criminal background or other security information.

The fact and scope of a background investigation

should be made known to the employee in advance. The fact that deliberate omissions or falsification of the employment application will result in immediate dismissal should also be made clear. A formal application form developed with the help of a personnel consultant should be used, even though the institution may not have a formal personnel program as such.

Security Responsibility

Responsibility for security, while properly assigned to one individual, should also be the responsibility of every individual employed by or otherwise connected with the operation of the institution. Employee job descriptions, whether written or verbal, should clearly assign to each position an appropriate security responsibility. For example, a paragraph in a secretary's job description which charges him or her with the responsibility to lock doors and windows of the office at the close of the day is appropriate. By the same token, a curator who exhibits an object in an insecure manner must accept responsibility for its loss or damage. The administrator of an archive, whose procedures for the usage of archival material are lax, must also assume responsibility for the loss of that material if and when it occurs. If security policy and procedures are prescribed in writing and security duties are assigned to specific positions in the job description, a system can be successful. Under such a system, if a loss does occur, identification of the weakness responsible can be more definitively made. If a failure to perform an assigned duty was the cause, responsibility may be fixed on an individual. Security responsibilities must be compatible with the principal duties and functions of the position. To assign a secretary the responsibility for responding to an alarm intended to detect unauthorized entry is certainly not compatible with the principal function of that position. All too often, however, such is the case.

What we have discussed so far is feasible for any institution and, except for the use of the services of credit agencies in the employment investigation, can be accomplished without expense. Without this administrative foundation, alarms and physical security measures, which are costly, will not be as effective as one might hope. For instance, how effective is a steel outer door alarmed with sophisticated equipment if no one is specifically charged to turn it on and lock the door?

Physical Security

This section will discuss alarm systems and physical security measures.

A word should first be said about construction. Any new construction should involve a security consultant who can work with the architect and later oversee the installation of the security measures. Building usage has an enormous effect on the design of physical security measures, and it is most important that the security consultant become intimately familiar with the particular idiosyncrasies of the institution involved. Building and life safety code considerations require the expertise of the consultant so that these requirements can be innovatively dealt with in the design of the physical security system. It is far cheaper and more effective to include security in the original design than to do the job after the building is completed.

Renovation of an existent structure is perhaps the most often encountered construction problem. Often the building which houses the institution was built at a time when aesthetic and other considerations were paramount and security problems were not as acute as they are today. Often the project involves renovating part of a building which will house the museum, library, or archive along with other activities.

The ideal situation is a building in which the museum, library, or archive is the only occupant, with the security dictates of the institution governing the overall security of the building. In the first instance of multiple occupancy, a pocket of security must be made encompassing the institution; however, regardless of what is done security is largely tied to what occurs in the other parts of the building. If an employee of an abutting occupant, for instance, is careless with a cigarette and a fire starts, the institution's well-designed fire detection system may do little. In multiple occupancy situations, one starts with a distinct security disadvantage which can never be completely overcome. Similarly, close proximity to abutting buildings can also represent difficult security problems. As a rule, multiple occupancy situations for museums, libraries, and archives should be avoided, as well as situations involving abutting buildings over which the institution does not exercise total control.

Alarm systems, in and of themselves, do not protect anything. Simply stated, an alarm notifies you *after the fact*. The advantage of an alarm lies in two areas; first, it is a relatively cheap way of continuously monitoring a security condition, and second, an alarm adds a degree of difficulty to an intended thief. A magnetic door switch, for example, wired to an anunciator will alert you only after the door is opened. How fast you are able to react (arrive at the door), vis-a-vis the length of time it takes a thief to steal and escape, is the effectiveness quotient of that alarm. If the door is metal and you install a capacitance type sensor, the thief activates the alarm when he disturbs the field (touches the door). In order to gain entry with this type of alarm, the thief must open the door, either by applying force or by manipulating the lock after the alarm has been activated. The physical security measures in this case come under attack after the alarm has activated, thus prolonging the time to effectively react.

The principal purpose of security is to prevent certain acts from occurring, and to increase the protector's ability to impede them if they do occur. A prerequisite to developing an effective defense is a survey to identify vulnerability. This survey should be peformed by a security professional, one who is intimately familiar with the criminal threat directed at museums, libraries, and archives, and who is also acquainted with the threat facing business and industry. Often, there is a preoccupation with the security of the collection, neglecting the security needs of other aspects of the institution's activities—parking lots, grounds, administrative areas, restaurant, bookstore, and so forth. The ideally designed security system is one that considers the total environment. A museum bookstore without adequate provision for cash handling or storage may act as a magnet, attracting a criminal element which would not otherwise be interested in the institution. A significant investment is made by museums, libraries, and archives. Inadequate protection of these assets can also result in a significant financial loss.

In assessing the threat, there are two aspects to consider: attack from outside and from within. Museums, libraries, and archives, for the most part, cater to the public, and have only limited control over those who visit. Therefore, it is highly possible that one who uses the front entrance during public hours as a visitor may hide and represent a threat after the building is secured for the night. As discussed earlier, employees represent a significant internal threat, and their activities must be both circumscribed and regulated. Not too many years ago, it was common to find major art museums keyed so that a single key issued to all employees fit all doors, including storage.

The internal threat is best addressed by compartmenting the building through a combination of doors and locks, so that access is based on need. A tight key control program is absolutely necessary to compartmentation. In large buildings, detection of the activity of the stay-behind or the criminally-minded employee may be difficult without some form of electronic surveillance system, motion detectors, case alarms, or closed circuit television. Protection against illegal entry from outside the building depends, to a large extent, on the materials used in constructing the building, the number and construction of doors and windows, roof construction, and openings. Skylights are frequently found in museum, library, and archive roofs and, depending on the height of the building or access to the roof from adjacent or abutting buildings, they may represent serious security problems. Vegetation, trees, and bushes which mask the view of basement and ground floor windows from the street or adjacent buildings offer protection while the thief cuts bars and forces entry through these openings.

The external threat is best met through physical security measures, not alarms. Earlier we mentioned that an alarm must be considered in terms of the length of time it takes to effectively react—in other words, to prevent a theft. An alarm, regardless of its type or design, provides intelligence only. It does not, in and of itself, stop anything.

A properly functioning alarm used in a motion detection role will indicate that someone is in the area covered by the sensor. To determine more, it may be necessary to investigate. It is common to use a motion detector in conjunction with CCTV, thus reducing the time needed to investigate and react. One can readily see how the effectiveness of an alarm is dramatically reduced if there is no policy regarding access by staff to an alarmed building.

An alarm used in conjunction with a physical security measure is best employed to advise that the assault has been initiated, rather than to advise it has been completed. The example used earlier of the capacitor type sensor on the steel door and its advantage over the contact switch illustrates the point. Window tape is frequently used on windows and unfortunately on the glass doors which some museums use as main entrances. A brick or sledge hammer applied to a glass window or door will shatter the glass and activate the alarm. If the thief can reach his objective before effective reaction is possible, the alarm has served no useful purpose as a defense. I would also say that the deterrent effect, often attributed to alarm systems intended to protect against the external threat, is overly optimistic if these systems are not used in conjunction with adequate physical security measures.

To the greatest extent possible, alarm wires and sensors must be hidden, shielded, and electronically supervised to detect tampering. The commercially available alarm can be purchased by anyone. Since alarm technology is basically simple, an alarm's weaknesses can be readily determined by a resourceful thief in advance. The defensive advantage of an alarm is therefore greatly enhanced if its parts are hidden from view. Alarm controls of any sort should not be placed so that they can be examined or manipulated by a thief. For example, it would not be wise under any circumstances to place the alarm switch outside the building it was protecting, regardless of the claims made by the manufacturer. All alarm wires which connect a system to the police station, private alarm company, or other reaction force should be buried, enclosed in conduit, and electronically supervised. All lines should exit the building below ground. An emergency power source should be provided to permit the system to operate during a power interruption. As a minimum, all doors, windows, and other openings such as skylights and other openings on the perimeter of a museum, library, or archive should be alarmed so as to detect intrusion. If possible, the alarm selected should detect the application of force before the assailant breaches the opening. In the case of windows or skylights, a back-up covering such as Lexan could be used in combination as follows: the window glass, alarm sensor, then Lexan, in that configuration.

The sources for alarms and their installers are quite extensive today. The yellow pages of the telephone directory attest to this. There is a significant mark-up in the alarm product field, and the principles behind most alarms are essentially basic technology. There are no mysteries. The application of computers in the alarm field has only now begun, and in a short time may well offer the degree of sophistication necessary to increase the difficulty of defeating an alarm. The advice here is to deal only with long-established firms, and to insist that the work paid for is in fact done. An alarm is not a substitute for other aspects of security—guards and physical barriers. An alarm is merely an adjunct, as it were, to a solid defense. As alarms are basic in their technology, they thus cannot stand alone, but all too frequently they are the only defense offered against the thief.

The museum, library, or archive must, to protect itself, develop, implement, and enforce a total security program. Discipline of the staff and management's support of the program are two of the most significant aspects of the program.

Fire Protection

Fire protection within a museum, library, or archive must take into consideration the nature of the materials which compromise the collection, but at the same time must provide adequate protection to the people who work within or visit the institution. Sprinkler systems are looked on by most institutions as being too risky. What if a vandal sets off one of the sprinkler heads, or a pipe breaks, or one of a dozen possible disasters occurs? Halon offers an alternative, since it will extinguish a fire and leave no residue. Halon is expensive, and while it will effectively extinguish a fire, there is limited protection against restart. Also, the Halon alternative may not be a practical solution for an institution housed in a combustible structure. Carbon dioxide is also used to a very large extent in restaurant kitchens. Its advantage is much the same as Halon; however, carbon dioxide will leave a residue.

Many institutions shy away from sprinkler systems unless forced to use them because of the nature of the structure in which the collection is housed and because of applicable fire and building codes. If the codes do not force the use of an extinguishing system, the general trend is either to have none or to employ Halon on a limited basis, and perhaps use a sprinkler system only in high risk areas, such as carpentry shops, boiler rooms, paint shops and the like.

Sprinkler systems are a safe and effective way of protecting collections. Today, there are options available to reduce the chance of accidental discharge. The Museum of Fine Arts in Boston uses a wet system to protect its Faneuil Hall Museum, employing the following options: heads which will automatically shut off if the heat source is removed; and shutoff valves (alarmed) on the stand pipe risers, permitting an immediate shutoff of the system in the event of a leak or pipe rupture, both relatively unlikely in a properly installed and maintained system.

The question that an administrator must ask when selecting a system is whether that system will do the job if the need arises. All too often, the least expensive and most effective system is not used because of a fear of the side effects. It is more practical to ensure the survivability of the collection than it is to avoid installing the necessary protection because of an inflated concern about accidents.

The Disaster

An important part of a museum's planning should be the disaster plan—the *what happens if* plan. If there is a fire in the institution, how will you protect the building during the fire, and most importantly, after the fire is out? In areas where tornadoes frequently occur, what plans have been made to protect the contents if the institution is involved? Disaster planning unfortunately is all too often ignored until too late, and there are not enough people or materials to protect the collection until the recovery process begins.

Guards

Today, professional guards play an important part in protecting the collections of museums, and to a lesser extent, archives and libraries. In the case of archives and libraries, the security role is more one of building security than it is one of actual protection of the collection. The trend today is to select guards who have retired, or others who are willing to work for a wage level generally paid to unskilled workers. It is not at all uncommon to entrust the entire assets of an institution amounting to many millions of dollars in value to a security guard who earns the minimum wage. Motivating and training such people is difficult and there is little wonder that they will let you down when the pressure is on them. Frankly, what can one expect of a night guard earning $5.50 to 8.50 an hour if he is approached by someone to assist in stealing from the institution? And how resourceful will this individual be in an emergency situation? There is an old adage, very true when it comes to security personnel—you get what you pay for.

Institutions employing guards should have, as a minimum, three documents. The first should be a guard's manual outlining the general duties of a guard and spelling out administrative procedures; the second document should be a set of guard orders for each guard position; and finally a set of emergency instructions covering steps to be taken under every possible emergency situation, such as: flood, fire, theft, stay-behind, bomb threat, medical emergency, tornado, hurricane, gas leak, break in a water pipe, fire in an adjacent structure, riot, etc.

Most institutions do not have personnel available to formally train guard personnel, nor can they afford to pay the guard's salary while sent off for formal training. The ideal situation would be to provide the guard with classroom-type formal training on all aspects of his duties. But in fact, even among the larger institutions training of this sort is seldom given. An acceptable training program would require the guard to read the guard's manual before being posted to the first assignment, followed by instruction from the person charged with management of the security program. During this instruction session, the areas of the manual dealing with the administrative aspects of the job should be covered along with the general responsibilities of the guard. This should be followed by two weeks of on-the-job training, interspersed with frequent discussions with the supervisor, during which time the supervisor questions the guard on the reactions to given situations. The idea is to teach the guard through recall of the principles in the manual as applied to specific situations. With the type of resources generally available for guard work, this approach has been proven to offer the most effective training process. On an annual basis, a retraining session for experienced guards is a necessity. During retraining, the salient points of the manual should be covered.

In addition to a set of written instructions, the guard force requires supervision. Frequently guards are assigned to their posts with no one designated to supervise their work or their compliance with policy.

Guards are relied on very heavily to protect museum collections. Because of this it is important to recognize that a human being is the least effective method of accomplishing this job and is also the most expensive. The guard's job is boring—he can only protect what is seen—and the guard can easily be distracted. Art museums particularly rely extensively on guards to prevent theft and vandalism. Yet in spite of rising salaries, training, and increased supervision, the effectiveness of the guard remains essentially the same, while the threat has increased and the number of incidents has multiplied. Art museums particularly would be wise to seek innovative ways of protecting their objects using a combination of physical security and alarms. A small painting in an isolated area of the museum is a prime candidate for vandalism or theft. If this painting is protected by den glass, and securely fastened to the wall, the probabilities of theft or vandalism occurring are lessened. This does not mean that guard protection is not required. Even the best physical security measures require supervision by a guard.

Transporting Collectables

A variety of safe, secure methods are available to transport collectables of high value. Four elements are fundamental to the security of all moves:

1. Maintaining secrecy concerning the arrangements
2. Reliability and resourcefulness of the courier or transporter

3. Proper packaging
4. Supervision during movement

The method selected must consider factors such as the size and weight of the object and its value. Large steel sculptures valued at several thousand dollars are frequently moved, employing an open-bodied semi trailer over the highway. The mails are frequently used to transport collectables, as are common carrier shippers and parcel post organizations such as United Parcel Service (UPS).

All moves of valuable objects, regardless of size, should include these security precautions:

- Whenever possible, have a courier representing the interests of the owner accompany the object.
- Use point to point transportation. Do not choose a method involving change of transportation mode or a period in a depot situation.
- Make sure that the object is insured during transit for its market value.
- Make sure that a point of coordination at the management level of the courier is established to facilitate the move.
- Do not mark packing crates in such a way that the contents may be deduced.
- Arrange an itinerary for the move, specifying times for all aspects of the move, i.e., pick-up time, time in transit, arrival time. Arrangements should be made to report any deviation in the itinerary to the owner or shipper.
- Do not use a method which will not permit accurate tracing or supervision of an object being moved. If using the mail, use certified or registered mail.

The best method for transporting very valuable collectables is through the use of couriers. If the shipment is a large one, the courier's equipment may include the use of escort vehicles, two-way radio communication, and surveillance by the law enforcement agencies through whose jurisdiction the shipment passes.

All to frequently, couriers selected to accompany valuable shipments between museums have consisted of an assistant curator and family following along behind a rented truck carrying a painting collection valued at many millions of dollars. In a practical sense such a courier is not going to prove effective in a crisis situation perhaps involving armed hijackers.

Chapter 10
High-Rise Office Building Security*

ROBERT BARNARD

An architectural firm is designing a twenty-story office building where the owners want to emphasize security as one of their standard building features. The security system will include intrusion detection, access control, television surveillance, and security guards. Based on the size of the building and the desired security functions, the most effective way to monitor and control these functions is with a computerized monitoring system. Besides managing the security functions, the computer can monitor the fire detectors and control the building's environmental conditions.

The computerized monitoring system consists of a communication processor, CPU, memory, CRT display, and line printer. In operation, the communication processor sequentially polls each office or office complex transponder. When addressed, the transponder reports to the communicator the fire alarm status, the identification of anyone requesting access, and the temperature and humidity in the area. The communicator then transfers this information to the computer's CPU. The processing unit retrieves the programmed instructions from memory and executes the commands accordingly. Any changes in the area's access/secure status or alarms initiated by either the intrusion detection sensors or fire detection sensors or monitoring devices are displayed on the CRT display and printed on the line printer. The identity of any authorized individual requesting access to a controlled area also is printed on the line printer. The systematic integration of the computer components with the individual subsystems will result in an effective central control system.

The individual subsystems that will be managed by the computerized monitoring system will be discussed in the following paragraphs.

Intrusion Detection System

The intrusion detection system for each office or office complex will be designed to satisfy the tenant's specific security requirements. At a minimum, however, the entrance door to each office or office complex will have a magnetic door switch concealed in the door assembly to detect anyone opening the door. Concealing the switch obviously eliminates any cosmetic distractions in the door's appearance and also provides physical protection against abuse and tampering.

Since many of the tenants will require additional security for their offices, plans are made to accommodate their needs. The offices requiring motion detection will be protected by either ultrasonic motion or passive infrared detectors. Provisions are being made for installing the detector transducers by installing junction boxes above the ceilings in all the offices. Detector transducer selection depends on the office layout; therefore, the selection and transducer placement will be decided after the tenant submits a drawing of the layout.

Infrared beam detectors will be used in those offices that require only a minimum level of detection. In this application they will detect anyone moving freely through the office or anyone approaching a specific area or object without concern. To reduce the detectors' chance to compromise by avoiding the infrared beam, the detectors will be concealed in the walls. The detectors will be installed to project an invisible beam across entrances and

---

*From *Intrusion Detection Systems*, 2nd ed, by Robert Barnard (Stoneham, MA: Butterworths, 1988).

normal traffic paths. Some applications might require redirecting the infrared beam to increase the length of the barrier, but the beam will be limited to three directions. Redirecting the beam requires the use of mirrors, which can create operational problems if a mirror should become misaligned.

Passive infrared motion detectors will be installed in the foyer in front of the elevator entrances on each floor. They will detect anyone getting off or on the elevator, or anyone moving through the foyer after the floor has been secured. Ultrasonic motion detectors were considered for this application, but it was feared that escaping air around the elevator doors when the elevator approached and passed the floor might cause the detector to false alarm. Microwave motion detectors also were considered for this application but were not selected because it was believed their energy might penetrate the elevator shaft and consequently detect the elevators when they passed the floor. If additional detection is needed to cover the corridors, microwave detectors will be used, but they will be directed down the corridors away from the elevators. A television camera will be installed in each foyer to cover the elevator entrances and adjoining area so that when there is an alarm, the monitoring operator can observe the corresponding television monitor and assess the cause.

The intrusion detectors in each office or office complex will be connected to separate monitoring terminals in the transponder that monitor the sensors and control devices in that area. When the transponder is polled after an alarm has been initiated, the detector that initiated the alarm will be identified on the CRT and recorded on the line printer. The detector location will be graphically displayed and alphanumerically identified for the monitoring operator. The graphic display will depict the outline of the office or office complex where the alarm originated and also identify the location of the particular detector that is in the alarm state. Since many of the offices and office complex outlines on the various floors are identical, the floor and office must be identified in English text. Any special instructions for the monitoring operator will be displayed, along with the alarm location. With a system designed in this manner, the monitoring operator can quickly dispatch the response guard to the appropriate office.

Access Control System

The access control system will provide the necessary level of security to accommodate each tenant's requirements. It will use encoded proximity cards and card readers to provide the basic level of access control. Although proximity cards are more expensive than magnetic-type cards, they are more difficult to counterfeit. They also can be read without having to insert them into or pass them through slot readers. This makes the proximity system more user friendly, reduces physical wear on the cards, and eliminates repairs to the readers caused by people jamming things into the slot. Over a period of time, the proximity card system is believed to be more cost-effective than magnetic card systems.

The proximity card alone will grant access to the parking lot and main entrance to the office building when the building is secure. But for the remainder of the building, the access control system will control who enters where, and where necessary, when they enter. Areas requiring medium level access security will be controlled by the access card, a keyboard with a five-digit memorized code, and perhaps a hand-geometry or fingerprint verification reader.

Areas requiring the highest level of access security will be controlled by a portal system, which will require card access and a five-digit memorized code to gain access to either an eye retina, voice, or signature verification reader.

Television Systems

Standard vidicon television cameras will be installed in strategic locations throughout the office building to assist the security guards in performing their surveillance function. Besides the cameras that view the elevator entrances in the foyer on each floor, additional cameras will be installed in the main lobby and parking areas. Two cameras will be installed near the ceiling, one on each side of the lobby, and positioned to view the main entrance and elevator entrances. Two articulated cameras with zoom lenses will be installed on each of the two building parking levels. One camera will be positioned near the ceiling to view the front elevator foyer entrance, and the second camera will be positioned on the opposite side of the parking lot to view the back foyer entrance. In these positions, the cameras can be articulated to view the entire parking area for general surveillance or for tracking anyone leaving the building until reaching one's automobile. Another camera will be positioned at the vehicle entrance on the first level to view the ramp and vehicle card access gate. Additional cameras will be installed as necessary to satisfy special tenant surveillance requirements.

The two lobby cameras will be connected to separate 9-inch broadcast-quality television monitors. These monitors will provide the monitoring operator with continuous surveillance of the lobby. The two articulated cameras and the fixed camera covering the entrance to the first parking level will be connected to a single 9-inch monitor through a sequential video switcher. A video switcher can automatically sequence the video signals from several cameras to one of several monitors for display. It also can be manually switched to one camera or from one camera to another, depending on the surveillance functions.

In operation, when the parking area cameras are not being used for selected surveillance, the video switcher will sequentially switch from one camera to the next. This arrangement will provide the monitoring operator with 10-second sequential views of the front and back entrances to the elevator foyer and the vehicle entrance to the first parking level. An operator wishing to monitor just one entrance or use one of the articulated cameras for surveillance can switch the video switch from automatic sequencing to the desired camera. When finished with the specific surveillance function, an operator can switch back to the automatic switching mode. A sync generator will be used to synchronize the picture from each camera to eliminate flicker or roll on the monitor when switching from one camera to the next. The two surveillance cameras on the second parking level will be monitored in the same way.

Twenty cameras covering the elevator entrances in the foyer on each floor will be monitored on six monitors. Video switching between the cameras and the six monitors will be computer controlled. In the normal operating mode the twenty cameras will be sequentially switched, and each camera will be periodically displayed on one of the six monitors. When there is an alarm, the computer will connect the camera covering the affected foyer and display it on the first of the six monitors. If another alarm occurs while the first scene is being displayed, then this foyer is displayed on the second monitor. It is very unlikely that six of the twenty foyers will be in alarm simultaneously. It is not planned to have video recording at this time, although it would alleviate the problem if more than six alarms should occur at any one time. Provisions are being made to add a video recorder if it is required by any of the tenants.

The twenty foyer cameras also will be synchronized to prevent picture flicker or roll when switching from camera to camera. When there is an alarm, the foyer where the alarm originated will be identified on the CRT display. But when the cameras are being sequentially displayed for general surveillance, the CRT will not be used in identifying the video. Instead, the monitoring operator can view a large number identifying the floor level. The number will be installed over the elevator door in full view of the camera in each foyer. Putting the numbers above the elevator doors will keep them from being blocked by normal personnel traffic entering and exiting the elevators.

Provisions are being made during the building design phase to accommodate almost any tenant video surveillance requirements. One such provision is the addition of video coaxial cables installed between the monitoring area and a central junction box located on each floor. Subsequently, if additional cameras are needed, new cables will only have to be routed through the ceiling from each camera to the central junction box. It is much easier to install spare cables when the building is being built than it is to add cables afterward. For this reason, spare transponder communications cables are being installed to accommodate other intrusion detection or building control systems.

The television monitors will be mounted above the computer console and CRT display. In this position the 9-inch monitors will be in easy viewing distance for the monitoring operator seated at the console. The video switcher control panel will be mounted just below the television monitors and to the side of the CRT display. The pan, tilt, and zoom controls for articulating the cameras in the parking areas will be located below the switcher control panel. Both the video switcher controls and the articulated camera controls will be within easy reach of the operator.

Another provision being made during the building design with respect to expanding the video surveillance capability is to allow adequate space in the monitoring room for additional monitors. The first ten additional monitors would be mounted above the existing monitors. If still more monitors are needed, the console will be extended on either side to accommodate them. Each time new television monitors or video controls are added, the console arrangement will be human-engineered to maintain a totally integrated system.

Security Guard Patrol System

One function of the security guards is to conduct scheduled fire and security patrols through the office building. The security guard patrol function uses an integrated computer-managed reporting and communicating system. The guards will patrol each floor at least once every four hours. With this schedule, each floor will be checked three times each night during the week and six times during the weekend and on holidays. To ensure that the patrols are being conducted, the desired patrol schedule will be programmed into the computer instructions. Each floor will be assigned a time window during which a patrolling guard must patrol the floor or area and report to the computer.

The report is communicated by activating a card access reader. These readers are located in strategic areas on each floor where the guard should check the building's fire and security status. When the guard reports, the transmitted message identifies his location to the computer. The information is printed on the line printer, along with the reporting time. With this information, the monitoring operator can observe the guard's progress. If the guard does not report during the programmed time window or reports in a sequence other than the programmed patrol route, an alarm will be initiated to alert the monitoring operator.

Communication between the patrolling security guard and monitoring operator in any security system is vital. For example, if a patrolling guard encounters any security or emergency hazard, he must be able to alert the monitoring operator or the proper authorities. Many installations provide their guards with portable radios, but radio communication will be limited in the high-rise office building because of all the steel and high-density building materials. Therefore, an intercom will be installed as part of the guard patrol system. The guard will be able to communicate with the monitoring operator by plugging a portable handset into the tour station telephone jack. Since it is often necessary for the monitoring operator to communicate with a patrolling guard, there will be a call-in lamp at each tour station, which can be activated on command from the control station.

Fire Control System

The purpose of this discussion is not to describe a fire control system for the high-rise office building. That task requires someone who is knowledgeable about fire and building codes and standards, as well as the operation and application of fire control equipment. Instead, the purpose here is to describe the types of fire control devices that can be monitored by the computerized monitoring system. These devices include fire detectors and sprinkler systems. The computer can monitor the sprinkler water flow detectors, water temperature, air pressure in tanks and dry-pipe sprinkler systems, and water tank levels. There are two basic types of fire detectors—smoke and heat detectors. The most important of these are the smoke detectors because they detect fires in the earliest stage of combustion, thus reducing the loss of life and property.

Smoke Detectors

Smoke detectors are required by Underwriters' Laboratories (UL), Inc., to respond to gray smoke in concentrations of from 0.2 to 4.0 percent obscuration per foot. (Obscuration is the proportional reduction in transmitted light caused by smoke particles interposed between the light source and point of measurement at a stated distance from the light source.)

The two most common types of smoke detectors that can comply with the UL requirement are ionization and photoelectric. Ionization smoke detectors contain a small amount of radioactive material safely sealed inside the detector's chamber. The radioactive material ionizes the air inside the chamber so that the air will be an electrical conductor. A precise electric current is passed through the ionized air and monitored by the detector. If smoke particles or particles of combustion enter the detector chamber, they mix with the ionized air and reduce the current flow. The reduction in the current flow is sensed, thereby initiating an alarm.

Photoelectric detectors sense the scattering of a light beam caused by smoke particles entering the beam. The detectors can either sense the attenuated light directly or detect the light scattered by the smoke particles. With both devices, a photosensitive cell detects either the attenuation of the direct light or the magnitude of scattered light impinging on the photosensitive cell. Detectors sensing the attenuated light illuminate the photocell directly and initiate an alarm when the smoke particles attenuate the light below a preset level. Detectors sensing the scattered light initiate an alarm when the light impinging on the photocell reaches a preset level.

Thermal Detectors

Thermal detectors sense the presence of fire when the ambient temperature rises above a preset temperature, usually about 135°F, or when the increase in ambient temperature exceeds a preset rate. The two detectors that sense these conditions are the fixed-temperature thermal detector and the rate-of-rise detector. Fixed-temperature detectors initiate an alarm when the ambient temperature reaches

135°F. They are available in both restorable and nonrestorable types. Restorable fixed-temperature detectors use replaceable fusible elements or are available with self-resetting bimetallic elements.

Rate-of-rise detectors initiate an alarm when the ambient temperature rises more than 15°F per minute. Most of these detectors also have a fixed temperature element in case the temperature rise is too slow to satisfy the rate-of-rise criteria.

Environmental Control System

The computerized monitoring system also can be used to monitor building temperature and humidity and to control the heating and air-conditioning systems to maintain a comfortable working environment inside the office building.

Emergency Power

The building will have an emergency generator to provide power to essential equipment and lighting when commercial power fails. The intrusion detection system, access control system, CCTV, console control and monitoring system, and fire detection system are considered essential equipment. Emergency lighting, intrusion detection equipment, and fire detection sensors have standby batteries to provide power during power interrupt. But the television cameras and access control system's electric door strikes will be without power during this time.

This means that the access-controlled doors will have to fail-secure or fail-open during power interrupt. Electrically controlled parking lot gates will fail-open so that vehicles can exit and enter the lot unimpeded. Anyone can enter the lot during the time the gate is open, but at least authorized individuals will not be denied access and the emergency generator should be on-line within 2 or 3 minutes.

All controlled doors inside the building will fail-secure but will have inside manual releases so that occupants can exit the areas. Master control keys will be stored in a safe in the security office so they can be used to override the electric strike and gain access to any controlled area in an emergency.

Summary

Computers have both a monitoring and a command and control capability that, when properly integrated, can provide a cost-effective way to manage security, access control, fire, and environmental control systems. There are, however, two important application considerations that could affect the computer system's operability and vulnerability. Operability can be affected if the maintenance plan is poorly conceived. There should be adequate spare modules available so that when there is a computer failure, making the computer operational is simply a matter of the technician replacing the failed module. The application might justify a spare computer that can assume control of the monitoring and control functions if the on-line system fails. Vulnerability is a prime consideration in high-security applications, so the number of functions performed by the computer should be limited to only those involved directly with security. Adding additional functions, especially functions that would involve other computer operators, will increase the vulnerability of the system to compromise.

Computerized monitoring systems, like all other intrusion detection sensors and monitoring systems, can improve security effectiveness only when they are properly designed and adequately supported by security personnel.

References

Books

Broder, James F. *Risk Analysis and the Security Survey*. Stoneham, MA: Butterworth Publishers, 1984.

Fisher, A. James. *Security for Business and Industry*. Englewood Cliffs, New Jersey: Prentice-Hall, Inc., 1979.

Gigliotti, Richard J., and Ronald C. Jason. *Security Design for Maximum Protection*. Stoneham, MA: Butterworth Publishers, 1984.

Healy, Richard J. *Design for Security*. 2d ed. New York: John Wiley, 1983.

Illuminating Engineering Society. *IES Lighting Handbook*. 5th ed. New York: Illuminating Engineering Society, 1972.

Purpura, Phillip P. *Security and Loss Prevention*. Stoneham, MA: Butterworth Publishers, 1984.

San Luis, Ed. *Office and Office Building Security*. Stoneham, MA: Butterworth Publishers, 1973.

Walsh, Timothy J., and Richard J. Healy. *Protection of Assets Manual*. Santa Monica, CA: Merritt, 1974.

Weber, Thad L. *Alarm Systems and Theft Prevention*. 2d ed. Stoneham, MA: Butterworth Publishers, 1985.

Reports

Barry, Joseph A. *Personnel Fatigue in Closed Circuit Television Assessment*. Boston: The Loss Prevention Institute, Inc., June 1981.

Carpency, Frank M. *Fiber Optic Cable Applications in Security*. Gaithersburg, MD: NUS Corporation.

Fite, Robert A. *Commercial Perimeter Sensor Evaluation*. Report 2209. Fort Belvoir, VA: MERADCOM, May 1977.

Garrett, William C. *Infrared Motion Sensor Evaluation*. Report 2237. Fort Belvoir, VA: MERADCOM, March 1978.

Moore, R.T. *Penetration Resistance Test of Reinforced Concrete Barriers*. NBSIR 73-101. Washington, DC: National Bureau of Standards, December 1972.

Moore, R.T. *Penetration Tests on J—SIIDS Barriers*. NBSIR 73-223. Washington, DC: National Bureau of Standards, June 4, 1973.

PART THREE
LOSS PREVENTION PLANNING AND CONTROLS

Chapter 11
Computer Security

JOHN M. CARROLL

Every day, the shape of the future becomes clearer: a cashless society with paperless offices, peopleless factories.

Eventually, the computer will become the principal repository of our personal and intellectual property, and our access to privileges and entitlements will depend upon how we identify ourselves to it.

Misappropriation of goods or services of any magnitude will necessitate defeating one or more computers. Often, the only way to do this will be to use another computer.

Computer crime is defined as any action not sanctioned in law or conventional business practice that is harmful to persons or property and which is either directed against or employs high technology information systems.

Why worry about computer crime? Usually, computer crime is nonviolent, although there have been at least three murders in the New York City area that stemmed from computer crime, and certainly the terrorist attacks on computer centers in France and Italy during 1979 and 1980 can scarcely be called nonviolent. In general, however, computer crime involves fraud and deceit rather than assault or violence.

There are at least two reasons to view an upsurge in computer crime with alarm. First, a successful computer crime can be extremely profitable to its perpetrator. Some observers have placed the average return in excess of half a million dollars. This can be a staggering loss to the victim; more often, it is shared broadly throughout society.

The second reason for concern is that the fabric of our society is held together by high-technology systems and the confidence of the public in the integrity of these systems is essential to the smooth functioning of society. Acts that threaten information systems strike at the heart of society.

What's so special about computer crime? Most computer crimes can ultimately be resolved in terms of theft, forgery, false pretenses, and fraud. When, then, should it be addressed as a special category? The answer is that successful investigation, prosecution and prevention of computer crime requires that criminal justice officials acquire technical knowledge not traditionally part of their training.

There are five reasons to believe that computer crime will continue to increase in years to come:

1. A maturing criminal population
2. Disaffection of the middle class
3. Increasing literacy
4. Concentration of assets within computers
5. Inadequate response from the criminal justice community

Maturation of the Criminal Population

As criminals grow older, they tend to abandon violent crime in favor of less risky and more profitable criminal pursuits. There is no criminal endeavor more lucrative than computer crime, and none with less risk to the perpetrator. We can expect that some of the demographic bubble that strained correctional services in the seventies will take to computer crime as they outgrow the 15-to-25 age bracket. Indeed, many offenders have been trained in data processing as part of well-meaning rehabilitation programs!

Middle Class Disaffection

It has been said that the rich get richer, the poor get welfare, and the middle class pays the shot. Inflation has pushed the good life beyond the reach of many middle class taxpayers. Moreover, the ranks of middle management are quickly filling up. Passage to upper management is becoming increasingly competitive and selective.

For all these reasons, and maybe more, we can predict a glut of highly-trained and critically-placed men and women increasingly frustrated by unfulfilled and unfulfillable expectations. These people, if they have access to the levers of computer power, may, if confronted with unsharable problems, resort to computer crime.

Rising Computer Literacy

The four thirteen-year-old New York schoolboys who penetrated a corporate computer in Montreal speak eloquently to the pervasive spread of computer literacy. Kids growing up with an intimate knowledge of the latest computers see nothing wrong with penetrating someone else's machine. Indeed, some of them see every new security safeguard as a challenge to be answered.

Response of the Criminal Justice Community

The fact that the criminal justice community has thus far failed to respond to the challenge of computer crime is perhaps rooted in its nonviolent, impersonal nature. There is no bleeding and battered victim voicelessly crying for retribution. This is perhaps why computer criminals incur less than a one-chance-in-a-hundred of going to jail. Judges may deal leniently with computer criminals because our legal system historically has assessed the seriousness of crime by its potential for personal injury or loss of life. The fact that computer criminals seldom come to trial may derive from the extreme difficulty or futility of building cases against them.

Increasingly, the onus will be on the private sector to safeguard its own computer-stored assets. The emphasis must necessarily be on security rather than detection; it takes an average of two years to establish a computer fraud case and may require the victimized firm to produce in court its detailed corporate records—with all its faults showing.

To a degree unequalled in nearly any other field, computer security is the only answer to protection of computer-stored assets.

Personnel Screening of Key Employees

There neither is nor can there be any effective protection against a well-qualified but dishonest person who can gain access to the computer system either at the computer room or at a remote terminal. The person who rips off a company's computer may very well be a member of staff, maybe even in charge of the whole operation.

Knowledge of computer systems is becoming more widespread all the time. Turnover in the industry is higher than in most others; and computer people are generally quite young. The point is that it is foolhardy to trust in the supposed ignorance of a possible intruder. A newly hired clerk or messenger may formerly have been an IBM software troubleshooter and may know your type of system inside out.

Computer crime is one of the least frequently prosecuted of all criminal activities. Fewer than 15 percent of all scams ever come to light. Furthermore, it is a crime that victims do not like to talk about. Banking, insurance, or trust company executives generally feel it is bad for business to reveal that one of their hand-picked officers was an embezzler.

It is still not unknown for victims to absorb their losses and let the culprit go with a golden handshake and a letter of recommendation.

Unfortunately, computer criminals are largely undistinguishable from the most creative of computer scientists, which, in a perverse way, they are.

In personnel selection, the computer center is the most critical area there is. The center director and deputy, the security coordinator or administrator, the operations manager, the manager of programming, the manager of systems analysis, the database administrator, all senior systems programmers, and your local software and hardware field representatives are all key persons who should be subjected to the same screening and security as any senior fiduciary or law enforcement officer.

Vulnerabilities of Computers

Computer systems have several unique vulnerabilities:

1. The coprocessing of two files can result in the production of a file deserving a higher level of security classification than that possessed by any of the input files; and the aggregation of records in a computer can often result in a file attracting a higher classification than that of any record in it.
2. The cost of EDP systems tends to centralize corporate data processing so that continuity of service becomes essential to operations.

3. Once an error is introduced into a system, it is difficult to extricate and tends to propagate rapidly.
4. Effective supervision is often lacking because company management and security officers do not understand computers sufficiently to know what the employees are up to; and the actions are difficult to trace to a specific individual.
5. A high labor turnover rate is common in computer facilities; also, personnel tend to demonstrate a greater loyalty to their profession and to each other than to their employers.

Computer files also have special characteristics that contribute to security concerns. Among these are:

1. The density of information is much higher than that of print media.
2. The nature and contents of files cannot be determined by visual inspection.
3. Stored information is more accessible at remote terminals than is information in conventional files.
4. When stored information has been modified in an unauthorized manner, such modification cannot be detected.
5. Computer storage devices, after having been erased, may retain images of data previously recorded on them.
6. Computers are fragile and their behavior can be subtly modified by changes in environment. A computer can be subverted to assume the role of a hostile penetration agent.

The points of vulnerability in computer systems are the central processors, storage devices, communication facilities, remote terminals, users, and systems personnel.

Central processors are vulnerable to failure of protection circuits, confounding of bounds and relocation registers, and misuse of privileged instructions. Their software is vulnerable to bypassing of file protection and access control programs or falsification of user identification.

Storage devices are vulnerable to unauthorized copying of stored information, theft of removable media, and hardware or software failure that could result in compromise.

Communications facilities can be compromised by undesired signal data emanations, crosstalk between secure and insecure circuits, and technical surveillance devices.

Users may misrepresent or forge identification or authorization, seek unauthorized access to sensitive material by browsing, and use debugging procedures to circumvent security mechanisms.

Remote terminals can produce undesired signal data emanations, are vulnerable to technical surveillance devices, and produce compromising text as hard copy or remnant images on platens or ink ribbons.

Systems personnel have normal access to supervisor programs, accounting files, systems files, protective features, core dumps, and files stored on removable computer media, and if not loyal and reliable, become security risks.

The most probable kinds of attack fall into these seven categories:

1. Subversion can result in destruction of equipment or facilities, disclosure of classified programs and data, interruption of service, improper modification of programs or data, loss of programs or data, theft of property, and misuse of resources.
2. Negligence can result in disclosure of classified information, corruption of programs or data, interruption of service, loss of programs or data, and destruction of equipment and facilities.
3. Accidents can result in interruption of service, corruption of programs or data, loss of programs or data, and destruction of equipment and facilities.
4. Covert attacks by nonemployees (stealth and deceit) can result in disclosure of classified information, interruption of service, corruption of programs or data, destruction of equipment and facilities, and theft of property.
5. Forced, overt attacks by outsiders can result in interruption of service and destruction of equipment and facilities.
6. Overt attacks by employees can result in interruption of service and destruction of equipment and facilities.
7. Input errors can result in interruption of service and corruption of data.

Summary of Security Approaches

Computer security involves measures relating to hardware, computer programs, (also called software), and communications. There are at least twenty cardinal points of protection:

1. Establish authority
2. Ensure loyalty and reliability of employees
3. Establish means whereby authorizing actions may be recognized as valid

4. Identify assets deserving of protection
5. Count the protected assets
6. Concentrate the valuable assets so they can be protected
7. Establish defined perimeters around the protected areas
8. Defend these protection perimeters
9. Maintain surveillance over protected assets
10. Control access to protected assets
11. Restrict knowledge of protected assets and protection mechanisms
12. Reduce exposure of protected assets
13. Limit privilege with respect to protected assets
14. Fix responsibility for protected assets
15. Document actions affecting protected assets
16. Double-check all actions affecting protected assets
17. Analyze documentation of actions affecting protected assets
18. Investigate all discrepancies
19. Punish deviations
20. Be prepared to fall back to prepared positions

Any protective mechanism should demonstrate *completeness*, in that it should function in every mode of operation, and *correctness*, in that it should provide a desired response to every stimulus.

The protective mechanism should be as simple as possible (least mechanism), and maintainable within its environment. It should be spoof-proof, offering maximal resistance to deception and producing the least possible number of false or nuisance alarms.

Survivability is essential. There must be a high probability that the protective mechanism will deliver a specified level of protection for a preestablished period of time under predefined operating conditions. It must be fail safe. If the mechanism is subjected to malfunction, loss of power or control, loss of communications capability, tampering or deception, it must revert to and remain in its most secure state.

Defense Mechanisms

There are seven defense mechanisms for computer security:

1. *Physical security* deals with physical access controls, such as fences and doors, locks and alarm systems, and environmental and emergency protection.
2. *Personnel security* concerns the selection, screening, clearance, and supervision of employees.
3. *Encryption of sensitive information* protects data communications and sensitive files by making them unintelligible to personnel other than the authorized recipients.
4. *Technical surveillance inspection* protects against intrusion devices that might be planted within secure premises.
5. *Suppression of compromising emanations* may be necessary because typewriters, terminals, line printers, card-readers, keypunches, and visual display units produce acoustical and electromagnetic emanations that can be intercepted, analyzed and interpreted.
6. *Line security* safeguard communications lines from the physical access needed to implant intrusion devices.
7. *Systems security* concerns protecting data within the system against unauthorized access.

Organizing for Security

The security coordinator is a key person in the computer security field and has at least six responsibilities:

1. Threat evaluation
2. Security procedures
3. Countermeasures
4. Backup and recovery planning
5. Intelligence (threat warning)
6. Security training

Threat evaluation identifies and places a value on hardware, programs, data, documentation and supplies. It should be updated at least annually.

The key to effective security precautions is to identify, by name, persons who will be held personally responsible for each and every specified asset. Accountable persons should maintain up-to-date documentary evidence relating to the security of assets and attesting to the acquisition, deployment, movement, modification, utilization, disposal, and authorized destruction of them. The next step in organizing for security is to control access to every specified asset. The security coordinator should maintain up-to-date records to help do this. These should include:

- Lists of persons holding passes or badges
- Control list of keys, cards, or combinations
- Passwords, lockwords, codewords, and who knows them
- Logical working names for hardware devices
- Unique names of all files and who can see or change them

- Unique names of all programs and who can run or modify them
- Security profiles of all center users; that is, what each user is permitted to do with every file or program to which she or he has access
- Project and programmer names and numbers
- Continuous custody histories of all confidential documents
- Local rules regarding where and how to lock away specified assets

Security Management

There are three management principles basic to computer security. The *never-alone* principle dictates that two or more designated persons should witness every security-relevant action and attest to it by signing a log.

The following illustrate security-relevant actions:

- Issue and return of access-control items or credentials
- Issue and return of card decks, tapes, and disk packs
- Systems initialization and shutdown
- Processing sensitive information
- Hardware and software maintenance
- Test and acceptance of hardware
- Modification of hardware
- Systems reconfiguration
- Design and implementation of databases
- Design, implementation, and modification of applications programs
- Design, implementation, and modification of operating systems
- Design, implementation, and modification of security software
- Changes to documentation
- Changes to emergency or contingency plans
- Declaring a state of emergency
- Destruction or erasure of important programs or data
- Reproduction of sensitive information
- Changes to computer operating procedures
- Receipt, issue, or shipment of valuable material

The principle of *limited tenure* suggests that no person should ever be left in any security-related position so long as to create a feeling of permanence or predictability.

Crews should be randomly rotated among shifts; individuals should be randomly rotated among crews; and mandatory vacation periods should be enforced.

The principle of *separation of duties* insists that no person should have knowledge of, be exposed to, or participate in any security-related functions outside one's own area of responsibility.

There are at least ten pairs of functions that should be performed by different individuals or groups. They are:

1. Computer operation and computer programming
2. Data preparation and data processing
3. Data processing and quality control
4. Computer operations and custody of tapes or disk packs
5. Receipt of sensitive or valuable material and transmittal of same
6. Reproduction, issue, or destruction of sensitive information and the granting of authorization for these acts
7. Applications programming and systems programming
8. Applications programming and database administration
9. Design, implementation, and modification of security software and *any other function*
10. Control of access credentials and *any other function*

There are four *administrative* rules necessary to implement the separation of duties:

1. Programmers shall not operate computers.
2. Operators shall neither write nor submit programs.
3. Implementation of modifications to computer operating systems that are intended to enhance security shall be a separate, distinct duty.
4. Quality control and audit shall exist as functions separate and distinct from dataprocessing production.

Physical and Environmental Security

Physical security can be based upon the onionskin concept of keeping one's most sensitive assets within concentric perimeters that become increasingly well-defended as one progresses inward. The onionskin concept is illustrated in Figure 11–1.

In suburban sites, the outer perimeter is the fence around the company compound. It should maintain a distance of at least 200 feet between any sensitive structure and a public access road.

In multistory urban sites, the outer perimeter is the building wall. Computer centers should not be

186 LOSS PREVENTION PLANNING AND CONTROLS

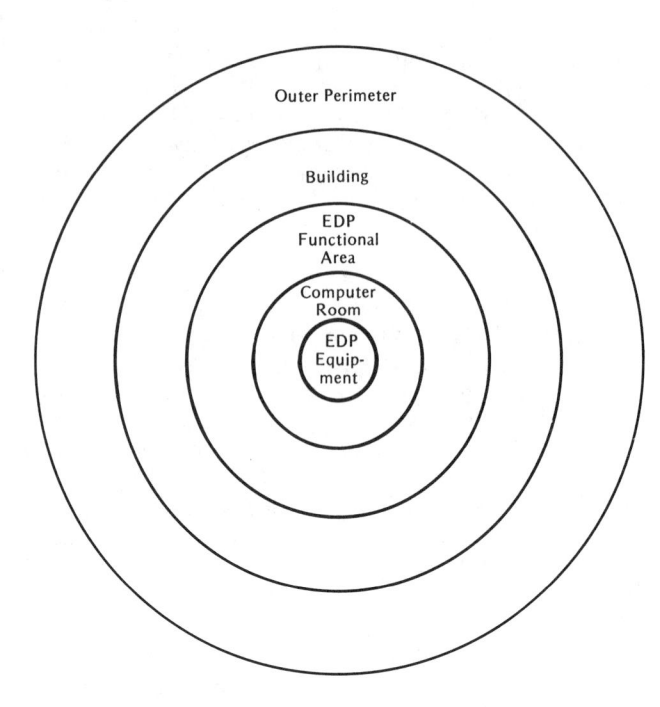

Figure 11-1. The onionskin concept takes advantage of existing features so that the most sensitive asset, the computer, lies within several concentric perimeters.

housed in basement, first floor, or top floor locations. There should be company occupancy on the computer room floor and on the floor above and below.

A screen of offices and/or a corridor should insulate the computer center from exterior walls. Building security should equal that provided in any high rise corporate setting.

Computer rooms should be windowless, lockable, and alarmed. Spaces above suspended ceilings and below the raised floor should be fire stopped and alarmed against intrusion. Air ducts should be secured against intrusion and equipped with fire dampers. Water pipes and wet columns should not cross the computer rooms, and effective underfloor drainage should be provided.

Computer rooms should be electrically shielded to fulfill manufacturer's recommendations in respect of radio frequency and electromagnetic interference, or to reduce undesired signal data emanations to an acceptable level, whichever is the most demanding.

There are six physical barriers essential to a secure computer center:

1. A media (tape) library must exist in a secure location contiguous to, but separate from, the computer room.
2. Data preparation must be done in a secure area close to, but separate from, the computer room.
3. Programmers' offices must be separate from the computer.
4. The security office must be a restricted area to all personnel except those directly connected with security.
5. The computer room must be a secure area restricted to operators actually on duty or maintenance technicians working under strict supervision.
6. Sensitive waste material awaiting destruction must be stored in a secure area well away from the computer room.

Figure 11-2 shows how dataprocessing areas surround the computer room affording defence in depth while furnishing support. Physical barriers protect and isolate critical areas. Figure 11-3 lists additional suggestions on protecting the computer room.

At any physical barrier, a security system must be able to dinstinguish between authorized and unauthorized persons. Discrimination is based upon one or more of the following:

- *Identification.* That is, confirmation of identity, visually, aurally, by signature, or by comparison with previously stored physical characteristics such as fingerprint impressions or palm geometry.
- *Passwords.* That is, a memorized combination or password.
- *Cards and Keys.* That is, possession of a key, card, badge, or other access item.

The simple possession of an access control item should not, in itself, grant access privileges to any

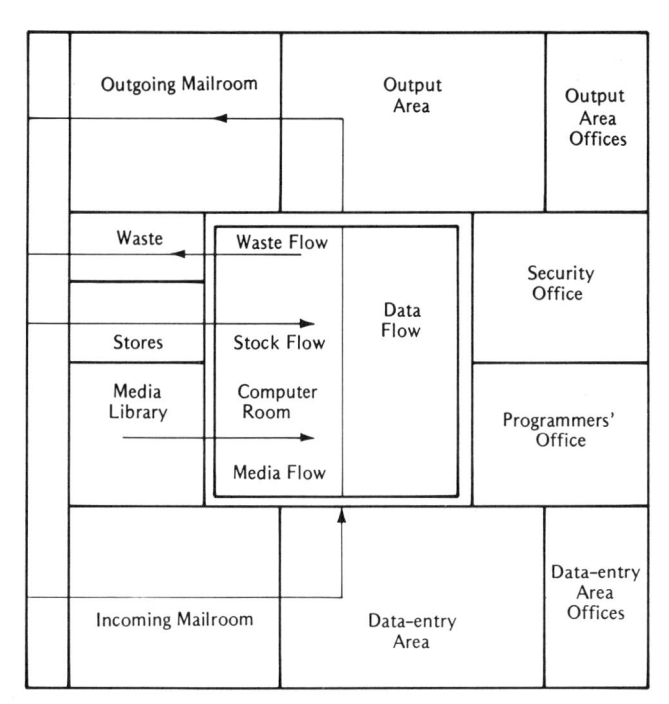

Figure 11–2. Data processing areas surround the computer room. The incoming and outgoing mailrooms are separated. Waste awaiting destruction is kept in a locked room. The media library is adjacent to, but separate from, the computer room.

sensitive asset because of the potential for forgery. The more sensitive the asset, the fewer individuals should have access to it. No person should ever be granted access to, custody of, or knowledge of a sensitive asset solely by reason of rank or position.

Interruption-resistant electrical power service should be provided to fulfill the service requirements of the center. A UPS (uninterruptible power supply) can deliver 45 minutes of emergency power, depending upon the size of its battery bank. It consists of a solid state rectifier working off the AC supply which continually charges a battery bank. This battery bank in turn drives a solid state inverter that powers the computer.

An installation using multiple independent UPS, each rated at 100 kilovolt-amperes, will be able to sustain continual operation inasmuch as it affords protection against UPS failure.

In some areas, it may be wise to install an on-site diesel engine or gas turbine to drive an emergency alternator. The prime mover should have its fuel supply stored on site. It is wise to provide a modular air conditioning system; for example, five 20-ton units to supply a 100-ton requirement.

Ionization-type smoke detectors should provide the initial input to the computer room fire emergency response system. They should be installed in:

- ceilings of the computer room and media library
- concealed spaces behind suspended ceilings
- spaces underneath the raised floor
- electrical panel boxes and telephone closets
- cable tunnels and air-handling ducts

It is not a good idea to have sprinkler systems in computer rooms because the inadvertent activation of a sprinkler head can create serious damage to equipment. A sprinkler system is essential, however, in media libraries and in data input areas where paper tends to accumulate. The best fire protection for a computer room is an underfloor Halon 1301 system.

If a general evacuation becomes necessary, at least one employee should be designated to preserve the security of classified, sensitive, or irreplaceable material. Documents requiring such care should be clearly marked or tagged.

Security of Data Communications

No persons should be able to gain access to a computer from a remote terminal unless able to identify themselves by an authorized programmer identifier (usually both programmer number and the number

- Tape and other magnetic media should be kept in a secure library separate from the computer room.
- Smoking, eating, horseplay, or socializing should not be allowed in the computer rooms.
- Communications lines and equipment should be aggregated within locked telephone closets.
- Equipment cabinets should be kept closed and locked unless access is needed for authorized servicing.
- There should be no accumulation of paper trash in computer rooms.
- Computer rooms should be windowless restricted areas with access controlled by the senior operator on duty.
- There should be an easily accessible portable CO_2 fire extinguisher of at least 15 lb. capacity within 50 ft. of any point in a computer room.
- Emergency "stop" buttons should be covered by spring-loaded housings to prevent accidental actuation.
- Logged-in terminals should never be left unattended.
- There should be no directional signs posted outside the computer room that might give assistance to possible intruders.
- Computer print-out should be handled in a secure manner until delivered to the intended user.
- Documentation, spare parts, and removable media should be locked away when not in use.
- Dust sources such as keypunches, chalkboards, and outer clothing should be kept out of computer rooms.
- Data-preparation areas should be located in secure zones separated from the computer room.
- Customer-service counters should have limited access and be separated from the computer room.
- No keypunches, photocopiers, cameras or other reproduction equipment should be allowed in computer rooms.

Figure 11-3. Computer room checklist. (Reprinted with permission of Cahners Publishing Co., from State and County Administrator, May 1978.)

of the project they are working on) and authenticate their identification by giving a password uniquely associated with their identification and known only to them.

Reauthentication should be required:

1. Periodically (to confirm user presence)
2. At random intervals (to forestall spoofing)
3. Subsequent to:
   (a) An attempt to use a forbidden computer command
   (b) An attempt to address a forbidden area of memory or otherwise exceed the programmer's authorized work area
   (c) Request for a highly classified program or data file
   (d) Request for excessive input or output service

It is necessary to test periodically for user presence at a remote terminal, because operators sometimes sign on and then monopolize a line all day, not even dropping it when they leave for lunch. Not only is this a waste of resources, it also facilitates *between-lines* entry. This occurs when an unauthorized person assumes the legitimate operator's position while absent or misuses the system.

In case of malfunction, loss of power, or intentional override of any security mechanism or error-checking arrangement, traffic over the line involved should be suspended until the fault has been corrected. No telephone number which affords dial-in access to a computer should be published in an unclassified telephone directory.

Every center which supports the ability of users of remote terminals to converse with each other, or with the computer operator, should establish detailed procedures for the exchange of such communications. The security coordinator should see that such procedures are observed.

Data communication messages should conform to a standard format. The message format should provide for indicating:

- Priority of service
- Identification of the originator
- Authorized recipients and whether for action or information
- Message routing
- Date and time of transmission
- Message group count
- Body of the message

- Unique identification (numbering) of the message
- Authentication of the message
- Date and time of receipt

The National Bureau of Standards and IBM have developed a system for encrypting computer communications and sensitive files. For the software representation, see *Federal Register* (March 17, 1975), Vol. 40, No. 52, pages 12067-12250. Its hardware representation is described in *Scientific American* (May, 1973), Vol. 228, No. 5, pages 15-23. The algorithm is described in National Bureau of Standards FIPS-PUB 46.

The algorithm published by NBS uses a 64-bit key (56 of which are active) and is known as the DES, or *Data Encryption Standard*. It is a block cypher in which sequences of plain text are taken 64 bits (eight characters) at a time. The bits are shuffled at random, albeit repeatable fashion by 16 consecutive, non-linear transformations. Latitude is afforded in selecting between alternative transformations by a 64-bit key.

In the Cypher Feed Back (CFB) Mode, input to the DES algorithm is not the data itself but rather a string of the output that was previously generated by the transmitter of the data. The 64-bit input is the last 64 bits of cypher that was previously transmitted. The receiver must independently generate the identical initial fill.

The transmitter must transmit sufficient cypher text to fill the receiver's input register with the same cypher text as is currently in the transmitter's input register. The actual plain text transmitted is irrelevant but must be discarded by the receiver.

The cypher protects against spoofing, misrouting, monitoring, and interference.

- *Spoofing* is falsely accepting a claim of identity.
- *Misrouting* of messages, in both message and circuit switching systems, can be either accidental or intentional.
- *Monitoring* of messages during data transmission can occur all along the transmission path by wiretapping or radio reception of the transmitted data.
- *Interference* is the activity of a penetrator who is actively disrupting the communications system or is receiving and then retransmitting data communications.

To counter storage threats, the key need be at only one location but must be retained for reuse when the data are to be retrieved.

Encryption of stored computer data provides protection against the use of stolen data. If each record is encrypted, beginning with a new initialization, the algorithm makes access to each record independent when encrypted in the CFB mode.

Erasing computer data on magnetic storage media may be time–consuming but destruction of the key destroys access to the data and the storage media may be released for general use. Addressing failures can lead to accessing the wrong, and possibly unauthorized, data; accidental destruction of one's own data may result. Encrypting the data by using the location of the data as part of the key will prevent unauthorized, accidental reading of data. Technical surveillance of the computer center can be productive of knowledge regarding the protective posture of the company and specific access-control items such as passwords. Intrusion devices can be planted within the computer or its peripheral equipment.

The methods of technical surveillance can be classified according to seven categories, based upon the physical principles involved:

1. Mechanical
2. Microphone and wire
3. Free-space active transmission
4. Free-space passive transmission
5. Carrier-current transmission
6. Visual or optical
7. Telephone

Defense against technical surveillance falls into three general categories:

1. Physical
2. Electrical
3. Radio frequency

Sweeping an area is the name commonly given to a technical surveillance countermeasures inspection. Rather than calling in a sweep team only when evidence suggests that an intrusion device has been installed, it is preferable to have regular sweeps performed, beginning at a time when the area can reasonably be presumed to be clean or free from intrusion devices.

Computer Systems Security

Systems security involves identification, isolation, access control, detection and surveillance, and integrity. The purpose of *identification* is to attach unique machine-readable identifiers to users, data files, computer programs, and items of hardware so that they can be brought under control of preestablished access rules.

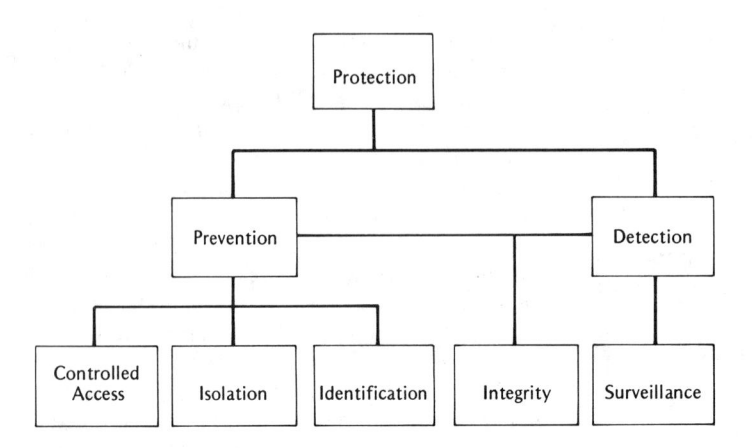

Figure 11-4. All five elements of systems security protect sensitive information assets. Access control, isolation, and identification are preventive measures. Surveillance facilitates detection of security breaches. Integrity applies to all security functions, ensuring that protective mechanisms are trustworthy.

Isolations erect barriers around the components so that access rules can be enforced in the computer environment.

Access control implements access rules by permitting isolation barriers to be penetrated in accordance with these rules.

Detection and surveillance ensure that access rules are observed. They provide the opportunity to take corrective action after security violations.

Integrity includes all techniques that guarantee that the mechanisms providing for identification, isolation, access control, and detection and surveillance are trustworthy.

Figure 11-4 depicts the interrelations among the five elements of systems security.

Identification

Identification means that a complete description uniform within each class of item must exist for every asset of the facility including the users. Every asset must be assigned a machine-sensible identifier unique within the facility.

All the active components of a transaction should be identified before it takes place. The user should know at least seven things:

1. The number of one's own terminal
2. The identification of its operating system if it has one
3. The communications routing, including the identification of all switchers or message concentrators
4. The computers in use
5. The operating system under which they are running
6. The programs called
7. The data files opened for use

The computer's operating system should be aware of the identification of the user at the remote terminal or the submittor of a batch job, the terminal identification, and the communications routing. Personnel should also be aware of the ID of any other computers involved.

Isolation

Isolation means that every user's data and programs, including those of the operating system, must be protected from every other user so that the other users cannot affect the present or future contents of the protected user's work space nor the present or future sequencing of instructions as explicitly stated in the access rules. Isolation erects logical barriers around every user's domain allocated within the system.

There are several nonexclusive ways to isolate systems and users. The oldest one is *dedication* of an entire machine, otherwise known as spatial separation. The second way is *sanitizing* or temporal separation of users.

In a multiprogramming environment, hardware *bounds registers* isolate concurrent users in main memory while *symbolic addressing* helps isolate them in virtual memory.

State switching is a technique used to protect the computer operating system from users.

Virtual machines are implemented by operating systems. They seemingly give each of several concurrent users a machine of their own. At one time, this was intended to ensure user isolation but subsequently teams were able to penetrate other user's address space by going through shared input/output facilities.

Rings of protection result from implementing multistate operating systems in which the most sensitive users possess the lowest state number (0) while the least sensitive users possess the highest state number (8). Naturally, users are never permitted to have read access downward (i.e., users with confidential clearance are never allowed to see material classified *secret*). It has been found, however, that within a computer process (i.e., a currently executing program) users must, in addition, never be able to write upward if global security is to be preserved (i.e., these users with secret clearance should never be permitted to create material classified as *confidential* for fear of leaking secrets). Nor should process be permitted to write downward for fear that they might plant a latent defect in data; in other words a user with confidential clearance should not be permitted to create secret material documents, although sometimes this is permitted by special provisions that allow them to append certain highly specific data to secret files.

The security kernel is kept within the most secure state of a multistate machine; it is reserved for security–relevant operating system programs and data.

Front end machines are dedicated to controlling systems access; *back-end* or *database machines* are dedicated to controlling access to data. *File encryption* is generally performed by hardware or firmware (read-only-memory) components.

The appropriate means to isolate user programs and data from each other depends upon the processing mode in use.

1. The environment of the system may be open or closed. In a closed environment, all users have security clearance at least to the level of the facility.
2. A local system is one in which all systems components and interconnecting telecommunications links lie within a common secure perimeter.
3. In a batch system, all program instructions, sequences of instructions, decision rules, and data files are predetermined when the job is submitted for processing.
4. In a serial system, jobs are run sequentially in the computer. A job is run to completion before control of the computer is relinquished to another job. The job currently being processed possesses exclusive rights to all on-line computer resources. The running of one job can in no way influence the running of any other job.

Serial processing can provide acceptable temporal isolation of users. Similarly, users possessing a common level of security clearance can be assigned scheduled time periods during which they may be granted exclusive access to a multiprogrammed system.

To achieve acceptable temporal isolation, no information classified higher than the job or jobs currently being processed may concurrently be stored in any computer system. Conversely, every job must attract a level of security classification equal to greater than that of the most highly classified program or data file used in its production. All programs and data having a lower level of classification than that of the job currently being processed must effectively be protected against writing to avoid leakage of classified data to nonclassified files.

Before any upward change in the security level of information is made, a fresh copy of the operating system and all permanent libraries should be loaded. These copies should first have been verified by line-for-line computer comparison with master copies maintained under secure conditions.

Each user, upon relinquishing the system, should be responsible for removing or overwriting storage areas as dictated by need-to-know considerations. Likewise, each successive user should be made responsible for the initialization of that workspace.

Spatial isolation is implicit in maintaining the local character of a system. All peripheral devices, especially terminals and storage devices, not under physically secure conditions, must be positively disconnected from a local system when it is processing sensitive information.

Before a computer environment can be considered secure, a local processing mode must be used, or cryptographic isolation must be provided on data communications lines; and users processing different levels of security classification must be isolated from each other.

State switching enables a computer to operate with its full capability when under control of its trusted operating system but to withhold its more sensitive capabilities when under control of a less trusted user.

The system has to be able to determine whether it is in the problem or supervisor state and to identify whether an instruction is privileged or not. If a user program issues an instruction containing a privileged operation code or memory location within the supervisor's area, interrupt is issued to impede processing. The upper limit of the supervisor area can be delineated by a single bounds register which is implemented in hardware. If users wish to have a privileged operation performed for them, they must summon the supervisor; the operating system will then perform the operation in a supervisor state.

If there are two or more users working independently, all user workspaces must be delineated. This can be done by describing each user's workspace in the contents of two registers—a *base register* holding the lower limit of workspace and a *bounds register* holding the address of the upper limit of workspace.

The main memory may be divided into partitions with one assigned to each user. It is then necessary only to identify a separate partition with a single user. The relative sizes of the partitions may be pre-established in the computer's architecture, allocated by the operator, or decided by the operating system. The limits of each user's partition are described by an entry in a memory allocation table.

Use of paged architecture implies that the main memory is subdivided into fixed-length blocks. Each user's workspace is described in terms of pages with one page corresponding to a block. In large systems, user workspace can be represented by a number of segments each consisting of several pages. In partitioned architecture, a hardware register called the program status word can be set to denote both the identity of the current user and the limits of that address space.

Segmented or paged architecture requires identification of the user by job number and the following of a trail through one or more tables to ascertain the limits of an address space. The operating system's address space in main memory and that of each user should be protected by hardware bounds registers. Before space in main memory is given to another user, it should be overwritten with zeros.

Virtual memory is secondary memory, usually on-line disk storage, which is used by the computer operating system as an extension of primary memory. User address space in virtual memory should be defined by either: pages of uniform size (often 1024 computer words), in which the page numbers imply the upper and lower bounds of user address space; or segments which are defined by upper and lower bounds (i.e., physical locations on the disk). It should be assigned only by the computer's operating system.

No user should be permitted to execute or have executed by system utility programs any computer operation that entails reading a portion of newly assigned virtual memory until the user first writes upon the space. This prevents a new user from seeing material created by the prior user. Secondary memory means on-line storage in a direct access storage device, usually a disk, that is utilized for the storage of frequently accessed programs and data files.

Every on-line file name should correspond to a physical location in secondary memory. The assignment of these locations should not be under the users' control, nor should they be permitted to gain any knowledge of the locations. Assignment of locations, writing in them, and retrieval of information from them should be handled totally by operating systems programs (symbolic addressing).

Checking access privileges, including those which govern access to storage devices, should be accomplished in supervisor state to preclude user interference. No file should be opened for reading until the user's access privileges and authority codes have been checked and verified.

Off-line memory means storage of information on tape or in any removable disk packs that are mounted by the operator upon request from the user.

All tape drives should be capable of recognizing and observing the use of write rings. The removal of these rings from magnetic tapes is a common safeguard against unauthorized writing on tapes. All disk drives should be equipped with write-lock-out switches.

Computer Access Control

Access control requires that isolating barriers permit information to flow only in accordance with pre-established access rules. Implicit in the existence of any access control system is a specification on what access privileges exist within the system (e.g., read, write, execute, delete, append); a security profile of each user specifying access privileges for each information asset; and an access list specific to every information asset specifying what access privileges each authorized user has over that asset.

When the set of access privileges recognized within a particular machine environment is reduced to binary form, these representations are called *authority codes* when they relate to users, and *protection bits* when they relate to information assets.

User security profiles, when stored within a machine, are known as *user directories*. There is an authority code conjoined with the identifier of each information asset. When user security profiles are conjoined with authority codes and passwords in a single data structure it is called an *authorization table*. The entire set of identification tables and access control tables is referred to as the *security tables*. Figure 11–5 shows the workings of the user-directory system.

Two basic principles should be reflected in access rules:

1. *Minimal privilege.* Only such information or processing capability that the user absolutely must have to carry out an assigned task should be made available to that user.
2. *Minimal exposure.* Once a user gains access to

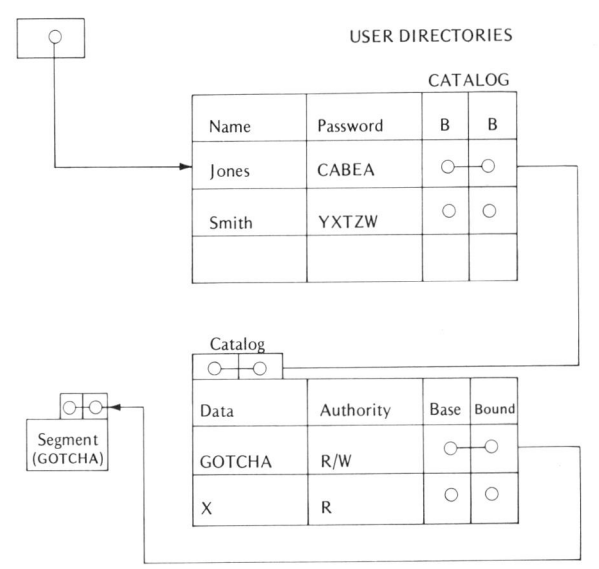

Figure 11–5. The user-directory system. When programmer Jones signs on and requests access to segment *Gotcha*, the computer searches the master directory for user ID, authenticates user password *CABEA*, and locates user's personal directory. It contains the name and location of segment *Gotcha* which has read/write authority bits set so the operating system returns the segment to Jones' work area.

sensitive information or material, that user has the responsibility of protecting it. The user should make sure that only persons whose duties require it obtain knowledge of this information while it is being processed, stored, or in transit.

Authorization tables exist in computer code; they reside within the computer's operating system along with the other control programs. They are to be accorded a level of protection equal to that of the most highly protected control program.

The security coordinator should maintain an access-control list for every computer program and every data file. These lists should reflect the intent of access rules laid down by department heads, security statements sent along with each program or data file, and the expressed desires of the originator. Access-control lists should contain:

- The identifier of the particular asset
- The unique identifier of every authorized user
- What each user is permitted to do with the asset
- Conditions under which access may be granted

The security coordinator should maintain a security profile of every authorized user. The profile should consist of:

- The identifier of the user
- The identifiers of all projects to which the user belongs
- The identifiers of all categories to which the user belongs such as security clearance number
- The identifiers of specific files to which each user has access, and what can be done to each

No access to a computer system should be granted until the user's password has been checked against the list of valid passwords.

When a user submits a job over the counter, the clerk should eyeball the person and require that a codeword be included in the program deck. In some systems, an officer sits at a security console and grants or denies all requests to enter the system or any of its files. In other systems, the security console only receives a warning of violations.

There are ten kinds of violations usually monitored:

1. The third consecutive unsuccessful attempt to log on
2. An improper response to a request for re-authentication
3. Improper use of a privileged instruction
4. Attempt to access memory outside of a user's allocated address space
5. Attempt to gain unauthorized access to classified files
6. Failure of any protective mechanism
7. Any attempt to read newly allocated memory without first having written upon it
8. Inability of the system to successfully verify a systems program
9. A similar inability to verify a security or relocation table
10. Improper attempt to enter a more highly privileged operating state

There is no generally agreed upon set of access privileges. For purposes of illustration one might consider the commands: EXECUTE, DELETE, APPEND, WRITE, READ, and READ/WRITE.

EXECUTE can most safely be implemented by having a systems utility routine run the desired program on behalf of the user.

DELETE means that the entry corresponding to the protected information asset is stricken from the user's directory.

APPEND means that the user can write in his or her own workspace the information later to be appended to the protected file.

WRITE without read access implies overwriting.

READ access transfers the protected information asset to the user's work space.

READ/WRITE access implies the power to alter and erase. The information asset is written in the user's workspace and can be replaced with data created there.

Some systems afford a higher level of access privilege than read/write access, namely CHANGE-PROTECT. This means the user can grant access privileges to other users with respect to any file with change-protect access.

Some experts assert that there are only three levels of access: read, read/write, and change-protect. They reason that EXECUTE can be extended to READ by executing the protected program with all possible combinations of input, and observing all corresponding output, and thus deducing the nature and contents of the program. DELETE can be considered a subset of WRITE as deletion can be achieved by overwriting; APPENDING is viewed as another case of WRITING. It is also asserted that the privilege to WRITE necessarily implies READ inasmuch as users can choose to write to a device which they are able to read.

Access in a computer environment consists of two steps: access to the system, and access to data residing within the system. Access to the system is controlled by the operating system, communications software, interface programs, or teleprocessing monitors. Sometimes, when the security provisions of a system has proved to be inadequate for some users, software suppliers have come up with security add-ons.

The user is identified by a user identification (USER ID) or project/programmer pair. This is authenticated by the personal password.

Access to data is controlled by one or more of three software mechanisms: (1) the file system which is a part of the operating system, (2) security programs that can be thought of as add-ons to the operating system, and (3) database management systems.

The file system uses the project-programmer pair to enter a user directory in which data resources privileged to the user are listed together with authority bits. Security programs such as RACF, ACF-2 or SECURE establish lists of USER-IDs containing their system privileges and records of their past behavior and match these entries with resource lists containing sets of USER-IDs and their access privileges.

Database management systems depend upon user subschemas to give each user only a suitably restricted view of the database. They may also use privacy locks and keys (usually passwords) to protect areas of the database against certain predefined user actions.

Detection and Surveillance

Surveillance requires that the activities of the users, the activities of computer routines acting on behalf of users, and changes made to the contents of security tables be monitored on a continuing basis. The system should possess the capability of detecting security infractions both immediately and retrospectively.

Surveillance activities that contribute to immediate detection of security infractions are called *threat monitoring*. Surveillance activities that contribute to retrospective detection of security infractions include both the analysis of systems activity profiles and the establishment of audit trails.

Activity profiles focus upon users of information assets, and acquire data concerning security-related actions or incidents. The analytic methodology employed involves establishing norms or base values and investigating perceived deviations from these norms.

Audit trails focus principally on data items. For reasons of economy, they are usually established for only a few items which are selected either by random sampling or when evidence suggests a specific area of investigation.

Detection and surveillance focus on terminals, classified files, sensitive programs, and users. Focusing attention on terminals helps determine *where* security violations are taking place. The parameters surveyed are: (1) time between log-ons;

(2) core kiloword-seconds per job; (3) connect time per job; and (4) time between changes to security tables.

The study of classified files may reveal the intended targets of penetration attempts. Parameters to observe are:

- Time between unsuccessful access attempts
- Time between successful *read* attempts
- Time between successful *writes* or *appends*

Study of the use of sensitive programs can give clues to the actions of intruders. Sensitive programs include those which modify security tables and access classified files. Key parameters are:

- Time between unsuccessful attempts to execute
- Time between successful executions
- Running time per execution
- Output per run or job

Violation profiles should be developed for users and jobs that are security significant.

User violation profiles can be used to establish responsibility for system misuse. They are generated by recording: user identifier and type, number and type of security violations, and identifiers of files involved in violations. A run book should be kept for every job that has security significance.

In addition to set-up and operating instructions, the run book should contain a record of each run. Norms should be established for key parameters and deviations duly noted. Key parameters include: user identifier, core kiloword-seconds, run time, programs called, data files accessed and created and the number of *reads, writes,* or *appends* on each, and output created (to ensure against unauthorized dumping of classified files).

Following are twenty events that should be investigated by the security coordinator:

1. Compromise or suspected compromise of classified information
2. Loss or inability to account for a valuable asset
3. Unexplained intervention of a computer operator in the running of a job
4. Presence or suspected presence of an intruder
5. Unexplained absence of a person possessing access to SPECIAL CONTROL information
6. Appearance of trap names on another firm's mailing list
7. Requests to see trapped records or files; or use of trapped or cancelled passwords
8. Unexplained increases in systems usage, especially during off hours or normally silent periods, by normally infrequent users, or from normally dormant terminals
9. Loss of identification, access control, or recognition items, or unauthorized use of same
10. Unexplainable customer complaints regarding improper or multiple billing, misaddressing, incorrect balances, zero-balance running, or omitted payments
11. Inability to balance any account
12. Inability to reconcile checks, invoices, or purchase orders
13. Unexplained inventory shortages
14. Unexplained increases in the frequency of unsuccessful attempts to gain service
15. Excessive demand for input or output service
16. Unexplained changes in data communications traffic
17. Unexplained appearance of new code in operating systems or programs; or of new names on access control lists, payrolls, vendor lists, or customer lists
18. Unexplained changes in job or user profiles
19. Unexplained access to classified files or increased activity in normally dormant files or accounts
20. Unexpected increases in the incidence of hardware faults, software failure, or abnormal program terminations, reruns, or fall-back recoveries.

To assist in investigations, the security coordinator will make use of logs, trend analyses, and profiles. In addition, she or he may require other information obtainable from computer operators, by dumping files, or by impounding program listings and terminal transcripts.

The security coordinator should periodically dump randomly selected samples of classified files to ensure that no unauthorized changes have been made. The dumping must be done under secure conditions, and the resulting print-out must be destroyed. Provision should be made for admission of auditors, after satisfactory identification, to the computer room, and for the secure running of audit programs.

A security audit trail consists of establishing an unbroken chain of evidence regarding the history of some security–significant event from its origin to its ultimate outcome. The security coordinator should be able to construct an audit trail of any incident falling into these:

1. Any visit to the computer center of a person not on staff or a regular visitor where such a visit deals with classified matters
2. The history of any access control item, including its current location and status
3. The history of any staff member with respect to

access privileges granted, altered, or revoked
4. A chronology of the specific experiences of any staff member with respect to having seen, handled, or had custody of accountable documents
5. A chronology of the specific incidents affecting any accountable document with respect to its having been seen by, handled by, or been in the custody of staff members
6. The transactional history of events preceding and following the creation of a valuable negotiable instrument
7. The evolution of a computer operating system program with respect to its initial implementation, actions taken in response to supplier's change letters or locally authorized modifications
8. The reconciliation of transactions affecting any record through three generations of a periodically updated, machine-readable file
9. The evolution of a sensitive applications program with respect to its creation, development and testing, implementation, modification, maintenance, and use
10. A chronological list of specific incidents experienced with respect to a security violation
11. A chronological record of incidents involved in the investigation of a security violation

These actions can be taken after a security violation:

1. Treat any departure from the prescribed log-on ritual as an unsuccessful attempt to log-on. Make a record of it, ask for the user to try again, and give no information that would inform why that attempt was unsuccessful.
2. After a third consecutive unsuccessful attempt to log-on, sever communications with the offending terminal. Print an alarm message at the computer operator's control console and/or the security console. Temporarily invalidate the user's access credentials and lock those files until privileges are restored. Notify the user's immediate supervisor, detain the user if possible, and investigate the incident.
3. Treat any unsuccessful attempt to reauthenticate or failure to reply to a challenge the same as a third consecutive and unsuccessful attempt to log-on; and in addition, dump the offending user's address space for subsequent investigation.
4. In the event of an attempt by the user to gain unauthorized access to a file or to use off-limits instructions, make a record of the incident, and send a message to the user stating that the file or instruction does not exist.

Since software mechanisms are more easily deceived than hardware mechanisms, it is imperative that as many system security features as possible be implemented in hardware. These should include:

1. Indication of the system's current operating state (e.g., Program Status Word) and the means for testing it
2. Limits of the current user's address space (e.g., base, bounds and relocation registers; partition protection lock and keys; and page and segment tables and registers) and means for testing them
3. Means for testing the level of privilege of operation codes
4. Access privileges of the memory block currently in use and means for testing it
5. Contents of the current linkage segment and means for testing it (e.g., Segment Linkage Stack Register)
6. Implementation of security-relevant systems interrupts
7. Algorithms for encrypting and decrypting
8. Algorithms for computing and comparing check digits
9. Algorithms for verifying the contents of security tables and similarly sensitive memory blocks
10. Identification of the user currently in control of the central processor and means for sensing this condition
11. Protective mechanisms for bulk storage and the means by which their protection is implemented (e.g., permit-to-write rings and write-lock-out switches)
12. Time-keeping mechanisms and means for sensing their contents and making decisions based on them

Integrity

Systems integrity requires that all hardware and software mechanisms that contribute to the security posture of a computer system function properly whenever sensitive information assets are likely to become exposed. The system must be fail-safe. If it becomes unsafe, it must fail. If it cannot be verified to be safe, it should not be allowed to start.

Integrity also requires that every security-relevant action should be validated before any part of it is carried out. Validation must establish a chain to a primary source of authority, usually authorization tables.

Integrity demands that none of the mechanisms for safeguarding systems security ever come under user control. No record which is needed for surveil-

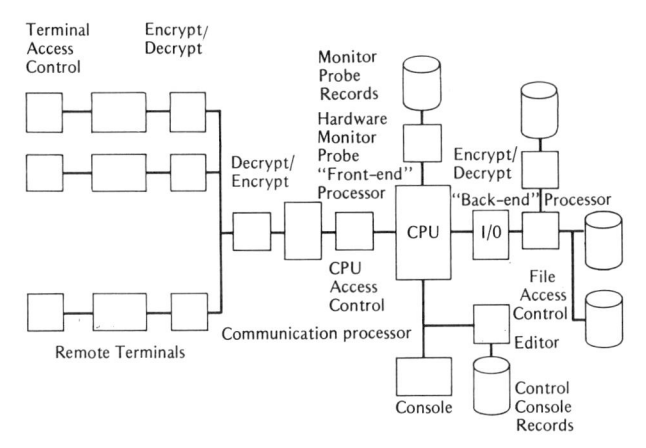

Figure 11–6. Several dedicated microcomputers (shaded symbols) protecting a computer. Terminal access control microcomputers authenticate user ID and test for user presence; encrypt/decrypt microcomputers protect data on communications lines; a front-end processor controls user access to the main frame; a hardware monitor probe carries out surveillance of the central processing unit (CPU); an editing microcomputer keeps audit-trail records safe from operator intervention; a back-end processor controls access to files; and an encrypt/decrypt microcomputer protects sensitive files.

lance should ever be exposed to unauthorized intervention by users or systems personnel. Security should be maintained at all times and under all conditions. Unusual conditions, such as severe data-processing overloads, call for enhanced security.

Security procedures must protect the integrity of the system when hardware malfunction occurs, software failure, or breakdown of support equipment.

Security programs should be written or modified locally only by competent programmers cleared to the appropriate level of security. The programs should be accounted for by the security coordinator and should be modified only with security's knowledge and consent. The logs used for auditing security-relevant actions can be protected against unauthorized operator or programmer intervention by recording such data on a tape handler whose rewind capability has been disabled.

To summarize, the strategy of securing a system depends upon six principles:

1. As many of the protective mechanisms as possible must be removed from the control of the user.
2. Every security-relevant action must be checked before it is permitted to be started.
3. No uncontrolled actions must be permitted after a check has been made.
4. Every security-relevant action must be checked.
5. It is not sufficient to check the storage protection keys on the memory block currently in use. It is necessary to go back to the authorization tables and find out which users possess access privileges to these data and what authority each one has.
6. In computer security, the critical function is performed by the password list, user directories, and segment access lists. It is essential to ensure that changes to them be made only by authorized persons and that no unauthorized changes are permitted.

The policy of selecting hardware protection over software protection wherever possible has become increasingly viable with the advent of cheap microprocessors. Figure 11–6 illustrates how several dedicated microcomputers could be used to protect a remotely accessed multiprogrammed computer.

Data Security

A database management system* (DBMS) is a core-resident program running in *problem* state under the control of the operating system. As part of its workspace it contains a DBMS buffer, and the database resides in secondary storage. The schema and subschemas are, like the DBMS itself, core-resident in problem storage. A schema is a list of all the data items contained in the data base. Each user has a particular subschema. This is a list of the data items that a user is allowed to see. The run-units are user programs that contain within their workspace program-work-area (PWA) locations to receive status information.

In a sequence of operations (Figure 11–7), the run-unit issues a call to the DBMS to read record (1). The DBMS consults the subschema appropriate to that run-unit for a description of the data requested, and then consults the schema for a logical description of the data types required to satisfy the request (2).

---
*This description of a data base management system is based upon recommendations of the U.S. government committee on Computer Data Systems and Languages (CODASYL) Data Base Task Group (DBTG).

198 LOSS PREVENTION PLANNING AND CONTROLS

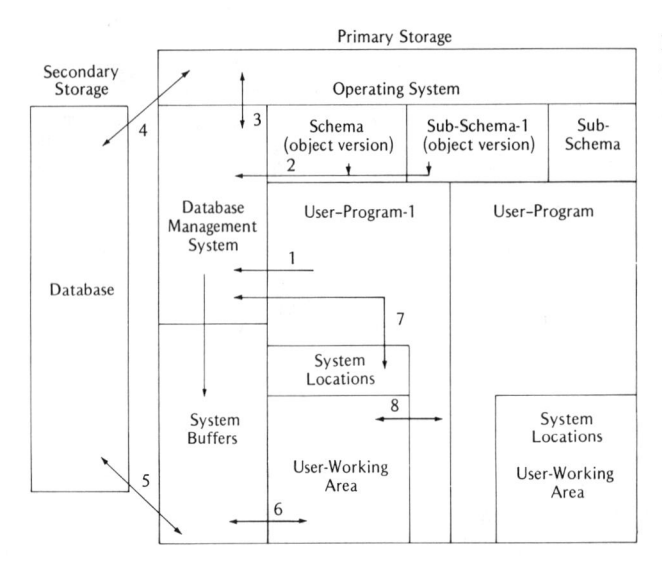

Figure 11-7. Main memory allocation for a CODASYL DBTG database management system. The numbers 1 to 8 correspond to the numbers in the conceptual description of DBMS operations.

The DBMS issues commands to the operating system to retrieve the data (3). The operating system consults its tables to obtain the physical description of the records sought and transfers the data from secondary storage to the DBMS buffer (4, 5).

Based on the data descriptions contained in the schema and subschema, the DBMS transfers from the DBMS buffer to the PWA, such data as are: (a) responsive to the run-unit request, and (b) appropriate to the privilege of the run-unit (6). The DBMS also provides status information to the run-unit on the outcome of its call including any error indications (7). The run-unit may now operate with the data in its PWA (8).

There are twelve commands that govern DBMS operations and concern database security; these do not include operations specific to database security mechanisms, which will be discussed later. The commands, DISPLAY, COMPILE, and ALTER, are specific to the schema and to subschemas. The command DISPLAY permits the schema or subschemas to be viewed. The command COMPILE permits the schema or a subschema to be used in compiling an applications program. The command ALTER permits a change to be made in the schema or a subschema.

The command OPEN is specific to areas. It can be qualified by the terms *retrieval* or *update* which correspond to *read-only* or *read-write* access. These terms can be qualified by the term *exclusive* which means that only the current run-unit can access the area, effectively locking out concurrent run-units.

The commands INSERT, REMOVE, STORE, DELETE, MODIFY, FIND, and GET apply to records. The command INSERT enters a record currently resident in the PWA into one or more sets of records. The command REMOVE cancels the membership of a specific record from a set. The command STORE writes a new occurrence of a record in the database. The command DELETE removes a specified occurrence of a record from the database. The command MODIFY replaces the values of specified data-items in a specified record occurrence with values currently in the PWA. The command FIND locates a specific occurrence of a named record type. The command GET transfers the contents of a specified record occurrence to the PWA.

Security Mechanisms

The security mechanism is a system of *privacy locks* and *privacy keys*. A privacy lock is a procedure, literal, or data-name that must be supplied by a run-unit as a privacy key to gain a specified kind of access to a specified data structure type or to specified occurrences of a data structure type.

The command LOCKS is specific to the schema and to subschemas. It permits a run-unit to see, change, cancel, or extend all privacy locks.

The various levels at which privacy locks may be specified are:

- Privacy locks may be established for the commands LOCKS, DISPLAY, COMPILE, and ALTER for the *schema* and *subschemas*
- Privacy locks may be established for the commands EXCLUSIVE, RETRIEVAL, EXCLU-

SIVE UPDATE, RETRIEVAL, and UPDATE
- Privacy locks may be established for the commands INSERT, REMOVE, STORE, DELETE, MODIFY, FIND, and GET for *records*
- Privacy locks may be established for the commands STORE, GET, and MODIFY for *data-items* (parts of records)
- Privacy locks may be established for the commands ORDER, INSERT, REMOVE, and FIND for *sets* (collections of records)

All DBMS share a common weakness: None is any more secure than the operating system under which it runs. In the case of any two-state operating system, a penetrator who can successfully and unlawfully access any other user's program can also dump all or any part of the DBMS or the database managed by it.

A defense mechanism used by database systems involves retaining *before* and *after* images of record or data-item occurrences that have been subjected to MODIFY transactions. Such a mechanism will not only permit correction of errors introduced accidentally or maliciously but also aid in reconstructing the circumstances surrounding a security infraction.

The protection of databases from destruction or corruption demands that the databases be reconstructable. In the batch update mode that requirement demanded preserving at least the two last files. These were called *father* and *grandfather* files; the current file was called the *son*.

To avoid having to repeat an entire run after a mishap, checkpoints should be built in so that batch totals and other transitory data values needed to reconstruct the file can be taken at short intervals during processing. In this way, a run can be restarted by falling back to the most recent checkpoint.

When records are processed in a randomly accessed mode, provisions for recovery may include keeping a continuous record of input transactions and taking periodic dumps of the file.

Not all data management systems follow the Data Base Task Group model. Three popular DBMS that do not do so and incorporate alternative security philosophies are ADABAS, System 2000, and IMS.

ADABAS security features are passwords and data encryption. Both retrieval and modification security are provided on file and field levels. When the user issues an OPEN to initiate database processing, he or she supplies a password. A security table is searched for that password. If found, it is converted into a file authorization profile. This profile shows the type of access (retrieval or update) on which fields on which files this user is allowed. Subsequent database activity is restricted accordingly. Data encryption is an optional feature that can be applied on a file basis.

ADABAS does not have its own communications processor to handle teleprocessing. It does, however, provide an interface to four popular communications systems. These systems control user terminals, and receive and send messages.

ADABAS has special commands used when updating records in the multiple-user environment. The application program issues a READ HOLD or FIND HOLD command on the record to be updated and then issues an UPDATE or DELETE command. ADABAS prevents other programs from accessing or modifying the record between the HOLD and the UPDATE or DELETE.

Data can be protected by ADABAS by using a dump utility and a journal file that has both before and after images. During processing, either a program or ADABAS can issue a CHECKPOINT command. When the CHECKPOINT is issued, every active program is given the opportunity to execute to a point at which it can be restarted and to save necessary data. If a failure occurs, the database can be restored to the checkpoint and programs restarted. A database can be restored by applying previous images back to a checkpoint. A database can be regenerated from a dump by applying after images. ADABAS can repair the database automatically after a system failure.

Passwords are the primary data security feature of SYSTEM 2000. There are a master password and other passwords (called *valid passwords*) for each database. The holder of the master password can perform all operations on any portion of the database and assigns the valid passwords. The database administrator holds the master password. Each valid password has permitted actions on specified portions of the database. Valid passwords are held by users or application programs.

The master password holder assigns authorities by specifying names of components and a list of permitted actions on those components. The actions may be:

R—Retrieve
W—Use
U—Update
V—Use the item in an update

SYSTEM 2000 can be run in a multiple-user mode that permits communication between users using a single copy of SYSTEM 2000 code. It can be run in the teleprocessing environment using a data communications package.

Users can concurrently process a SYSTEM 2000

database. To ensure data integrity, a user can temporarily hold or lock portions of the database. Holding means the user wants update control of the data, but allows other users to retrieve it. Locking means no other users can access the data in any way.

SYSTEM 2000 data protection facilities consist of a dump utility and an update log file. Any time after a database loading, the master password holder can request SYSTEM 2000 to dump the database. When doing this the user can define a file to be used for update logging. If so, SYSTEM 2000 automatically maintains a record of updates. Update records can go directly to tape, or a batch of update records can be saved on disk and later transferred to tape. The direct mode protects the user in the case of serious system failure.

As another option, update records can be written immediately at the time the database is changed. This causes each update to be saved. The dump and update tape can be used to restore the database when it is lost due to machine failure.

IMS provides data security in several ways. The first is by segment sensitivity. When a *logical data base record* (that is, a list of the data items that a specific user is allowed to see) is defined, it is described as being sensitive to segments that belong to the underlying *physical data base record* (that is, a list of all the data items that the data base contains). A segment in the PDBR to which an LDBR is not sensitive will be unknown to the user of the LDBR. Consequently, access to certain data can be denied to some users by not including definitions of certain segments in the LDBR to which the users refer.

To prevent unintentional deletions and other security problems, IMS allows the database administrator to specify processing options for each segment in the LDBR. If the user attempts to perform an operation that is not specified in the LDBR description, IMS rejects the attempt.

IMS allows the user to encrypt data before storage. In the teleprocessing environment, IMS provides for terminal and password security. Terminal security allows certain transactions and commands to be entered only from specific terminals. Password security forces the user to give a password before requesting IMS to process specified transactions or commands.

IMS allows two or more users to process a database concurrently. If a task (run-unit) updates a segment, that segment is locked from access by other tasks until the updating task terminates. If the task terminates abnormally, all of its updates are automatically rolled back. Thus, the database cannot contain modifications made by a task that failed to complete successfully.

A series of utility programs provides data protection for IMS databases. IMS has a checkpoint/restart capability whereby the system can be brought to an orderly termination and all unprocessed requests and undelivered messages saved. When restart is initiated, the system picks up at the checkpoint and begins processing without loss of data or messages.

To provide for situations where IMS is terminated unexpectedly, such as because of machine failure, IMS databases are protected using dump/restore utilities. IMS records database modifications on a system log. When a database is lost, it can be restored from a previous dump, and all modifications applied in reverse. A utility is available to read the log backwards and remove changes made to the database.

Logs

Logs provide a chronological record of events that can ascribe causation to breaches of security. Backup files make it possible to recover from catastrophic errors and malfunctions. Documentation describes in detail the intellectual processes that went into the development of procedures and safe-guards. And inventory records establish facts regarding the movement of critical resources.

Logs should be kept of activity taking place at the computer operator's console, at the security officer's console (if one exists), at the media library counter, and at remote terminals or customer service counters.

Backup Files

For every information resource, sufficient backup should exist so that any such program or data file that is inadvertently or maliciously erased or lost can be reconstructed. Copies of backup files should be stored in secure, off-site facilities.

Information for backup and the information and material needed to implement backup should never be subjected to the same hazards as the original. It should be regularly updated.

When restart and recovery is initiated after a hardware malfunction or software failure, data should be assembled to aid in any subsequent investigation:

1. Take a hard-copy image of the contents of primary memory (dump) remembering to preserve its security.
2. Reload as much of the operating system as is

needed to reestablish service.
3. Determine the condition of all files. If open, could the contents of classified files have become compromised?
4. Close protected files found to be open and recover copies, or take other measures to account for leakage of classified information.
5. Sanitize the machine (clean out files) before turning it over to hardware or software maintenance personnel lacking appropriate clearance or need-to-know.

When processing is interrupted because of a security violation, collect:

1. Identify of the user responsible
2. Identity of all classified files or programs involved
3. Type of security violation

Next, dump the offending user's address space for purposes of investigation. Retrieve and impound any print-out delivered to the offending user. And take measures to prevent the offending user from reentering the system.

A formalized record-retention policy should exist. Any record that does not fall within an established retention schedule should be marked for erasure and destruction after some specified date. This policy helps preserve security by reducing the amount of material to be safeguarded, and releases space for other users.

A computer center should keep lists and inventories to maintain accountability for its assets. Continually updated lists should be maintained of personnel; access-control, identification, and recognition items; and locally created negotiables such as checks, warrants, and insurance policies.

Unissued access-control items and stock to be used for printing negotiables should be serially numbered. Stock should be verified by actual count periodically and before and after every withdrawal.

All software defects should be corrected as soon as possible after they become evident. All active copies of the computer operating system or other programs for software maintenance should be updated to reflect these actions. Whenever changes are made to the computer operating system or other critical programs, before and after copies should be kept. Whenever a change is implemented, all hardware, systems software, and user programs affected by it should be modified so that they are compatible before normal production is permitted to resume. All programs, documentation and operating procedures reflecting a change should be updated. Archival copies of the originals should be retained.

Emergency planning requires providing for emergency operation at the primary site, providing for off-site storage of records, and arranging for an alternate site in case the primary one is destroyed or otherwise denied.

Here are some situations that should be considered in emergency planning:

1. Work overload necessitating overtime
2. Catastrophic failure of hardware or support equipment
3. Catastrophic failure of the control program
4. Reduced workloads entailing shortened work hours
5. Operation with one or more hardware components inoperable
6. Strike, absence of staff, adverse occupation, riot, or civil insurrection
7. Fire, flood, windstorm, bombing, earthquake, building collapse.

The most important part of emergency planning is designating the person who has the power to say that an emergency exists; and an assistant who can step in if the designated person is unavailable or incapacitated.

Decide which personnel will be required to sustain a minimal level of essential service and treat them as members of your first team. Inform them of their emergency duties and give them training.

If your contingency plans call for handing off jobs to other centers, make sure they can provide at least as high a level of protection as would have been provided at the primary site; then send somebody along to see the jobs get the protection.

At least once a year, every computer center should carry out an exercise in which staff demonstrate the capability of restoring a predefined minimum level of essential service using only resources designated for backup purposes. Security should be preserved during this exercise.

Appendix 11a

Computer Security Checklist*

VICTOR HAROLD

1. Do you forbid the use of shared passwords?
2. Have you made sure you don't have a common password?
3. Are you using a long password, i.e., a *pass sentence*?
4. Is your password hard to memorize?
5. Have you made sure that users must log off the system when leaving the terminal area?
6. Is your location secure and difficult to enter?
7. Do you have a password concealment procedure during operation?
8. Have you avoided installing an override procedure that permits a user to bypass internal central checks and other system protection designs?
9. Is your error display clear and does it permit the user only to correct?
10. Is access to files and accounts restricted?
11. If a file or account must be updated, is permission to amend restricted?
12. Is authorization necessary prior to changing a program or file?
13. Can reentered data be verified?
14. Are critical programs and files duplicated and stored at another location?
15. Does your log indicate which items were sent to another location and when they have arrived?
16. Have you back-ups of central processing and communications?
17. Do you frequently test back-up procedures and programs?
18. Do you have a disaster recovery program?
19. Do you periodically test your back-up and disaster recovery programs?
20. Are all systems interruptions, errors, and transactions logged?
21. Do you have a variable timetable for checking, monitoring and reporting errors, unauthorized systems use, and unauthorized systems transactions?
22. Do you have a good error correction procedure?
23. Can programmable controls be implemented easily?
24. Do you check control figures manually?
25. Are employees constantly reminded to avoid carelessly leaving items lying around?
26. Overall, do you know who used the computer, when they used it, what was done with it, and how they did it?
27. Are you certain no one person has absolute control over the computer?
28. Is the owner of the business or a very high level manager the only persons permitted to run the audits and tests?
29. Has management clearly stated that it will prosecute anyone who steals corporate data?
30. To be sure that computer systems theft does not have a chance to occur:
    a. Are your computer-generated reports properly discarded?
    b. Do your auditors have computer operations expertise?
    c. Are EDP operations after business hours secure?
    d. Are employee relations good?
    e. If your computer generates negotiable instruments, is every document accounted for by a responsible executive?
    f. Is access to the computer area secure?
    g. If the computer must be used to transfer funds, get credit ratings, and handle loans, are those functions limited to very senior and responsible executives?
    h. Are all computer functions separated and handled by specialists with expertise in a particular area?
    i. Is the data preparation area secure?

*Reprinted with permission of Victor Harold.

31. If a computer operator leaves the firm, are passwords changed?
32. Are two people present when EDP equipment is operating?
33. Is the computer operation closed to the general public?
34. Are errors and malfunctions investigated?
35. Have you performed an in-depth screening of all EDP personnel?
36. Are rejected transactions thoroughly checked?
37. Do you check out all systems errors and malfunctions?
38. Are customer complains about their transactions investigated?
39. To make sure your computer security program is most effective, do you:
    a. Have cards or badges which are machine readable and can identify terminal users?
    b. Periodically change security codes and passwords?
    c. Permit authorized personnel only to have access to confidential files?
    d. Use scramblers?
    e. Safeguard user files by employing lockwords?
    f. Have a record made of unsuccessful attempts to penetrate the computer?
    g. Have a program which disconnects the computer at certain times?
    h. Require terminal users to log in and log out when using the computer?
40. Are there frequent counts of all original and duplicate computer inventory items such as tapes, disks, programs, and documents?
41. Is the computer library staffed and kept locked?
42. Do you keep track of all outputs and keep them logged and stored?
43. Have you a system for reporting and investigating errors?
44. Are your input documents stored and accounted for?
45. Are source document corrections made only by the originators?
46. Are your source records retained and kept secure until it is determined they can be destroyed?
47. Are the duties and responsibilities of computer personnel clearly defined?
48. Is there a division of duties and responsibilities?
49. Is the EDP department kept independent from all others?
50. Are the EDP staff vacations generally taken during the month's end wind-down?
51. Are you certain after-hours processing is strictly regulated?
52. Do you frequently review and audit the security program?
53. Are programmers always given written instructions?
54. Do all your programs contain a statement of ownership?
55. Have all your programs sufficient controls to prevent being overridden?
56. Have you divided the responsibility for writing and modifying your programs?
57. When a program changes, is an audit trail maintained?
58. Is there a division of responsibilities for program production and debugging activities?
59. Are the programs maintained between two or more employees?
60. Are you preventing computer screens from being seen from a distance?
61. Are accidental hazards such as magnetism and sunlight scrupulously avoided to prevent damage to disks and tapes?
62. Have you limited the number of attempts that can be made by anyone logging on to your system?
63. If telephone lines are used for data transmission, have you installed signal scramblers to inhibit wiretaps and to prevent unauthorized entry?
64. Do you periodically change the access telephone number?

Chapter 12
Planning for Emergencies and Disasters*

EUGENE FINNERAN

A well-planned security system must include plans and procedures to allow for the efficient handling of any possible emergency situation which might arise. This would include fires, explosions, floods, earthquakes, hurricanes or tornados, volcanic eruptions, riots and other civil disturbances, or any life-threatening condition that would require response by emergency teams and evacuation of employees.

These plans must provide for the designation of key personnel who will be authorized to make decisions as to when an emergency exists and to implement the appropriate plan of action. The plan must include the emergency evacuation of the affected area, a method of identifying personnel, formation of emergency teams, establishment of a communications capability, coordination with outside agencies, availability and location of medical services, authority to shut down the facility, procedures which will be followed, and training of all personnel in the duties or actions they are to take in the event an emergency situation is declared.

Among the items to be included in the written plans, or manual, will be the following**:

1. Statement of policy.
2. Description of potential hazards, with risk assessment.
3. Description of the facility, including size, construction, location, access roads or other means of transportation, entry points, type of operations, hours open, numbers of personnel on hand at each shift or stage of operation, building plans, utilities lines, etc.
4. Emergency organization, showing the chain of command, responsibilities of each position, with succession list.
5. Emergency facilities, including the emergency command center, evacuation routes, assembly points, communications and alarm systems and their locations.
6. Emergency equipment and supplies, including medical and first aid, fire fighting equipment, salvage equipment, food and water supplies, with their locations.
7. List of mutual aid agreements.
8. List of outside agencies, with emergency phone numbers.
9. Shutdown procedures.
10. Physical security procedures.
11. Evacuation procedures.
12. Other related items applicable to the specific organization.

The person in overall charge of operations within the facility also will be primarily responsible for the formulation and implementation of the emergency procedures plan. It will be necessary for this individual to delegate authority to implement the plan in the event an emergency arises during that person's absence. She or he will designate key personnel and assign specific duties to each. For the most part the personnel and their designated duties will follow the organizational structure of the company.

Emergency Team

The emergency team will be made up of personnel from each shift who possess skills necessary to complete a plant shutdown, shutting off or reestablishing the power supply; contacting and coordinating out-

---

*From *Security Supervision: A Handbook for Supervisors and Managers*, by Eugene Finneran (Stoneham, MA: Butterworths, 1981).

**From *Industrial Security*, by David L. Berger (Stoneham, MA: Butterworths, 1981).

side assistance, such as the bomb squad, fire or police department or medical triage teams; cutting off or restoring the water supply; fire fighting; and administering first aid and any other skill which might be required during an emergency situation. An example of how these teams are normally composed and how responsibilities would be divided is listed below.

Team Leader

Maintenance supervisor, facility engineer or whatever the designation, the person responsible for the mechanical operation of the facility will be in charge of the emergency team. The individual selected is usually more familiar with the power sources, facility layout and mechanical operations than any other person.

The team leader will receive instructions from the facility manager or the team leader designated representative. Based on decisions made by top management, the team leader will coordinate all activities to be conducted by the team. Although this supervisor is a permanent member of the team, the team leader must have an alternate on each shift in the event she or he is unable to respond to the emergency. These alternates will normally be shift supervisors from the team leader's department.

Assistant Team Leader

The facility security or risk manager is the next permanent member of the emergency team. The risk manager too will be familiar with the layout of the entire facility but probably will not be familiar with the mechanical aspects. The risk manager will receive briefings and orders from the team leader.

The assistant team leader must also have on the emergency team designated alternates who will normally be the shift supervisors of the security force. It will be the responsibility of this member to make and coordinate all requests for outside assistance. The risk manager will deploy the security force to prevent any unauthorized entry to the area during the emergency by potential pilferers or looters. During such emergencies the security force's communications network will be used to supplement other emergency communications systems.

Technicians

Technicians assigned to the team will consist of electricians, mechanics, maintenance personnel, medics, security officers, and coordinators from each department within the facility. The team designated as the facility fire brigade will normally form separately and stand by for instructions from the team leader.

Station assignments and specific duties of each technician will be spelled out in the emergency plan. However, they are not to take any action without clearance from their team leader. Most of these technicians will be involved in the shut down of facility operations should that become necessary. Medical personnel would normally be assigned to an emergency medical station, and their primary purpose is to treat members of the emergency team who are injured while carrying out their mission. A security officer will normally be assigned to the team leader as a runner and coordinator for the security manager.

Emergency Evacuation

The purpose of the evacuation of personnel from a facility is to remove them from a potentially dangerous area or structure to a place of safety. The evacuation also gives members of emergency teams unobstructed access to the area affected by the cause of the emergency. During an actual emergency, tension runs high and some personnel may be prone to panic. This potential for panic can be greatly reduced if all personnel are thoroughly familiar with the evacuation plan and have confidence in its ability to see them safely out of the danger area.

To achieve this will require many factors, beginning with the designation of routes which will be used for evacuation. It will be necessary to obtain a plot plan of the facility which indicates the location of all departments and operations. It will be necessary to record the number of personnel employed in each department or area, broken down by the shifts when they are present. The plot plan must also show all aisles throughout the facility which can be used for emergency evacuation routes, and it must indicate the location of each emergency exit from the facility.

An analysis must be made of the number of persons who can be safely evacuated through each emergency exit, the time required to move personnel of a given department to the nearest emergency exit, and how many of the facility's personnel would be required to leave through each of the designated exits. Following this analysis it may be necessary to establish additional exits to provide for the safe evacuation of all personnel. Once these factors have been established, specific routes and exits can be designated to be used in the event of the necessity of evacuating any given department or section within the facility. The selected routes must be capable of moving the maximum number of personnel present in the facility at one time, should that eventuality arise.

The selected routes and designated exits should be marked on the plot plan, using a color coding method, with a different color being assigned for each department or section to be evacuated. If a color coded identification system has been adopted utilizing different colors for each work unit, the same color codes can be applied to the emergency evacuation plan. The completed plot plan should be reproduced in sufficient quantity to permit its posting throughout the facility and its use in the indoctrination of all personnel.

The next step to be taken is to provide for the safe evacuation of personnel by marking the actual routes using the same color coding used on the plot plan. There are several methods which can be used and are adaptable to any facility which might be encountered. The three most common methods presently being used are:

1. Printed signs bearing the words *Emergency Exit* and having an arrow pointing in the direction of travel. These signs would be posted at any turning point along the route. This method is particularly adaptable for office areas where the floors are carpeted and which would not lend themselves to the other methods described below.
2. Color coded arrows painted on structural support beams along the evacuation route with the words *Emergency Exit* beneath the arrows. These arrows should be closely spaced and must be marked at any location where a directional change is required. This method is normally used in large open areas such as warehouses and storage areas.
3. A continuous stripe, in the appropriate color, painted on the floor starting at the department designated by each color and continuing to the emergency exit that is to be used by that department. This method is most often found in use in plant production areas.

It will not be sufficient to merely move your personnel out of the area or structure being threatened. They must be moved to a staging area which is located a safe distance from the facility so that any possible explosion or fire would not harm the assembled personnel. These staging areas must be clearly marked using the same color coding as was used in marking the evacuation routes. All personnel not assigned to an emergency team will be required to report to the staging area so that a head count may be made, by supervisory personnel, for the purpose of determining that all personnel have been safely evacuated.

Personnel who have been assigned to an emergency team will report to a central predesignated location where they will join their team and pick up the equipment and the supplies required to accomplish the duties that have been assigned to their team.

A method must be established for alerting all employees to leave their workplace and evacuate the facility. This will normally take the form of a siren or bell accompanied by flashing emergency lights. Whatever form is used, it should be electrically operated with an automatic switching device to back-up battery power in the event of a power failure. All evacuation routes should be provided with sufficient emergency lighting to permit the safe evacuation of all personnel during a power failure. This emergency power source should be extended to all lights used to mark the location of emergency exits.

Once the evacuation plan has been completed, routes have been designated and marked, and exit doors and staging areas assigned, consideration must be given to identifying and designating alternate routes. Should primary evacuation routes be blocked by fire, structural collapse or explosion, personnel affected will have to be preassigned to alternate evacuation routes. This may be accomplished by assigning each department a secondary color for the route to follow should it become necessary. The alternate routes should be close to the primary route and will have to be capable of handling the additional personnel.

At this point training sessions should be initiated within each department or evacuation unit to fully familiarize all personnel with the evacuation plan. Initially, drills should be held by individual departments walking through their evacuation route and their alternate route. When this has been accomplished it would be desirable to hold an actual drill of the entire facility and system. This can be most economically accomplished at the end of a shift. Employees can be released directly from their staging area to clock out for the day. These drills should then be scheduled periodically, with the frequency determined by the percentage of employee turnover.

Protecting Vital Records*

Vital records—those necessary to insure the survival of business—constitute a small part of a company's records, normally no more than two percent. Therefore, it is important that vital records be given maximum protection from every disaster, including nuclear. Indeed, it is not unusual to find that for some businesses the information contained in their records is the single most valuable asset. The loss of processing and trade secrets, drawings, formulas, and the like, could easily end a prospering business.

What Records Are Vital?

What records *must* a company have in order to function? This varies, depending upon the type of business, but there are certain fundamental records vital to any corporate organization. For instance, the incorporation certificate, the bylaws, the stock record books, the board of directors, minutes, and certain corporate finance records would be vital to any corporation.

A manufacturing organization would, in addition, require engineering drawings and specifications, work processes and procedures, lists of employee skills required, and similar information. Without these, it is impossible to produce a product. The task of recreating all this information would be virtually insurmountable, especially for a complex product. A banking institution, on the other hand, would require current information on the status of depositors' accounts, accounts with other banks, loan accounts, and similar information.

Selecting Records to Be Protected

The vital records protection program is an administrative device for safeguarding vital information, not for preserving existing records. Management begins to protect vital records by systematically determining what information is vital and which records contain this information.

Management need not waste time and energy identifying records. A record either contains vital information or it does not. If a record contains vital information it should be protected against every possible peril, but, if a particular record does not contain vital information, it has no place in the company's vital records protection program.

The following is a suggested procedure for analyzing a company's vital records:

A middle management project team analyzes a company's vital records information needs in four steps. (The logical team leader is the company records manager.)

Step 1. The project team classifies company operations into broad *functional* categories. These categories will be different in each company, but in general they should include at least:

- Finance—bill payment, account collection, and cost accounting.
- Production—research, engineering, purchasing, and related activities.
- Sales—inventory control and shipping activities.
- General administration—personnel, legal, public relations, and similar staff activities.

Step 2. The project team determines the role of each function in an emergency. Not every current company function and activity will be essential to prompt postdisaster recovery. Some activities must be suspended during the recovery period and some can be eliminated completely. If elimination or curtailment of an activity after a disaster will restrict the company's ability to restore some essential aspect of its operations, then that activity is vital. The information needed to maintain it is also vital and should be protected.

Step 3. The project team identifies the minimum information that must be readily accessible during a postdisaster emergency to assure that vital functions perform properly. For instance, to stabilize customer account collections, it may be necessary to know, from the most recent account statement, the outstanding balance at the time of the disaster and to have a record of subsequent purchases and payments. To clarify field parts inventory conditions, it may be necessary to have access to a copy of the most recent sales agents' reports. (This step may disclose that some records that would be needed in an emergency are not created in the routine of daily business, and that a system should be developed to assure that these records are available for possible postdisaster use.)

Step 4. Finally, the team identifies the particular records that contain this vital information and

---
*From *Disaster Planning Guide for Business and Industry* (Baltimore: Defense Civil Preparedness Agency, 1974), pp 17–22.

the departments in which they are, or should be, maintained.

Safeguarding Vital Computer Information and Records

Much of the vital company information that management seeks to protect is not captured on paper records and kept in filing cabinets. Instead, this information is processed by a computer and captured on the distinctive media associated with electronic data processing.

Effective protection of this vital information is more complicated than safeguarding paper vital records for the following reasons:

1. Formerly dispersed information is consolidated, which intensifies its exposure to possible destruction or compromise. Paradoxically, this consolidated information becomes, through remote data transmission equipment, more accessible to more people. Greater accessibility to this information means greater chances for compromise and the introduction of error.
2. The data processing medium is extremely vulnerable to a wide variety of perils. Even a sharp blow through careless handling can render worthless the magnetic data impulses that it records.
3. A nuclear detonation also produces an electromagnetic pulse (EMP) which has been described as generally similar to a lightning strike. Experiments show there is a significant difference between the two phenomena and their effects on electrical components. Although standard lightning protection devices do provide some measure of protection, they cannot be relied upon for protection against EMP. Until well-tested EMP protective devices are available, the most effective technique for minimizing damage to sensitive computer components involves isolating them electrically and/or magnetically from the EMP environment. In many cases, satisfactory isolation can be achieved by temporarily disconnecting equipment containing the sensitive components from power sources, antennas, or other input/output leads that enter the computer area.

The company is almost totally dependent on the continued use of the computer to process information. There is no option in the event of emergency to revert to manual processing of this data. The manual records, the machines on which they were processed, and the proper personnel may no longer be available to the company.

Other reasons why effective protection of the vital information is complicated are:

4. Information transmitted over a distance for remote computer processing or handled by a computer service bureau is out of the company's direct control and custody for an extended period of time.
5. The computer and the physical area in which it is located must be protected along with the vital information which is so closely linked to it.
6. The adequacy and validity of the programs used to process this information and related computer operations documentation must be safeguarded to assure the usefulness, currency, and accuracy of the basic information.

Much of what already has been said about selecting and protecting vital records applies to records processed by a computer. Many of the general measures taken to protect vital data processing operations and records are measures that ordinarily should be taken to assure the general efficiency of the computer and its use by the company. However, procedures used to protect vital data processing records must be compatible with the information system design policies and the computer programming concepts used by the company.

In protecting paper or microfilm vital records, it is necessary to safeguard only the record itself. In contrast, in protecting vital data processing records, there are three distinctive elements: (a) the computer facility, (b) the physical data processing media, and (c) the inherent integrity of the information itself.

Controlling the Computer Facility

It is preferable prior to computer installations to build many protective features into the area in which it is to be operated. However, where these features have been omitted or function inadequately, it is possible to remedy some defects even when the computer has been operating for some time. The following are some recommendations for improving computer facility security. They are not meant to cover every aspect of computer operations area design and layout.

1. Make the facility as inconspicuous as possible. Remove door and direction signs that identify the computer's location. It may be preferable to leave the computer facility entrances unmarked. Block off or otherwise eliminate display windows originally installed to exhibit computer operations to visitors or those passing the building on the outside.
2. Strengthen controls over access to the computer facility. Provide 24-hour security guard sur-

veillance of the area. In some companies this may be supplemented by installation of closed circuit video equipment to permit monitoring of computer facility approaches by the company security center duty officer. Install security hardware on all computer facility entrance doors. Maintain a record of all visitors, both outsiders and nonassigned company employees, and require that they wear a distinctive badge while they are in the computer facility. Encourage assigned employees to question visitors about their right to be in the facility, even when they are wearing visitor badges. Prohibit delivery by vendors of stocks of forms and other supplies directly to the computer facility. Do not permit vendor representatives to stack these materials in halls or corridors.

3. Review possible computer facility exposure to water and fire damage. Determine if the walls, ceiling, and air conditioning system are sufficiently watertight to prevent possible damage in an emergency. Be sure that drainage under the facility's raised floor is adequate to avoid water accumulation wherever flooding may occur or water might be used to extinguish a fire. Reduce exposure to arson attempts by moving air conditioning and ventilation air intake ducts from the ground floor to high up on the building side—possibly even up under the eaves. Wherever local fire codes permit, remove water-sprinkler fire extinguisher systems and replace them with carbon dioxide or Halon 1301 systems. Make prompt removal of continuous form carbon paper accumulation and similar waste a part of computer facility work routines. To further reduce fire hazards from accumulating waste, prohibit smoking or eating within the computer facility. Include a test of the heat/smoke detection system in the periodic check of computer facility emergency action procedures.

If the computer facility is located in an urban highrise controlled-environment building, consider the possible fire effects of two basic building construction features. Heat circulation openings between the exterior and interior building walls may encourage fire to spread by breaching fire walls. The sealed windows may intensify heat buildup by offering no openings through which it might dissipate naturally.

4. Provide sufficient emergency power generation capacity to compensate for voltage surges or extended voltage reductions, and otherwise maintain uninterrupted power input to the computer. A battery pack with converter, possibly integrated into a motorized generator, will be required.

Minimum power failure reserve needs: from 15 to 18 minutes, sufficient to reach the nearest restart point and to shut operations down in an orderly manner. Attach an audible warning device to the emergency power source to assure that the computer operator and others are notified promptly of the shift to reserve power.

5. Alternate computer facilities to be used in an emergency must be truly compatible with the company's computer and the work performed on it. The alternate facility should have sufficient reserve processing capacity and mainframe schedule time to permit it to handle the company's work and its own in an emergency. Be sure that the two computers are comparable in memory size, hardware options available, resident operating system supervisory program, and peripheral equipment. Determine the impact of any differences on the program itself and the data preparation routines used in the material to be processed on this alternate computer.

6. Consider a Farraday Cage to shield vital information containing company secrets from external monitoring while it is being processed.

7. Make sure that all unused wiring, including telephone cables, is removed from the computer facility. Make sure that heating and air conditioning ducts and water pipes are grounded as close to the computer facility as possible. Both practices will reduce the likelihood of a tap being placed on the data processing system.

Safeguarding Data Processing Media

Data processing tapes and disks containing vital information must be protected by placing them in specially insulated records containers which have been approved by Underwriters Laboratory for temperatures up to 150°F. Underwriters Laboratory does not certify a fire door that will protect the contents of a records vault or file room at 150°F during a fire. Thus, data processing media containing vital information can only be physically safeguarded by placing them in specially insulated records containers.

Vital records magnetic tape files and the computer programs used to process them can be provided with additional protection by copying them in card image form onto duplicate tape reels to be dispersed for storage at a geographically remote company operating location. The principles behind protecting paper vital records by dispersal apply here as well. To avoid possible fraud, these dispersed tapes should be placed in the custody of the internal auditor at the location where they are stored. Where data files are updated periodically, three generations of the file can be retained; the current version and two previous ones. The "grandfather" (oldest) version of the vital records file is stored off-site until replaced by a later version. When the vital information files are maintained on a disk file, a snapshot (complete copy of its content at a specific point in time) is made periodically by copying file content onto a tape. The two oldest snapshot tapes are stored off-site until replaced in turn by later snapshots. In both instances, these off-site stored tapes can be used with related transaction records to reconstruct files damaged in a disaster or by computer processing error.

Assuring Information Integrity

The integrity of computer processed vital records is maintained by careful control over data input and user file access, program content revisions, and computer facility operating practices. In each of these areas some protection will be provided by normal data processing management practices. However, special attention should be given to their impact on computer processed vital records.

1. *Input and file access.* Input data editing routines can be designed to detect and automatically reject spurious information. Vital records processing programs also can be designed to limit selectively user access to key file segments and to restrict user ability to modify certain types of information in the file. In addition, the resident supervisor or operating system program should maintain a log inaccessible to assigned computer operators. This log should routinely record programs processed, files used, computer operator assigned, and travel use rate and elapsed time. Where the computer facility services a data transmission network, this log also should record user terminal identification and the type of inquiry made. Computer facility supervisors and company security officers should review the log jointly at least once weekly and investigate questionable inquiries and other apparent irregularities.

2. *Program content revisions.* Computer programs used to process vital information should be fully documented. A current copy of this documentation should be stored off-site with the dispersed file copy tapes. Programs purchased or leased from another company should receive protection equal to that given to company-developed programs. Programs from outside sources may have been adapted in some way to the company's specific data processing needs. Documentation incorporating all the necessary features for such an adapted program may be difficult or impossible to obtain from the supplier on short notice. Company computer facility operating policies will determine full documentation file content, but the program segment of it should encompass at least the following: A plain English narrative description of what the program does; definition of transaction content; block or program logic diagrams; decision tables; source coding; assembly listings; a register of checkpoints, error messages, interruptions, together with restart and recovery instructions; and a description of input, output, and transaction processing controls.

As company operating policies and procedures evolve, the programs used to process computer vital records must be altered to reflect these changes. Program changes should be fully documented. Programmers involved in these changes must be clearly identified in the program documentation, and both the user department and computer facility supervisor must review and approve these changes before they are implemented. Programmers must not be permitted on their own initiative to make even minor changes in production programs that they are running.

3. *Computer facility operating procedures.* Assign computer operators to work in pairs at all times, even on weekends and holidays, and especially when vital records are being processed. Normally a supervisor or senior operator is assigned to work with a less experienced person. Teamwork will improve overall operations quality and make it more difficult for operator errors or data alteration to go undetected. The possibility of file destruction through computer operator error can be reduced further two other ways. Place the operating or executive supervisory program in read only memory. This will safeguard the program's memory protection feature which prevents accidental file segment destruction and blocks illegal file use. Also, include intensive job completion condition checks in every program. Instruct computer operators not to terminate or dismount any program until a satisfactory end of job message from the program has been logged on the console terminal.

Computer operator team assignments and work shifts should be rotated periodically to maintain separation of duties and operating controls. As part of these same controls, an operator should not process the same programmer's programs for an extended period of time.

Encourage operators to remain alert to facility physical condition changes. They should check periodically during each shift for such things as magnets, screwdrivers, files, and other small potential sabotage tools. Disengaged security and fire alarm equipment, and open doors to operational disk units and other peripheral equipment should also be noted.

Careful scheduling of computer time ensures more efficient use of the facility's capacity and makes it easier to spot diversion of computer time for unauthorized copying of vital information files. Vary the schedule for processing particular vital records by not running the program at the same time each day or on the same day each week.

Certain information can be identified as vital only as long as it remains uncompromised by industrial espionage efforts. Extra steps must be taken to prevent compromise of the company's computer processed vital information. Pulverize used continuous form carbon paper, impact printer ribbons, discarded forms and printout copies, and used tabulating cards in a document disintegrator located in the computer facility. The disintegrator's product does not require additional protective handling and may be disposed of with the building's regular trash. Write streams of random digits over disk packs and tapes that have contained vital information before they are released for use with other files and programs.

Special operating procedures for remote data transmission terminals will reduce the possibility of compromising vital information in connection with terminal use to add information to or process files. Data sent over common carrier lines is exposed to interception through line taps. It can be reproduced in a tap attached printer compatible with the data transmission system used. Addition of store and forward capabilities to the tap terminal makes it possible to hold the intercepted information and to modify it before releasing the information back into the system. Programs used with data transmission systems make it possible to disguise vital records message content during transmission. Decoding at either end of the transmission link makes the input processible or the program response understandable at the remote terminal.

Testing and Evaluating the Vital Records Program

The vital records protection program is designed to protect and provide the information needed by the company for survival in a disaster or emergency. But will it? Management must know. Periodic vital records protection program tests will provide the answer.

The company security officer, records manager, and internal auditor should test/evaluate the program at least once a year and note any program defects or problems in a joint test results report to be sent to the proper company officer for information and remedial action.

1. *Determining test objectives.* The test determines if the currently operated program will provide needed information under circumstances simulating disaster or emergency conditions. Every effort should be made to make test conditions as realistic as feasible.

Broadly, the tests should verify that vital records needed after a disaster are: (a) current; (b) protected sufficiently against natural disasters, nuclear detonation and other possible perils; and (c) retrievable as needed in useable form.

More specifically, the tests should determine that the company's various vital information needs can be satisfied in a typical emergency situation. For example:

- Employees can be paid and proper deductions made for taxes, the retirement fund, and other payroll accounts.
- The company's cash position and the location of its banked funds can be determined.
- The company assets and accounts receivable and payable ledgers are all current.
- The order entry, engineering, production, and customer account information needed to resume production and sales activities are available and current.

2. *Preparing and conducting the test.* After scheduling the test, determine where it will be held, who will participate, and how long it will last. Restrict advance knowledge of the test date to as few people as possible. Keep the test period as short as possible. The amount of time required for the test will vary. A large company may need several days to complete it. A small company may need no more than an afternoon. Reduce to a minimum participants' absence from their regular duties. The test should be located off company premises if possible. It may be held at a motel, an executive conference center, the company alternate headquarters site, or in a conference room made available by the local civil preparedness agency.

Well in advance of the scheduled test date, arrange for necessary test participant team support, working space, couriers, microfilm, copying equipment, and access to data processing equipment. Also, arrange for several company executives not scheduled to participate in the test to act as judges. They should be familiar with the records used in the test. They must be able to determine if the problems posed by the test have been answered successfully.

After the participants arrive at the test site, explain the test conditions. Emphasize that the problems posed must be answered by records currently included in the Vital Records Protection Program. Give the questions to the test team.

3. *Typical vital records test problems.* Assume that the plant has been completely destroyed during the night, with nothing salvageable. Demonstrate the company's ability to perform tasks such as the following:

- Notify all managers to report to an emergency center for reorganization planning.
- Notify all other employees not to report to work until further notice.
- Continue paying plant personnel on time.
- Send alternate shipping instructions to vendors with whom orders have been placed.
- Prepare a list of sources of supply for a specified product.
- Produce engineering drawings and the bill of materials for a small number of specific products.
- Prepare an insurance claim statement covering the complete destruction of the manufacturing buildings.
- Prepare a list of vendors in order to replenish operating departments.
- Produce a current statement of assets and liabilities and a statement of income and expense.
- Produce a list of commission balances for each manager and salesperson by a unit number.

Allow time for test team members to determine what records will be needed to answer the questions.

As the requested records arrive by courier from their various locations, they must be reproduced or reconstructed in useful form. If the records are on microfilm, prints must be made of the first 10 images on each reel. These prints must be inspected by the test judges to determine that they are sufficiently legible for use in performing the specified test task. If the records are on computer magnetic tape, the test must print out successfully the first 100 records on each reel. The judges must determine that the printout adequately reproduces these records.

Records used in the test must be returned by the couriers to their protected locations as soon as the test is over.

4. *Testing vital computer records and operations.* Vital computer records will be used to satisfy many of the basic test problems, but a comprehensive evaluation of the Vital Records Protection Program requires supplemental testing of vital records computer processing. Devise tests that will:

- Compare a clean copy of the tapes for selected vital records processing programs against a copy of the programs currently in use to determine that protected computer program documentation is up-to-date.
- Demonstrate that computer audit trails are being maintained in vital records computer programs
- Determine that the alternate computer and its associated supervisory programs are still compatible with the company computer facility.

Additional aspects of computer facility operations should be tested. These will be determined by company data processing management policies.

Public Information

Any information concerning the emergency that is to be released to the news media or the general public must be cleared through the industrial relations manager, or in case of absence the personnel manager or the security manager. The emergency plan should provide that no information is to be released without the express approval of the facility manager or designated representative.

Notification of key personnel of a facility, in the event of an emergency, will normally be the responsibility of the security force shift supervisor after being advised that such notifications are to be made by the risk manager or the industrial relations manager.

Emergency Equipment Location

A complete listing of all emergency equipment should be made and attached to the emergency plan. The location of each item of equipment should be exact, and provisions must be made for frequent periodic inspections and necessary maintenance of this equipment.

Training

The need for employees to be trained in emergency evacuation procedures, routes, and staging areas already has been discussed. It is also necessary for special training to be given to all persons who have been designated as members of the emergency teams. As mentioned in a previous chapter, the fire brigade normally will be able to obtain their training from the local fire department. Medical personnel, if not registered nurses or trained paramedics, should receive advance training in first aid treatment, including CPR, from the American Red Cross or other qualified agencies or sources.

All members of the emergency team should be given orientation tours throughout the facility to make sure that they are sufficiently familiar with the layout to be effective during an emergency situation. During an actual emergency, the security force will be called upon to perform their normal duties of providing protection to the facility. Additionally, they will be expected to provide such services as controlling traffic exiting the facility, and making sure that routes are kept open for the entry into the facility of emergency vehicles such as fire, police, or ambulances. They also may be called upon to control crowds which gather near the facility. Security officers who are off duty may be called in to augment the working shift and to provide for extra patrols around the perimeter of the facility.

A security force is put to the test daily but never to the extent that they face during an emergency situation. It is at this time that the professionalism, dedication, and training of a guard force will decide whether people will live or die and whether property will be saved or lost. Some of their duties will get boring and they will have to be reminded that the training and strict devotion to duty must be maintained and developed if they are to be effective when the emergency situation arises.

Appendix 12a

Checklist: Preparedness for Data Center Disaster*

IRA S. SOMERSON

Part I—The Facility

1. Is the computer facility constructed from noncombustible, nonexplosive materials?
   a. Concrete or other noncombustible subflooring? Does it have positive drainage?
   b. Aluminum or other noncombustible raised flooring (including supports)?
   c. Is underfloor cabling channeled through conduits?
   d. Tiling used on floor—is it nonpetroleum based (including asphalt, rubber, vinyl, linoleum, high pressure plastic laminates, or carpeting)?
   e. Are there water-based paints on walls or trim?
   f. Are walls and trim themselves noncombustible (fire resistance of at least one hour)?
   g. Are metal doors and partitions used, including metal framing?
   h. Is glass that is used of steel-web or reinforced type?
   i. Is tile from any dropped ceiling made or noncombustible or high-melting-point-materials?
   j. Are cables that connect ceiling lighting properly placed in conduits?
   k. Are all electrical connections properly grounded?

---

* Reprinted by permission of Ira S. Somerson, CPP, President of Loss Management Consultants, Plymonth Meeting, PA.

1. Is all furniture of metal construction?
2. Is the upper ceiling constructed to conduct water from higher levels away from the computer facility? Has floor above been made watertight? Pipe and wire openings?
3. Are the computer facility and the supporting facilities separated sufficiently, by distance or by fire-resistant materials, to prevent a fire in one area from affecting the other?
   a. Tape or disk libraries?
   b. Paper or card storage?
   c. Backup files?
   d. Copies of source decks?
   e. Copies of source listings?
   f. Supporting operating facilities?
      (1) Alternate computing facilities?
      (2) Punch-card processing facilities?
      (3) Remote job entry facilities?
      (4) Customer engineer facilities?
   g. Copies of operations procedures?
   h. Copies of control procedures?

   Certain facilities may be located far apart yet linked via cable-routing tunnels or vertical cable chases.
4. Are the facilities that house the activities listed above constructed to the same noncombustible, nonexplosive standards as the main computer facilities?
5. Are there sources of light, strategically located, that do not depend upon the main power source?
6. Have the facilities been constructed to permit easy access for fire-fighting personnel and equipment?
7. Have the facilities been constructed to permit distribution of detection sensors and extinguishing agents?
8. Have the air-conditioning facilities been constructed to permit only authorized access, including:
   a. High placement to restrict access?
   b. Protection of the source of water supply?
   c. Protection of fan or cooling mechanisms?
9. If a fire were to occur in one of the data-center facilities, would other offices of the business be "knocked out" as well?
10. What is the proximity of the data center to the organization's medical facilities?
11. Have the main power control boards been placed in a remote or restricted access position?
12. Have communications devices and equipment relative to the data center been placed in a remote or restricted access position?
13. Has closed-circuit television equipment been installed:
    a. To cover critical equipment?
    b. To cover access routes?
    c. To cover critical data storage locations?
    d. To cover air-conditioning and power sources?
    e. Beneath false floors and above false ceilings?
    f. To cover critical communications equipment and devices?
    g. To permit monitoring by safety or security officials?
14. Are master controls for detection and suppression systems located outside the data center?
15. Is the data center located in a nonhazardous area?
16. What hazards may be removed from the immediate surrounding area?
17. Are caustic or flammable cleaning agents permitted to be present in the data center?
18. Do elevator doors open on the data-center floor? Or near the data center?
19. Do hallways have false floors which could permit unauthorized access?
20. Are those hallways covered by closed-circuit television?
21. Is it possible for someone to externally access your communications lines?
22. Have you removed identification markings from power rooms, communications closets, etc.?
23. Is access to communications equipment, such as junction boxes, switching mechanisms, terminal outlets, etc., freely available? It should not be.
24. What is the position of the data center relative to sources of potential fire, such as cafeterias, power cabling, rubbish storage, caustic chemicals, fumes, odors, etc.?
25. What is the proximity of the data center to steam lines?
26. Is the data center located near places where hazardous processes are employed?
27. Is the space beneath the access flooring used for storage of Class A materials, e.g., IBM cards? It should not be.
28. Is sub-floor space cleaned regularly?

Part II—Disaster Protection

Detection Equipment

1. Do the facilities listed above have one or more of the following:
   a. Smoke detection equipment? (specifically recommended by the National Fire Protection Association)
   b. Heat rise detection equipment?

c. Optical or ultraviolet detection equipment?
   d. Humidity control equipment?
   e. Thermocouple detectors?
   f. Products-of-combustion detectors? (recommended by NFPA)
   g. Air temperature detection?
   h. Are any of these detection units mounted *inside* the cabinets of critical system components?

Alarm Mechanisms

1. Do the facilities listed above provide alarm mechanisms, such as automatic alarming upon the detection of fire(light), smoke, or inordinate heat rise?
2. Are there several strategically located, manually operated alarm systems?
3. Does the alarm device used report the location of the fire to a centralized fire or security position?
4. Does the alarm device used also report to municipal fire or security positions?
5. Does the alarm system provide any prealarm audible tone?
6. Does the alarming mechanism contain automatic shutdown of critical equipment? (particularly required with sprinkler systems)

Protection Equipment

1. Do the facilities listed above have one or more of the following:
   a. Automatic dispersal of a fire extinguishing or retardant agent, such as:
      (1) Gas—Halon 1301? (above and beneath floors and ceilings)
      (2) Foam? (not recommended by NFPA)
      (3) Water (last resort), including hoses and/or sprinkling systems?
      (4) Fixed flooding systems?
      (5) Dry suppressants? (not recommended by NFPA)
   b. Manual Equipment, such as:
      (1) Portable extinguishers for electrical and other fires?
         (a) CO_2
         (b) Halon 1211
         (c) Halon 1301
      (2) Water or other extinguishing agent for nonelectrical fires?
         (3) Are these strategically located *throughout* the facilities?
   c. Automatic and/or delayed interruption of power sources where electric fires have been discovered?
   d. Automatic shutdown of air-conditioning systems (particularly where Halon 1301 is used)?
   e. Automatic shutdown of heating or humidity systems?
   f. Automatic close-off of air ducts?
   g. Automatic illumination of emergency lighting upon the interruption of the prime power source?
   h. Automatic sealing of firebreaks or firedoors between sections of the facility?
2. Are there any fire-suppressant outlets mounted *inside* the cabinets of critical system components?
3. Is there a means to manually activate an automatic system?
4. Are the automatic devices used "rate-compensated" to allow for sudden increases in temperature?

Part III—People and Procedures

1. Are there specific procedures for use in the case of fire, such as:
   a. Power-down procedures to prevent destruction of critical data and devices?
      (1) Does each employee have a copy?
      (2) Is a copy prominently posted?
      (3) Do security officials have a copy?
      (4) Do organizational fire officials have a copy?
      (5) Do municipal fire officials have a copy?
   b. An emergency plan that will be invoked, to include:
      (1) Assigned responsibilities for each staff member?
         (a) Who will power-down the equipment?
         (b) Who will remove critical data files?
         (c) Who will attempt short-term fire-fighting measures?
         (d) Who will oversee the evacuation measures?
         (e) Who will notify fire or other responsible officials?
         (f) Who will cut the power?
         (g) Who will cut the air conditioning?

(h) Who will have responsibility to attend to any injured personnel?
   (2) Assigned evacuation procedures for each staff member or zone within the facility?
  c. Strategically located and prominently displayed:
   (1) Extinguishing agents?
   (2) Emergency telephone numbers:
     (a) Fire?
     (b) Police?
     (c) Ambulance?
     (d) Security?
     (e) Authority in company?
   (3) Direct-line telephone handsets?
   (4) Evacuation plans?
   (5) Responsibility assignments?
   (6) First-aid equipment and supplies?
   (7) Egress route (arrows or colored lines on the floor) and lighted exit signs?
2. Have the people involved read and understood these procedures?
3. Is there a periodic review of these procedures with the people involved?
4. Are new employees familiarized with these procedures as part of their initiation to the organization?
5. Do you conduct periodic fire drills to test and improve the plan?
6. Is there a periodic inspection to highlight needed improvements?
7. Is the use of incendiaries, such as smoking within the facilities, controlled?
8. If your facilities have electrically operated doors, is there a manual opening ability if the power source is interrupted?
9. Is access to the computer or other related facilities controlled?
10. Have all personnel been trained in the use of manual fire-fighting equipment?
11. What procedures exist to re-arm any fire-prevention equipment?
12. Are the temporary passes numbered to permit control of the pass as well as the person using it?
13. Is there a procedure for returning and accounting for these temporary passes?
14. Can temporary passes be easily duplicated?
15. Are the pass and access rules consistently enforced?
16. Is there metal-detection equipment available? Is it used?
17. Is there a means to inspect parcels and other articles moved in and out of the data center?
18. What guarantee do you have that boxes or crates containing products or equipment received at the data center actually contain their specified contents?
19. Do you restrict the entrance of radios and other electronic devices from the data center?
20. Are there *alert* mechanisms for the summoning of security personnel?
21. What mechanisms exist to ensure that the person is carrying his or her own badge?
22. Are there electric eye or proximity warning indicators positioned in infrequently used rooms or hallways?
23. In addition to closed circuit television, are there sound monitoring systems that permit security personnel to listen in when the facilities are unused?
24. Do the facilities exist to lock all external doors on command?
25. Are vendor personnel allowed to roam freely on the strength of their apparent vendor affiliation? They should not be.
26. If you have a closed circuit television setup, have you assigned someone to watch the monitors at all times? As a sole duty?
27. Can anyone who wants to see the data center do so on request?
28. Is there, at the data center, a current photograph of every person who has legitimate access to the area?
29. Are there double-door arrangements that will lock an intruder in-between them?
30. Are your security precautions the same at every entrance, including the loading dock?
31. How about access from stairwells? Are they restricted or in any other way controlled?
32. Floor access to communications cables—how are they monitored or controlled?
33. Are plans and blueprints for the data center and other important areas controlled or restricted? Where are they available outside the organization?
34. What restrictions are there on the introduction of camera or other photo recording equipment within the confines of the data-center area?
35. What restrictions are there on the introduction of sound or other magnetic recording equipment within the confines of the data-center area?
36. Do you have closed-circuit television on the outside of entrances or covering closely available parking spaces? How about the building periphery?
37. Is self-contained breathing apparatus available to fire-fighting personnel?

Part IV—Disaster Prevention and/or Recovery

1. Is there a regular inspection by qualified personnel of all automatic detection and protection systems?
2. Is there a regular inspection of the facilities by qualified fire prevention officials?
3. Are simulated fire drills conducted at random intervals? Is there a specific plan for such drills?
4. Do watchmen or other security officials perform an inspection of the facilities at regular intervals when not in use?
5. Have contingency plans been established to permit recovery and resumption of operations in the event of an emergency?
6. Is there an emergency power source to energize emergency lighting?
7. Is there an alternative power source that will permit resumption of operation if the prime power source is destroyed?
8. Are there arrangements made for backup computer service in your own or other facilities?
9. If there are backup arrangements made, have you ever run your systems on that hardware to ensure compatibility?
10. Is there a mandated and enforced housekeeping procedure that ensures that flammable materials, such as paper, inks, corrugated boxes, and ribbons are kept at an absolute minimum?
11. Are your critical files backstopped by keeping current copies in fireproof safes or remotely located facilities?
12. Could you recapture the last n days' processing with copies of your files and transactions remotely located?
13. If, as the result of a disaster, your automated systems were to become nonfunctioning, are there up-to-date manual backup systems that will permit your organization to continue operation?
14. Would your backup systems or procedures permit you to cover your short-term future needs?
15. Has your organization identified *all* the costs that would be incurred in such a disaster and adequately insured against such losses, including:
    a. Replacement of hardware?
    b. Replacement of critical software?
    c. Replacement of critical applications programming?
    d. Replacement of physical facilities at their current costs?
    e. Temporary replacement for injured personnel?
    f. Opportunity costs due to business interruption?
    g. Potential for natural disaster?
16. Have you fully evaluated your insurance coverage in light of the costs detailed above?
17. Has there been designated, within your department, a person responsible for review of the items contained in this checklist?
18. Does this person have the authority to cause changes to be made, or sufficient leverage to convince funding authorities to underwrite necessary improvements?
19. Are procedures drawn on this checklist reviewed periodically by insurance, safety, and fire prevention officials?
20. Are key personnel trained at the minimum in first-aid procedures to treat burns and smoke inhalation (hydrogen chloride gas from insulation)?
21. Is there first-aid equipment available within the facilities, whose location and contents are known by at least key personnel?
22. Have all personnel been trained in the use of fire-fighting equipment?
23. Is the location of additional equipment for such use known to all personnel?
24. Have you made arrangements with your suppliers for the speedy replacement of:
    a. Forms and card stock?
    b. Magnetic storage media?
    c. Air-conditioning equipment?
    d. Electrical equipment?
    e. Detection and prevention equipment?
    f. Security equipment?
    g. Construction materials?
        (1) Flooring?
        (2) Ceiling?
        (3) Partitioning?
        (4) Other related materials?
    h. Computing equipment?
    i. Data entry devices?
    j. Communications equipment?
25. Have arrangements been made for freight capabilities for such replacement and to move backup equipment, files, documents, etc.?
26. Have arrangements been made for specialized repair and/or clean-up services?
27. Do only authorized personnel have access to the computer facility and to programs, data, documentation, and procedures?
28. Have you established a system of signatures to control access to critical data and documents?
29. Have you classified your data, programs, and documentation in terms of their criticality to your organization?

30. Have you instituted a marking system whereby the most critical items are evacuated first?
31. Are data files of the highest classification grouped together in racks to permit easy removal?
32. Are data file storage racks mounted on wheels to permit easy removal?
33. Have you instituted a system of visitor control?
34. Would it be useful to install magnetic sensors in your access doorways?
35. Would it be useful to consider the use of security guards at data center accesses?
36. Would it be useful to put critical files "under lock and key," limiting the access to certain people only?
37. Should you conduct a periodic security check of all personnel:
    a. Spot inspection under operation?
    b. Complete background investigation on hiring?
    c. Thorough investigation of all personnel with access to the data center?
38. How are casual workers evaluated—such as maintenance and repair personnel from *outside* the organization?
39. Where there is sound-deadening material installed—on walls, in cabinets, or around desks and operating positions—has it been sprayed with fire-retardant chemicals? Foamed cellular plastics should never be used.
40. Is there wooden or other combustible furniture in the area? All furniture should be of metal construction.
41. Does the data center have a devoted power system? The source of power should not tie to other parts of the organization.
42. Does the data center have a devoted air-conditioning system? Ducting should not carry fumes and smoke to other offices.
43. Are you absolutely sure that even though your hardware and software systems are compatible with those of your backup source, its time available would be sufficient to absorb your work?
44. And if it can accommodate the load, could it do so for the days, weeks, or months necessary to get your system functioning again?
45. Have you installed self-closing mechanisms on all internal doors?
46. Are internal doors and passageways free of all obstructions? Including wedges?
47. Are internal aisles wide, straight, and free of obstructions, or are they "mouse in a maze" passageways?
48. Is all equipment positioned so that access doors open fully and freely?
49. Once open, do the cabinet doors permit room for a person to work on the device?
50. Is there sufficient room between units to permit free airflow and heat dissipation?
51. Have you considered the appointment of a "security task force" to:
    a. Perform random inspections?
    b. Perform random audits?
    c. Verify the use of protection procedures?
    d. Verify the currency and protection of programs and data?
    e. Verify the disaster prevention techniques?
    f. Call and monitor disaster training exercises?
52. Have you eliminated external walls and windows that permit easy access to a saboteur?
53. Have you eliminated the casual visitor by stopping organization executives from including the data center in a facilities tour?
54. If your organization may be subject to a civil disturbance, have you filed your disaster plan with local police, fire, civil defense, and/or national guard officials?
55. Do you have an annual service contract with a qualified fire and security systems maintenance and service organization?

Part V—Disaster Strikes: What to Do When; The Aftermath

1. Has a disaster team been designated?
2. Have there been procedures established for the sequence of events to be followed during and after a disaster?
3. Has there been established a master plan for disaster contingencies? Is it kept current?
4. What personnel notification procedures exist?
5. Are all team members provided with copies of home or other telephone numbers wherein team members or other department members may be notified?
6. Has a notification precedence been established?
7. Would it be useful to develop a pyramid structure notification procedure?
8. What if the telephone system is not functional as in the case of a natural disaster? Is there an alternative plan for notification?
9. Once these procedures have been established, have you *dry run* them to determine if they work and then modified them accordingly?
10. In establishing notification procedures, have you established precisely:
    a. Whom each person is to call?
    b. What to do if the notification chain breaks down?

c. What each person is to do externally to the organization?
   d. Where each person must report, and to whom?
   e. The sequence of events once these people are on-site?
11. How will the user be notified? As part of the pyramid structure? Separately? By a specifically responsible official?
12. How will the responsible company officials be notified?
13. Have you canvassed, located, or otherwise arranged for an alternative site within the organization complex to reestablish the base of operations?
14. If your facility is sufficiently destroyed as to make alternative location necessary, have you located such a facility?
15. In evaluating this facility, have you determined that the proper site facilities are available or can be expeditiously arranged:
   a. Power and lighting?
   b. Air conditioning?
   c. Humidity control, where necessary?
   d. Alternative disaster proceedings, in case of a secondary disaster?
16. Who will be responsible for notification of the news media? Who is the official spokesperson? In what form will this official notification take place—press releases, telephone calls, interviews?
17. Have you arranged for duplicate or sufficiently similar hardware to be installed on short notice to the vendor?
18. How will that vendor be notified in case of disaster? To whom will be assigned the responsibility to contact account representatives, customer engineers, or systems engineers?
19. To begin functioning again:
   a. Have arrangements been made to install a copy of the working operating system on the backup hardware? This, of course, implies that you had stored such a copy in safe-keeping, kept it maintained, and tested it periodically. Has this been done?
   b. Any communications lines to reestablish? How will the communications services and communications equipment vendors be notified? What will be their role?
   c. Once the recovery hardware is in place and functioning, have you sought additional backup for problems or overflow?
   d. Have you taken steps to duplicate your programs and decks (assuming you are using those you had put back for just such a disaster), so that you again have the proper disaster protection?
   e. Have you available a software or other control specialist to ensure that any job control specifications are valid and functioning? Have you duplicated such job control language to again provide backup?
   f. Have you obtained an emergency distribution of forms and supplies? If you used backup forms and supplies, have you taken steps to replenish that supply?
   g. Have you taken steps to duplicate the critical files, library programs, software, etc., necessary to reestablish backup?
   h. Have you duplicated any operator run books or documentation for backup?
20. What priority scheme have you established for recovery? What runs will be done first in order to get the company functioning again? If that is not part of your disaster plan, who will make the decision at this time?
21. What personnel realignments, additions, etc., will it be necessary to make during the recovery period?
22. Must you make use of specialized support services during this period:
   a. Commercially available payroll services?
   b. Commercially available timesharing?
   c. Use of specialized consulting services during the period?
   d. Specialized and short-term training for new personnel?
   e. Commercially available distribution services, such as mail trucks, courier services, security services?
23. What about media resupplies? Can you get additional tapes, disks, etc., on short notice? Are the arrangements made?
24. What about temporary furniture, equipment racks, storage for cards, tapes, disks, etc.?
25. What about office supplies and equipment? Could you on short notice obtain sufficient support equipment to regain operation?
26. Have you contacted the organization's legal representatives to determine liability for data, services, equipment, etc.?
27. Have you contacted the organization's insurance representatives to determine coverages, liabilities, and to arrange for coverage for the new site?
28. Have you contacted the accounting department to determine what manual records exist to assist you to reconstruct any data records? Could

the accounting department assume any manual backup operation during the interim period so that critical receivables can be obtained, processed, and accounted for, payroll services can be maintained, and accounts payable systems can continue to receive trade discounts?
29. Have you contacted other departments to which you supply information with similar requests?
30. Have you contacted real estate representatives to determine the availability of reconstruction sites if it is determined that you will not rebuild on the destroyed site?
31. Have you obtained and duplicated all written documentation and placed that copy away for safekeeping?
32. Have you assigned supervision to see that any manual or backup systems are functioning properly during the interim period?
33. Restoring the original site:
    a. Have you assigned responsibilities for clean-up of the destroyed site, or otherwise obtained that service?
    b. Have you established contracts for reconstruction or refurbishing of the data center, to include electrical, air conditioning, humidity control, fire prevention, heating, security, etc.?
34. Have you begun the short-range and long-range planning to return the installation to normal conditions and operations?
35. Have you done the public relations tasks?

Chapter 13
Bomb Threats and Search Techniques*

REX D. DAVIS

Bombing and the threat of bombing have created a need for practical knowledge to cope with the increasingly violent activities of people who represent segments of unrest in our society. Repeated criminal acts which use or threaten to use explosives against educational institutions, industry, law enforcement, and the general public, place a most urgent responsibility on law enforcement agencies. However, the protection of life and property is a responsibility that cannot be delegated to law enforcement alone. Every citizen must be prepared to accept responsibility if we are to enjoy a safe place in which to live and work.

Information for the preparation of this document was obtained from a wide range of official and private sources, including the actual experience of Alcohol, Tobacco, and Firearms special agents. The ideas and methods suggested reflect the most current information available to help you.

One suggestion in this chapter should be emphasized; it is preparedness. When one is equipped with an organized plan, most bomb threat problems can be resolved with minimal personal injury and property damage.

Purpose of Calls

The only two reasonable explanations for a call reporting that a bomb is to go off in a particular installation are:

1. The caller has definite knowledge or believes that an explosive or incendiary has been or will be placed and wants to minimize personal injury or property damage. The caller may be the person who placed the device or someone else who has become aware of such information.
2. The caller wants to create an atmosphere of anxiety and panic which will, in turn, possibly result in disruption of the normal activities at the installation where the device is purportedly located.

When a bomb threat call has been received, there will be a reaction to it. If the call is directed to an installation where a vacuum of leadership exists or where there has been no organized advance planning to handle such threats, the call will result in panic.

Panic

Panic is one of the most contagious of all human emotions. It is defined as a "sudden, excessive, unreasoning, infectious terror." Panic is caused by fear —of the known or the unknown. Panic can also be defined in the context of a bomb threat call as the ultimate achievement of the caller.

Once a state of panic has been reached, the potential for personal injury and property damage is dramatically increased.

People may be tempted to join a fleeing crowd; the fright of those in motion is enough to suggest the presence of something to fear. When this stage is reached, it becomes difficult to control the group. Attempting to reason with such a crowd may be futile, but it is possible to control the group by assuming leadership or distracting key members of the group. In any case, corrective action should be taken before the movement stage, if possible.

---
*The bulk of this chapter was originally published as a pamphlet by the Bureau of Alcohol, Tobacco, and Firearms (ATF), 1976.

1. *Panic Deterrents.* An effective preemergency program of informing personnel what is expected of them in an emergency coupled with the example of strong, competent leadership by officers of the organization will go far toward preventing panic. To reduce the likelihood of panic, the physical causes should be eliminated in an emergency. The organization should be prepared to remove the injured and the dead from general view; clear away debris which appears to cut off escape; quickly control fire; and approach any disturbance with calmness. Preemergency preparations should include arrangements to facilitate routes to be taken in evacuating the building or going to shelter, and locating organization personnel where they can take command and give calm, decisive instructions at places where groups are likely to congregate.
2. *Antidotes for Panic.* In certain circumstances, it is conceivable that, despite preemergency preparations, an unorganized group may be on the verge of panic. Organization personnel should be prepared to deal with this in terms of the following principles:
   - *Provide assurance.* Exert positive leadership. Reassure the group by giving information and instructions calmly.
   - *Eliminate unrest.* Dispel rumors. Identify troublemakers and prevent them from spreading discontent and fear.
   - *Demonstrate decisiveness.* Suggest positive actions. Indicate what to do, rather than what *not* to do.

Leaving facilities unattended can lead to destruction of the facility and the surrounding area. Large chemical manufacturing plants, power plants, unattended boilers, and other such facilities require the attention of operating personnel.

Other effects of not being prepared or not having an organized plan to handle bomb threat calls can result in a lack of confidence in the leadership. This will be reflected in lower productivity or reluctance to continue employment at a location that is being subjected to bomb threat calls.

Preparation

Lines of organization and plans must be made in advance to handle bomb threats. Clear-cut levels of authority must be established. It is important that each person handle each assignment without delay and without any signs of fear.

Only by using an established organization and procedures can you handle these problems with the least risk. This will instill confidence and eliminate panic.

In planning, you should designate a control center or command post. This control center should be located in the switchboard room or other focal point of telephone or radio communications. The management personnel assigned to operate the control center should have decisionmaking authority on the action to be taken during the threat. Reports on the progress of the search and evacuation should be made to the control center. Only those with assigned duties should be permitted in the control center. Make some provision for alternates in the event someone is absent when the threat is received.

Evacuation

The most serious of all decisions to be made by management in the event of a bomb threat is evacuation or nonevacuation of the building.

The decision to evacuate or not to evacuate may be made during the planning phase. Management may pronounce a carte blanche policy that in the event of a bomb threat, evacuation will be effected immediately. This decision circumvents the calculated risk and gives prime consideration for the safety of personnel in the building. This can result in production down-time, and can be costly, if the threat is a hoax. The alternative is for management to make the decision on the spot at the time of the threat. There is no magic formula which can produce the proper decision.

In the past, the vast majority of bomb threats turned out to be hoaxes. However, today more of the threats are materializing. Thus, management's first consideration must be for the safety of its people. It is impossible to determine immediately whether a bomb threat is real.

Investigations have revealed that the targets for terrorist bombings are not selected at random. The modus operandi for selecting the target(s) and planting the explosive appears to follow this pattern: The target is selected because of political or personal gain to the terrorist; it is then kept under surveillance to determine the entrances and exits most used, and when. This is done to determine the hours when very few people are in the building. The idea is that the intent is not to injure or kill people, but to destroy the building. Reconnaissance of the building is made to locate an area where a bomb can be concealed and do the most damage, and where the bomber is least likely to be observed.

A test, or dry run, of the plan is often made. After the dry run and at a predetermined time, the building is infiltrated by the bomber(s) to deliver the explosives or incendiary device. The device may be fully or partially preset prior to planting. If it is fully set and charged, it is a simple matter for one or two members of the group to plant the device in a pre-selected concealed area. This can be accomplished in a minimum of time. If the device is not fully set and charged, one member may act as a lookout while others arm and place the device. Most devices used for the destruction of property are usually of the time-delay type. These devices can be set for detonation to allow sufficient time for the bomber(s) to be a considerable distance away before the bomb-threat call is made and the device is detonated.

The terrorists have developed their plan of attack and the following procedures are suggested to business and industry for coping with bomb threats.

How to Prepare

1. Contact the police, fire department, or other local government agencies to determine whether any has a bomb disposal unit. Under what conditions is the bomb disposal unit available? What is the telephone number? How can you obtain the services of the bomb disposal unit in the event of a bomb threat? Will the bomb disposal unit assist in the physical search of the building or will they only disarm or remove explosives?

2. Establish strict procedures for control and inspection of packages and material entering critical areas.

3. Develop a positive means of identifying and controlling personnel who are authorized access to critical areas.

4. Arrange, if possible, to have police and/or fire representatives with members of your staff inspect the building for areas where explosives are likely to be concealed. This may be accomplished by reviewing the floor plan of the building.

5. During the inspection of the building, you should give particular attention to elevator shafts, all ceiling areas, rest rooms, access doors, and crawl space in rest rooms and areas used as access to plumbing fixtures, electrical fixtures, utility and other closet areas, space under stairwells, boiler (furnace) rooms, flammable storage areas, main switches and valves, e.g., electric, gas, and fuel, indoor trash receptacles, record storage areas, mail rooms, ceiling lights with easily removable panels, and fire hose racks. While this list is not complete, it can give you an idea where a time-delayed explosive or an incendiary device may be concealed.

6. All security and maintenance personnel should be alert to suspicious-looking or unfamiliar persons or objects.

7. You should instruct security and maintenance personnel to make periodic checks of all rest rooms, stairwells, under stairwells, and other areas of the building to assure that unauthorized personnel are not hiding, reconnoitering or surveying the area.

8. You should assure adequate protection for classified documents, proprietary information, and other records essential to the operation of your business. A well-planted, properly charged device could, upon detonation, destroy those records needed in day-to-day operations. Computers have also been singled out as targets by bombers.

9. Instruct all personnel, especially those at the telephone switch-board, in what to do if a bomb threat call is received.

As a minimum, every telephone operator or receptionist should be trained to respond calmly to a bomb threat call. To assist these individuals, a bomb threat call checklist of the type illustrated at the back of this pamphlet should be kept nearby. In addition, it is always desirable that more than one person listen in on the call. To do this, have a covert signalling system, perhaps a coded buzzer signal to a second reception point. A calm response to the bomb threat could result in getting additional information. This is especially true if the caller wishes to avoid injuries or deaths. If told that the building is occupied or cannot be evacuated in time, the bomber may be willing to give more specific information on the bomb's location.

10. Organize and train an evacuation unit consisting of key management personnel. The organization and training of this unit should be coordinated with other tenants of the building.

a) The evacuation unit should be trained on how to evacuate the building during a bomb threat. You should consider priority of evacuation, i.e., evacuation by floor level. Evacuate the floor levels above the danger area in order to remove those persons from danger as quickly as possible. Training in this type of evacuation should be available from police, fire, or other units within the community.

b) You may also train the evacuation unit in search techniques, or you may prefer a separate search unit. Volunteer personnel should be solicited for this function. Assignment of search wardens and team leaders can be employed. To be proficient in searching the building, search personnel must be thoroughly familiar with all hallways, restrooms, false ceiling areas, and every location in the building where an explosive or incendiary device may be concealed. When the police or firefighters arrive at the building, if they have not previously reconnoitered the building, the contents and the floor plan will be strange to them. Thus, it is extremely important that the evacuation or search unit be thoroughly trained and familiar with the floor plan of the building and immediate outside areas. When the room or particular facility is searched it should be marked, or the room should be sealed with a piece of tape and reported to the group supervisor.

c) The evacuation or search unit should be trained only in evacuation and search techniques and not in the techniques of neutralizing, removing, or otherwise having contact with the device. If a device is located it should not be disturbed but a string or paper tape may be run from the device location to a safe distance and used later as a guide to the device.

CHECKLIST WHEN YOU RECEIVE A BOMB THREAT

Time and Date Reported: _____

How Reported: _____

Exact Words of Caller: _____

_____

_____

Questions to Ask: _____

1. When is bomb going to explode? _____
2. Where is bomb right now? _____
3. What kind of bomb is it? _____
4. What does it look like? _____
5. Why did you place the bomb? _____
6. Where are you calling from? _____

Description of Callers Voice: _____

Male _____ Female _____ Young _____ Middle Age _____ Old _____ Accent _____

Tone of Voice _____ Background Noise _____ Is voice familiar? _____

If so, who did it sound like? _____

_____

Other voice characteristics: _____

Time Caller Hung Up: _____ Remarks: _____

_____

_____

Name, Address, Telephone of Recipient: _____

_____

_____

Figure 13–1. Bomb threat checklist.

When a Bomb Threat Is Called In

1. Keep the caller on the lines as long as possible. Ask the caller to repeat the message. Record every word spoken by the person. Figure 13-1 is a form which can be completed during the phone call to record all vital information.
2. If the caller does not indicate the location of the bomb or the time of possible detonation, you should ask for this information.
3. Inform the caller that the building is occupied and the detonation of a bomb could result in death or serious injury to many innocent people.
4. Pay particular attention to peculiar background noises such as motors running, background music, and any other noise which may give a clue as to the location of the caller.
5. Listen closely to the voice (male, female), voice quality (calm, excited), accents, and speech impediments. Immediately after the caller hangs up, you should report to the person designated by management to receive such information.

Emergency Telephone Numbers

Police

Sheriff

Fire

Ambulance

Telephone Security Department

Bureau of Alcohol, Tobacco and Firearms

FBI

Army Explo. Ord. Disp. (EOD)

Civilian Defense Unit

Figure 13-2. Emergency telephone number list.

Since the law enforcement personnel will want to talk first-hand with the person who received the call, he or she should remain available until they appear.
6. Report this information immediately to the police department, fire department. ATF, FBI, and other appropriate agencies. The sequence of notification should have been established during coordination in item 1 above.

Written Threats

While written messages are usually associated with generalized threats and extortion attempts, a written warning of a specific device may occasionally be received. It should never be ignored. With the growing use of voice print identification techniques to identify and convict telephone callers, there may well be an increase in the use of written warnings and calls to third parties.

Save all materials, including any envelope or container. Once the message is recognized as a bomb threat, further unnecessary handling should be avoided. Every possible effort must be made to retain evidence such as fingerprints, handwriting or typewriting, paper, and postal marks which are essential to tracing the threat and identifying the writer.

Bomb Search Techniques

1. Do not touch a strange or suspicious object. Its location and description should be reported to the person designated to receive this information.
2. The removal and disarming of a bomb or suspicious object must be left to the professionals in explosive ordnance disposal. Who these professionals are and how to contact them for assistance is something that you should include in any bomb threat plan.
3. All requests for assistance should be directed to one or more emergency numbers. See Figure 13–2. Be sure that the telephone numbers for these agencies are included in your plan.
4. If the danger zone is located, the area should be blocked off or barricaded with a clear zone of three hundred feet until the object has been removed or disarmed.
5. During the search of the building, a rapid two-way communication system is of utmost importance. Such a system can be readily established through existing telephones. CAUTION—the use of radios during the search can be dangerous. The radio transmission energy can cause premature detonation of an electric initiator (blasting cap).
6. The signal for evacuating the building in the event of a bomb threat *should not* be the same as that for a fire. In the bomb threat, where possible, all doors and windows should be opened to permit the blast wave to escape in the event of an explosion. Also, evacuation routes will have to be determined if a bomb is found so as to lead people away from the bomb.
7. If the building is evacuated, controls must be established immediately to prevent unauthorized access to the building. These controls may have to be provided by management. If proper coordination has been effected with the local police and other agencies, these may assist in establishing controls to prevent reentry into the building until the danger has passed.
8. Evacuate the persons to a safe distance away from the building to protect them against debris and other flying objects if there is an explosion. If the building is evacuated, all gas and fuel lines should be cut off at the main valve. All electrical equipment should be turned off prior to evacuation. The decision to cut off electrical power at the main switch should be made by management with consideration given to lighting requirements for search teams.
9. During the search, the medical personnel of the building should be alerted to stand by in case of an accident caused by an explosion of the device.
10. Fire brigade personnel should be alerted to stand by to operate fire extinguishers.
11. Preemergency plans should include a temporary relocation in the event the bomb threat materializes and the building is determined to be unsafe.

Room Search

The following technique is based on use of a two-person searching team. There are many minor variations possible in searching a room. The following contains only the basic techniques.

First Team Action

When the two-person search team enters the room to be searched, they should first move to various parts of the room and stand quietly, with their eyes shut, and listen for a clock-work device. Frequently, a clock-work mechanism can be quickly detected without use of special equipment. Even if no clock-work mechanism is detected, the team is now aware of the background noise level within the room itself.

Background noise or transferred sound is always disturbing during a building search. In searching a building, if a ticking sound is heard but cannot be located, one might become unnerved. The ticking sound may come from an unbalanced air conditioner fan several floors away or from a dripping sink down the hall. Sound will transfer through air-conditioning ducts, along water pipes and through walls, etc. One of the worst types of buildings to work in is one that has steam or water heat. This type of building will constantly thump, crack, chatter, and tick due to the movement of the steam or hot water through the pipes and the expansion and contraction of the pipes. Background noise may also be outside traffic sounds, rain, wind, etc.

Second Team Action

The person in charge of the room searching team should look around the room and determine how the room is to be divided for searching and to what height the first searching sweep should extend. The first searching sweep will cover all items resting on the floor up to the selected height.

Dividing the Room. You should divide the room into two equal parts or as nearly equal as possible. This equal division should be based on the number and type of objects in the room to be searched, not the size of the room. An imaginary line is then drawn between two objects in the room, i.e., the edge of the window on the north wall to the floor lamp on the south wall.

Selection of First Searching Height. Look at the furniture or objects in the room and determine the average height of the majority of items resting on the floor. In an average room this height usually includes table or desk tops, chair backs, etc. The first searching height usually covers the items in the room up to hip height.

First Room Searching Sweep

After the room has been divided and a searching height has been selected, both searchers go to one end of the room division line and start from a back-to-back position. This is the starting point, and the same point will be used on each successive sweep. Each person now starts searching around the room, working toward the other, checking all items resting on the floor around the wall area of the room. When the two meet, they will have completed a *wall sweep* and should then work together and check all items in the middle of the room up to the selected hip height. Don't forget to check the floor under the rugs. This first searching sweep should also include those items which may be mounted on or in the walls, such as air-conditioning ducts, baseboard heaters, and built-in wall cupboards, if these fixtures are below hip height. The first searching sweep usually consumes the most time and effort. During all searching sweeps, use the electronic or medical stethoscope on walls, furniture items, and floors.

Second Room Searching Sweep

The person in charge again looks at the furniture or objects in the room and determines the height of the second searching sweep. This height is usually from the hip to the chin or top of the head. The two people return to the starting point and repeat the searching techniques at the second selected searching height. This sweep usually covers pictures hanging on the walls, built-in bookcases, and tall table lamps.

Third Room Searching Sweep

When the second searching sweep is completed, the person in charge again determines the next searching height, usually from the chin or the top of the head up to the ceiling. The third sweep is then made. This sweep usually covers high mounted air-conditioning ducts, and hanging light fixtures.

Fourth Room Searching Sweep

If the room has a false or suspended ceiling, the fourth sweep involves investigation of this area. Check flush or ceiling-mounted light fixtures, air-conditioning or ventilation ducts, sound or speaker systems, electrical wiring, structural frame members, etc.

Have a sign or marker posted indicating "Search Completed" conspicuously in the area. Use a piece of colored scotch tape across the door and door jamb approximately two feet above floor level if the use of signs is not practical.

The room searching technique can be expanded. The same basic technique can be used to search a convention hall or airport terminal.

Restated, to search an area you should:

1. Divide the area and select a search height;
2. Start from the bottom and work up;
3. Start back-to-back and work toward each other; and
4. Go around the walls then into the center of the room.

Encourage the use of common sense or logic in searching. If a guest speaker at a convention has been threatened, common sense would indicate searching the speaker's platform and microphones first, but always return to the searching technique. Do not rely on random or spot checking of only logical target areas. The bomber may not be a logical person.

(For comparison of search systems, see Figure 13–3.)

Suspicious Object Located

It is imperative that personnel involved in the search be instructed that their mission is only to search for and report suspicious objects, not to move, jar or touch the object or anything attached thereto. The removal/disarming of a bomb must be left to the professionals in explosive ordnance disposal. Remember that bombs and explosives are made to explode, and there are no absolutely safe methods of handling them.

1. Report the location and an accurate description of the object to the appropriate warden. This information is relayed immediately to the control center who will call police, fire department, and rescue squad. These officers should be met and escorted to the scene.
2. Place sandbags or mattresses, not metal shield plates, around the object. Do not attempt to cover the object.
3. Identify the danger area, and block it off with a clear zone of at least 300 feet—include area below and above the object.
4. Check to see that all doors and windows are open

	SEARCH SYSTEMS			
	SEARCH BY: Supervisors	ADVANTAGES	DISADVANTAGES	THOROUGHNESS
S U P E R V I S O R Y	BEST for Covert search POOR for thoroughness POOR for morale if detected	1. Covert 2. Fairly rapid 3. Loss of working time of supervisor only	1. Unfamiliarity with many areas. 2. Will not look in dirty places 3. Covert search is difficult to maintain 4. Generally results in search of obvious areas, *not* hard to reach ones 5. Violation of privacy problems 6. Danger to unevacuated workers	50-65%
O C C U P A N T	SEARCH BY: Occupants BEST for speed of search GOOD for thoroughness GOOD for morale (with confidence in training given beforehand)	1. Rapid 2. No privacy violation problem 3. Loss of work time for shorter period of time than for evacuation 4. Personal concern for own safety leads to good search 5. Personnel conducting search are familiar with area	1. Requires training of entire work force 2. Requires several practical training exercises 3. Danger to unevacuated workers	80-90%
T E A M	SEARCH BY: Trained Team BEST for safety BEST for thoroughness BEST for morale POOR for lost work time	1. Thorough 2. No danger to workers who have been evacuated 3. Workers feel company cares for their safety	1. Loss of production time 2. Very slow operation 3. Requires comprehensive training and practice 4. Privacy violation problems	90-100%

Figure 13–3. Comparison of search systems.

to minimize primary damage from blast and secondary damage from fragmentation.
5. Evacuate the building.
6. Do not permit reentry into the building until the device has been removed/disarmed, and the building declared safe for reentry.

Part of your responsibility to the public necessitates maintaining good public relations. This responsibility, however, also includes the safety and protection of the public. We may well be approaching the point, when in the interest of security and protection of people, some inconvenience may have to be imposed on persons visiting public buildings.

Perhaps entrances and exits can be modified with a minimal expenditure to channel all personnel through someone at a registration desk. Personnel entering the building would be required to sign a register showing the name and room number of the person whom they wish to visit. Employees at these registration desks could contact the person to be visited and advise her or him that a visitor, by name, is in the lobby. The person to be visited may, in the interest of security and protection, decide to come to the lobby to meet this individual to ascertain that the purpose of the visit is valid and official. A system for signing out when the individual departs could be integrated into this procedure. There is no question that such a procedure would result in many complaints from the public. If it were explained to the visitor by the person at the registration desk that these procedures were implemented in the best interest and safety, complaints would be reduced.

The following are other factors for consideration:

1. During the inspection of the building, particular attention should be given to such areas as elevator shafts, all ceiling areas, rest rooms, access doors, crawl space, and other areas which are used as a means of immediate access to plumbing fixtures, electrical fixtures and the like, utility and other closet areas, areas under stairwells, boiler (furnace) rooms, flammable storage areas, main switches and valves, e.g., electric, gas, and fuel, indoor trash receptacles, record storage areas, mail rooms, ceiling lights with easily removable panels, and fire hose racks. While this list of areas to be noted with particular emphasis is not complete, it is sufficient to give an idea of those areas where a time-delayed explosive or an incerdiary device might be concealed.
2. Establish and enforce strict procedures for control and inspection of packages and material going into critical areas.
3. Develop and enforce a positive means of identifying and controlling personnel who are authorized access to critical areas and denying access to unauthorized personnel.
4. Instruct all security and maintenance personnel to be alert for suspicious looking and acting people. All personnel should be alert for foreign or suspicious objects, items, or parcels which do not appear to belong in the area where such items or parcels are observed.
5. Instruct all security and maintenance personnel to increase surveillance of all rest rooms, stairwells, areas under stairwells, and other areas of the building to insure that unauthorized personnel are not hiding in or reconnoitering these areas.
6. Insure that doors and/or access ways to such areas as boiler rooms, mail rooms, computer areas, switchboards, elevator machine rooms, and utility closets are securely locked when not in use.
7. Check key control procedures to see that all keys to all locks are accounted for. If keys are in possession of persons no longer employed at the facility, or keys cannot be accounted for—*locks should be changed.*
8. Check fire exits to be sure they are not obstructed.
9. Check fire hose racks and fire extinguishers regularly to assure they have not been tampered with, i.e., hoses cut or exposed to acid, nozzles damaged.
10. Increase patrols/surveillance of receiving and shipping areas, garage and parking areas.
11. Assure adequate protection for classified documents, proprietary information, and other records essential to the operation of your plant. (A well planted, properly charged devide could, upon detonation, destroy records which are vital to day-to-day operations.)
12. Check perimeter fences/walls/barriers to assure a good state of maintenance and adequate clear zones. Post with "No Trespassing" signs.
13. Check all exterior and protective lighting for proper operation and adequate illumination.
14. Protect ground floor windows with heavy mesh, grill work, or protective glass.
15. Conduct daily check for good housekeeping and proper disposal or protection of combustible material.
16. Have on hand, or arrange for immediate procurement of, sand, sand bags, or mattresses to be used as shielding in the event an explosive device is located in the building.

230 LOSS PREVENTION PLANNING AND CONTROLS

17. Have flashlights or battery-operated lanterns on hand, in the event electric power is cut off.
18. Install closed circuit television to monitor areas where a bomb might be placed.
19. Install metal detecting devices.
20. Post signs indicating the use of closed circuit televisions and other detection devices.

Letter and Package Bombs

Background

Letter and package bombs are not new. While the latest incidents have involved political terrorism, such bombs are made for a wide variety of motives. The particular form of these bombs varies in size, shape and components. They may have electric, non-electric, or other sophisticated firing systems.

Precautions

Mail handlers should be alert to recognize suspicious looking items. Mail should be separated into personal and business categories. Although there is no approved, standard detection method, the following precautions are suggested*:

1. *Place of origin.* Note the delivery's postmark. This may be from a country directing a terrorist campaign or from an area where such mail devices are popularly associated. If the arrival of such mail is uncommon, it should be treated as suspect.
2. *Sender's writing.* Any mail should be treated with caution if it features an unusual style of writing, not normally received, on the address. This should be considered in relation to No. 1, above.
3. *Balance.* Any letter should be treated as suspect if it is unbalanced, has loose contents, or is heavier on one side than the other.
4. *Weight.* If a package or letter seems excessively heavy for its volume, it should be treated as suspect.
5. *Feel.* If an envelope has any feeling of springiness at the top, bottom or sides, but it does not bend or flex, this is a key sign of an explo-

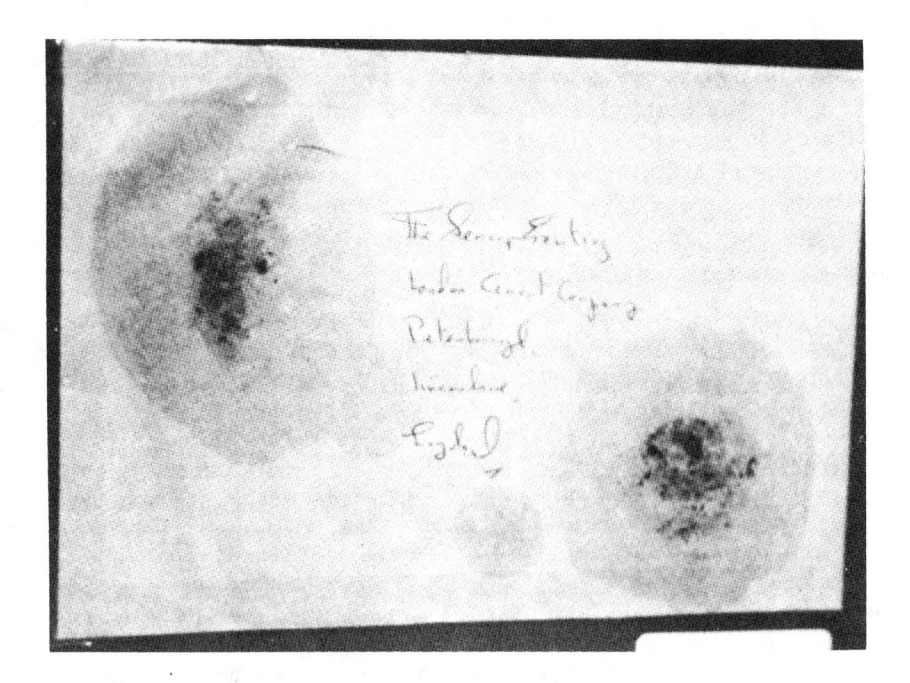

Figure 13–4. Example of a letter bomb. Greasy patches on envelope and tape all around edges are clues to concealed explosives. (From Graham Knowles, *Bomb Security Guide* (Stoneham, MA: Butterworths, 1976).)

---

*From *Bomb Security Guide*, by Graham Knowles (Stoneham, MA, Butterworths, 1976).

sive device. WARNING—EXAMINE MAIL GENTLY!
6. *Protruding wires.* Mail devices are often loosened or damaged by rough mail service handling. It is possible that fuse or electrical wires and connections will become loose and penetrate the wrapping or envelope. Any such device is unstable and highly dangerous. It must not be touched.
7. *Holes in the envelope or wrapping.* An explosive mail device which has been handled roughly may show wire or spring holes in its outer wrapping. This, by itself or in combination with the other clues described, should alert mail handlers to a suspect device.
8. *Grease marks.* Certain types of explosives leave greasy black marks on paper, a certain clue to the presence of a suspect device. It can also mean that the explosive has become old and unstable, making it very dangerous. (See Figure 13–4.)
9. *Smell.* A smell suggestive of almonds or marzipan, or any other strange smell, is a good clue to a suspect device.
10. *Unrequested deliveries.* Unrequested deliveries, especially packages, should be screened and treated with caution. A book or other thick brochure discovered upon opening a delivery should be reported to the security department or examined for any of the above clues. Any mail over which there is the slightest suspicion should *not be handled.* Remember the catch phrase—TOUCHING TRIGGERS TRAGEDY. (It is of benefit to both parties if senders place their name, organization address and telephone number on packets. Then, in case of suspicion, they may be contacted for an explanation of the contents.)
11. *Suspicious packaging.* If an envelope is taped down all around, instead of having a normal opening flap, it may contain a booby trap spring. Such letters should be handled very carefully and treated as suspect.
12. *Letter stiffness.* Feeling *very gently* should reveal whether an envelope contains folded paper or a device. The presence of stiff cardboard, metal, or plastic should alert the handler to a possible suspect device.
13. *Inner enclosures.* If, after opening a letter or package, the mail handler encounters an inner sealed enclosure—whether or not it fits any of the above descriptions—the item should be treated as suspect.

If You Have a Suspicious Looking Letter or Package

1. Do not try to open it.
2. Isolate it and evacuate everyone in the vicinity to a safe distance.
3. Notify local police and await their arrival.

Buildings—Their Problems

The physical construction of buildings and their surrounding areas varies widely. Following are a few of the problems search teams will encounter.

Outside Areas

When you search outside areas, pay particular attention to street drainage systems, manholes in the street and in the sidewalk. Thoroughly check trash receptacles, garbage cans, dumpsters, and incinerators. Check parked cars and trucks. Check mailboxes if there is a history of placement in your area.

Schools

School bombings are usually directed to nonstudent areas. Find out which teachers or staff members are unpopular and where they work. The problem areas in schools are student lockers and the chemistry laboratory.

Student lockers are locked; no accurate record of the combinations is available because students change lockers at will. Every other locker seems to "tick." Alarm clocks, wrist watches, leaking thermos jugs and white mice, all make "ticking" sounds. Have the school authorities or police cut off the locks; then search teams should open the lockers. If you cut off the lock, though, you may end up paying for it.

Chemistry labs should be treated with caution. Each year some student tries to make an explosive mixture or rocket fuel in the classroom, gets scared, and phones in a bomb call. The best procedure is to get the chemistry teacher to inspect the classroom, lab and chemical storage area with you. The teacher will know 90 percent of the items in the lab; that leaves only 10 percent to worry about.

If repeated bomb threats are received at schools in your area, recommend that the school board hold makeup classes on Saturday. This tends to cut down the number of bomb scares.

Office Buildings

The biggest problem in office buildings is many locked desks. A repair of desk locks is an expensive item. There will be many other items to keep you busy, such as filing cabinets, storage closets, and wall lockers. Watch out for the company's security system if they deal in fashions of any type, the automotive or aircraft industry, defense contracts, or the toy industry. Electrical leads, electrical tapes, electrical eyes, electrical pressure mats, electrical microswitches, will all ring those huge bells that no one knows how to turn off.

Auditoriums, Amphitheaters, and Convention Halls

Here, thousands of seats must be checked on hands and knees. Look for cut or unfastened seats with a bomb inserted into the cushion or back. Check out the stage area which has tons of equipment in it, the speaker's platform, and the microphones. The area under the stage generally has crawlways, tunnels, trap-doors, dressing rooms, and storage areas. The sound system is extensive and the air-conditioning system is unbelievable. The entire roof area, in a theater, frequently has one huge storage room and maintenance area above it. Check all hanging decorations and lighting fixtures.

Airport Terminals

This structure combines all problems covered under schools, office buildings, and auditoriums, plus outside areas and aircraft.

Aircraft

The complexities of aircraft design make it unlikely that even the trained searcher will locate any but the most obvious explosive or incendiary device. Thus, detailed searches of large aircraft must be conducted by maintenance and crew personnel who are entirely familiar with the construction and equipment of the plane. In emergency situations where searches must be conducted by public safety personnel without the aid of aircraft specialists, the following general procedures should be used:

1. Evacuate the area and remove all personal property.
2. Check the area around the craft for bombs, wires, or evidence of tampering.
3. Tow the aircraft to a distant area.
4. Starting on the outside, work toward the plane's interior.
5. Begin searching at the lowest level and work up.
6. Remove freight and baggage and search cargo areas.
7. Check out rest rooms and lounges.
8. Be alert for small charges placed to rupture the pressure hull or cut control cables. The control cables usually run underneath the center aisle.
9. With special attention to refuse disposal containers, check food preparation and service areas.
10. Search large cabin areas in two sweeps.
11. Check the flight deck.
12. Simultaneously, search the baggage and freight in a safe area under the supervision of airline personnel. If passengers are asked to come forward to identify and open their baggage for inspection, it may be possible to quickly focus in upon unclaimed baggage.

Elevator Wells and Shafts

Elevator wells are usually one to three feet deep with grease, dirt and trash and must be probed by hand. To check elevator shafts, get on the top of the car with two six-volt lanterns, move the car up a floor (or part of a floor) at a time and look around the shaft. Be prepared to find nooks, closets, storage rooms, false panels, walk areas, and hundreds of empty whiskey bottles in paper bags. Don't forget that as you go up, the counterweights are coming down—check them, too. The elevator machinery is generally located on the roof. A word of caution: Watch for strong winds in the elevator shaft. Don't stand near the edge of the car.

Handling the News Media

It is of paramount importance that all inquiries by the news media be directed to one person appointed as speaker. All other persons should be instructed not to discuss the situation with outsiders, especially the news media.

The purpose of this provision is to furnish the news media with accurate information and see that additional bomb threat calls are not precipitated by irresponsible statements from uninformed sources.

Chapter 14
Applicant Screening*

PHILIP P. PURPURA

This chapter focuses on one of the most important assets of an organization—its personnel. Job applicant screening is the process of obtaining the most appropriate individual for a particular job. This can be accomplished through interviews and testing, among other methods. Screening is a major loss prevention tactic.

Before specific applicant screening strategies are pointed out, two important areas are elaborated upon to assist the reader in understanding present-day applicant screening. These concerns are (1) the polygraph and the psychological stress evaluator (PSE) and (2) laws affecting the employment relationship and the screening and hiring process. A sound knowledge of these topics is vital for loss prevention practitioners, personnel administrators, and others involved in screening and hiring. In fact, new laws and court decisions are constantly changing this process. Legal assistance is a wise request when a personnel administrator or loss prevention practitioner is implementing or modifying a screening and hiring program.

The Polygraph and the Psychological Stress Evaluator (PSE)

There is an obvious need for truth verification devices; however, at present, these devices can use improvements in instrumentation, regulation, examiner's training, and research. Briefly, the polygraph and PSE can measure stress in the human body resulting from attempts at deception. The polygraph measures changes in the nervous and respiratory systems. The PSE measures changes in the inaudible characteristics of the human voice.

*From *Security and Loss Prevention*, by Philip P. Purpura (Stoneham, MA: Butterworths, 1984).

The History of These Devices

In 1895, Cesare Lombroso used the first scientific instrument to detect deception. Referred to as the hydrosphygmograph, it measured changes in pulse and blood pressure when suspects were questioned by police. Vittorio Benussi, in 1914, successfully detected deception with the use of a pneumograph that measured inhalation and exhalation. A few years later William Marston conducted research with the sphygmomanometer, which analyzed blood pressure during questioning. Marston is also credited with helping to develop the pneumograph and galvanometer; the former analyzes breathing changes while the latter analyzes skin electrical resistance changes. In 1921, a psychiatrist, Dr. John A. Larson, developed the polygraph. His original machine measured blood pressure, respiration, and pulse. By 1949, Leonard Keeler added galvanic skin response (electrical changes) measurement and developed the Keeler Polygraph.

The PSE is considerably newer than the polygraph. Many say that this newness has caused its limited acceptance. In 1964, Robert McQuiston, Allan Bell, and Wilson Ford developed the PSE for U.S. Army intelligence, but it was rejected by the Army. In the 1970s McQuiston patented a civilian version and its usage increased in the private sector.

How These Devices Work

The proper name for the present-day polygraph is the Pneumo-Cardio-Sphygma Psychogalvanic Reflex. Usually when a person tells a lie while being polygraphed, physiological changes occur because of the fear of being exposed. Consequently, the blood pressure, breathing, and galvanic skin response change. The actual painless hookup of an individual to the machine involves a blood pressure cuff attached to the upper arm. This also measures pulse. A rubber tube placed around the chest area measures breathing. Electrodes are connected to a hand for the galvanic (electrical) skin response changes. While questions are being asked during the examination, bodily changes are recorded on graph paper. This record corresponds to the questions, and the examiner can see which questions caused considerable bodily changes. Interspersed between important questions are less important questions such as name and address. The purpose is to compare changes in bodily reactions to more important questions. The examination usually takes less than an hour, but it depends on the number of questions.

One argument over the polygraph is whether or not an individual can fool this device. Many say that fooling the polygraph is difficult, but it can be done. Because the polygraph records changes in bodily functions, it is possible to produce inconclusive results by causing bodily changes. This sometimes can be accomplished by thinking of sex, crossing the eyes, tensing muscles, biting the tongue, or pressing a toe on a thumbtack previously hidden in a shoe.

The PSE does not require being hooked up to a subject. There are a few variations of the PSE now on the market. One type records the voice and then is played back at a reduced speed while a traveling pen moves over a data strip. Another type instantaneously produces a readout of the stress levels in the voice. More specifically, the PSE is a spectrum analyzer that measures frequencies of voice stress. The examiner studies the relationship of specific questions to the level of stress when each question was answered. This process is similar to the polygraph examination, but different characteristics of the subject are analyzed. Furthermore, as with the polygraph, the PSE does not measure truth or deception. An evaluation is made indirectly from the examiner's interpretation of the subject's responses to questioning. With this in mind, there is a general consensus that truth verification depends on the training and experience of the examiner.

The Types of Training Required

Unfortunately, there are no uniform, enforceable standards for polygraph examiners. The American Polygraph Association (APA) accepts members only after graduation from 1 of 23 polygraph schools accredited by them. Since only 23 states have licensing laws for examiners, there are operators in other states who call themselves "experts" and possess limited, if any, training. The APA requires members to have a college degree. Law enforcement officers are required to have two years of college. The polygraph schools regulated by the APA must comply with stringent rules. A minimum of 250 hours plus field work is required. The John E. Reid school in Chicago requires six months of training besides a college degree upon entrance. This school offers a master's degree for graduates.

The International Society of Stress Analysts has over 300 members and requires graduation from one of three PSE schools. These schools offer programs ranging between 120 and 150 hours long. Regulation is a problem for schools not part of the International Society. Consequently, other PSE operators are being trained in less time.

The Problems and Criticisms of Each Test

Besides weaknesses in some polygraph training programs, there are other problems. For instance, several techniques for using the polygraph are being taught. If one uniform technique were used there would be less confusion, especially during court testimony. Another problem is the disputes between private and public sector police. For example, the accuracy of examinations by each group is, at times, criticized by the other.

A major criticism directed at the PSE is inadequate training, especially since PSE training programs are often of shorter duration than those of polygraph programs. Regulation and uniformity are also a problem, as with polygraph usage. Many feel that because the PSE measures only one bodily function, it is not as good as the polygraph. In addition, many believe that because the PSE has not been in existence for very long, it has not yet proven itself. More research is needed. Another criticism is that the PSE can be used on an individual without his/her knowledge. Rights groups, besides polygraph advocates, have attacked the PSE on this issue. A major criticism of both devices is that they invade individual privacy.

The Validity of the Results of Each Test

This question has generated considerable controversy. However, because of continued research, more is being learned about these devices. Research on the polygraph has been extensive. One of the more popular studies was conducted at the University of Utah where researchers concluded that the polygraph can be over 90 percent accurate in detecting truth or deception in criminal cases.[1] Other studies have shown similar findings. These positive results from scientific studies have helped to eliminate negative attitudes toward the polygraph.

Considerable criticism has been leveled at the PSE. Since the development of the PSE, several validation studies have been conducted without any degree of scientific vigor. In addition, testimonials have been downgraded by PSE's detractors. Conversely, some researchers have supported the accuracy of the PSE—to 90 percent.[2] Before both PSE researchers and examiners can justify PSE accuracy, additional scientific research is necessary.

The Extent and Limitations of Usage

Anywhere from 200,000 to 500,000 polygraph tests are estimated as being administered yearly in the private sector.[3] This figure does not include the thousands of law enforcement agencies that use the polygraph. The polygraph serves useful purposes in aiding criminal investigations and in screening job applicants in both the private and public sectors. Furthermore, both insurance and bonding companies have used the polygraph to screen certain claims; this has saved these companies money that would have been paid for false claims. Attorneys sometimes use polygraph tests to verify prospective clients' stories before accepting cases.

Besides the extensive use of the polygraph, there is a continuing controversy about it. On the one hand, business people feel that the test is low in cost ($40 to $100 per test) and quick. Valuable time is saved during either background or criminal investigations. Because billions of dollars per year are lost due to employee crimes, many business people feel that polygraph tests are necessary for their survival. But there is a movement in the U.S. Congress to ban the use of the polygraph in employment settings. The APA, the examiners' professional association, is battling the proposed ban. The American Civil Liberties Union (ACLU) and the AFL–CIO are opposed to the polygraph. Former Senator Sam Ervin (D–NC), who during the Watergate hearings referred to the polygraph as "20th Century witchcraft," introduced a number of bills in Congress that would have halted polygraph usage by industry. These bills failed, but other senators are working on similar bills. Opponents of the polygraph feel that these tests are an invasion of privacy and that traditional methods of applicant screening are adequate.

John A. Belt and Peter B. Holden from Wichita State University conducted a study of "Polygraph Usage Among Major U.S. Corporations."[4] Of the 400 major corporations surveyed, the response rate was 143 or 35.7 percent. Twenty percent of the respondents signified that the polygraph was used in the firm's personnel program. Usage was most common among commercial banks and retail companies. The most frequent use of the polygraph was to verify employment applications; next, to periodically survey employee honesty and loyalty; and next, during criminal investigations. The firms that did not use the polygraph did so mostly because it was considered unnecessary and inappropriate in the business setting and because of reluctance due to the legal implications.

Additional research is needed on the use of the polygraph and the PSE. Independent university researchers are a source of high-quality research.

State laws affecting the use of the polygraph and PSE are varied. It is an up-hill struggle for acceptance for both devices. Note this smattering of reports:

> Decisions in the State courts indicate polygraph results have been admitted in 36 States, not admitted in 13 States, and no case of record in 1 State. Of the 36 States that have admitted results, 22 limited the admissions to stipulated results.[5]

The PSE has been admitted in court in Maryland, Florida, California, Louisiana and West Virginia. The technique has been banned in Illinois and other states and the PSE forces blame the polygraph forces in that state for its banishment. ... It has taken the polygraph profession 50 years to break the ban in court. PSE forces are only starting.[6]

Besides states prohibiting polygraph and PSE usage, these devices are not used in certain industries because of union contracts. Legislation in some states stipulates that no job applicant may be denied a job nor an employee fired for refusal to submit to a polygraph test. About 23 states require licensing of polygraph examiners; regulation is also expected for PSE examiners.[7]

Laws Affecting the Employment Relationship and the Screening and Hiring Process

A competent personnel administrator has no choice but to maintain an up-to-date knowledge of laws affecting the employment relationship. This task is not easy, especially since most personnel administrators do not have a law degree and changes are taking place rapidly. Legal assistance is periodically necessary. Other aids can include workshops or labor reports such as the *Retail/Services Labor Report* (Bureau of National Affairs, Inc., 1231 25th St., NW, Washington, DC 20037).

The major laws of concern can be divided into four categories.

TYPE OF LAW	EXAMPLE(S)	CHARACTERISTICS
Antidiscrimination	The Equal Employment Opportunity Act of 1972, amending Title VII of the Civil Rights Act of 1964	Prohibits discrimination in all matters relating to employment because of race, color, religion, national origin, or sex.
Wages, benefits, and working conditions	Federal Insurance Contribution Act of 1937	Social Security
	Occupational Safety and Health Act of 1970	Attempts to ensure a safe and healthy work environment.
Privacy	Fair Credit Reporting Act of 1971	Protects a person from misuse of personal information (e.g., credit information that may jeopardize employment opportunities).
	State laws affecting polygraph usage	See previous discussion.
Miscellaneous	National Labor Relations Act	Controls the relations between management and labor.
	State criminal laws	Covers crimes (e.g., larceny, committed by an employee).

The above laws, which comprise a brief outline, illustrate the loss prevention practitioner's overlapping responsibility with personnel administrators. Each organization differs, as do individual responsibilities. State criminal and polygraph laws and OSHA laws are obvious areas of concern for loss prevention practitioners.

The following information emphasizes laws pertaining to the screening and hiring process. However, at times, laws pertaining to the employment relationship and laws of interest to loss prevention practitioners are also pointed out. *Knowledge of these laws can prevent government investigation, litigation, and losses.*

Antidiscrimination

The Civil Rights Act of 1964 originally applied to the federal government and companies doing business with the federal government. Later, many state legislatures passed similar laws which became applicable to both local governments and businesses. In essence, Title VII of this act states that it is unlawful for an employer to discriminate against an individual in hiring, firing, or compensation because of an individual's race, color, religion, sex, or national origin. In 1972, Congress expanded the coverage of Title VII to most government activities. In addition, businesses involved in government contracts or work were to actively seek minorities for hiring and/or promotion. Employers not complying with affirmative action are subject to federal court suits entered by the Equal Employment Opportunity Commission (EEOC), which enforces Title VII. Affirmative action is a policy whereby an employer ceases to discriminate against particular racial, ethnic, religious, or sexual groups, and also tries to undo damage caused by past discrimination by actively seeking out and assisting those subjected to discrimination. Affirmative action programs stipulate that standards (e.g., testing) for selection or promotion must be job-related. Because of this stipulation, many employers mistakenly curtailed their testing programs for hiring and promoting. The famous case of *Griggs v. Duke Power Co.*, 915 Sup. Ct. 849 (1971), was also in the minds of many employers. In 1968, several employees of the Duke Power Company in North Carolina were given a pencil and paper aptitude test for manual labor. Willie Griggs and twelve other black workers sued their employer with the charge of job discrimination under the Civil Rights Act of 1964. Their contention was that the pencil and paper aptitude test had little to do with their ability to perform manual labor. The Supreme Court took this case and decided that a test is inherently discriminatory if it is not job-related and differentiates on the basis of race, sex, or religion.

As stated above, many employers curtailed testing because of fears of discrimination, litigation, and losses. However, the federal government's *Uniform Guidelines on Employee Selection Procedures* does not oppose testing. The federal government is opposed to any discriminatory selection procedure such as testing that is not job related. As these guidelines became clearer, many employers who previously feared testing began to test again.

Lester Mood, a consultant on affirmative action and EEO, discusses "Coping with Anti-discrimination Laws" in *Administrative Management*.[8] He points out that a charge of discrimination begins with a complaint, next with a government investigation, and finally the findings determine the case. In half of the cases the findings favor the employer; in those cases that favor the employee, an attempt is made at negotiations; a very small percentage of cases result in actual legal action; employers are most concerned about the investigation because of the expense (e.g., time and resources); attorneys are rarely involved unless voluntary cooperation is not obtained.

Lester Mood divides employment discrimination charges into three categories and then recommends preventive strategies. The first is the "Evil Motive," which is the typical discriminatory action by a person against another person because of race, sex, and so forth. This charge is difficult to prove, requires substantial evidence, and damage must be shown. Proof can consist of documented racial statements and statements by witnesses. Prevention can involve written rules that prohibit racial, ethnic, and sexual language and actions. By facilitating communications, management can prevent external redress of grievances. Documentation is also vital, especially to support your contention. The second charge is "Differential Treatment," which means that a person was treated differently from a person of another class under similar circumstances. This is the most frequent charge filed; it places a burden on the respondent (employer) because the charges depend on comparative evidence; the investigator seeks "comparative evidence" as to what happened to the charging party as opposed to other employees. When employers and supervisors make decisions based upon objective criteria "Differential Treatment" can be prevented. If subjective decisions are necessary, a comparison with other workers can insure consistency. In the third charge, known as "Disparate Impact," a particular employment mechanism (e.g., test) adversely affects a racial, sexual, or ethnic group, even though all groups were treated the same. The Griggs case illustrates this type of discrimination, where employment testing was not job-related. Lester Mood describes preventive strategies as consisting of analyzing the employment criteria for validity. In other words, are the employment or test criteria related to the actual job to be performed? Is there any adverse impact? A key point is that, according to law, preemployment screening must be job-related.

Privacy

Since 1966 Congress has passed several bills outlining privacy rights. Changes are constantly taking place and employers often become confused as they attempt to keep in tune with new developments. But what is the purpose of all this privacy legislation? Essentially, these laws protect citizens' privacy from government and business. Because a number of these bills have created confusion, eight specific bills are briefly presented below to aid the reader in developing a clearer picture of privacy legislation:

Freedom of Information Act of 1966 (amended in 1974). This law allows you to obtain your records from government agencies such as the FBI, CIA, IRS, and others.

Fair Credit Reporting Act of 1971. An individual has the right to know about and correct his/her file at a credit reporting agency. If credit is denied, an explanation is necessary. Also, if a job applicant is subject to a credit bureau check, his/her written consent must be obtained.

Equal Credit Opportunity Act of 1974. This legislation essentially prohibits credit discrimination based on sex, marital status, religion, race, national origin, or receipt of welfare. This law also forbids questions that do not specifically relate to credit ratings.

Privacy Act of 1974. Federal agencies must maintain only accurate and relevant data. An individual can ask that information outside of these criteria be removed. Information is prohibited from being released unless an individual's written consent has been received. Files can be passed to other government agencies in certain criminal proceedings if approved by a court.

Tax Reform Act of 1976. The IRS is required to inform a person, three days in advance, of intentions to subpoena his/her financial records from banks or any other source. The person has two weeks to notify the financial institution, in writing, not to release records. During this period, the person may take legal action to block the IRS from acquisition of the records. The IRS is prohibited, during this period, from examining the records or taking relevant testimony unless a court order has been obtained.

Fair Debt Collection Practices Act of 1977. This law limits collection agency harassment of debtors. An agency may ask a person's employer only for that person's address.

Right to Financial Privacy Act of 1978. Under this law an individual's bank records are private property. No federal agency can examine the records without written consent. A person can go to court to challenge action to force exposure of the records. In some criminal cases, court authority can release the records.

Bankruptcy Act of 1979. This law further restricts collection agencies in locating debtors and repossessing goods. Debtor repayment requirements are also eased.

Besides these eight bills, a ninth bill is expected to pass that will add more privacy rights in insurance and credit. Several insurance companies have already made changes to comply.

Although some citizens view "privacy laws" as unnecessary, others see these laws as important for protecting individual privacy. A Louis Harris survey released in May 1979 showed that 64 percent of the people surveyed "said they feel a sense of real concern about threats to their privacy—up from 47 percent a year ago. Fifty percent of the police and 64 percent of the business leaders questioned agree. Fully 84 percent of the general public—and 71 percent of law enforcement officers—support the Privacy Act of 1974 provision that permits an individual to see and correct records about himself or herself held by federal agencies."[9]

Robert Ellis Smith, editor of *Privacy Journal*, points out in a *Security Management* article that "You *Can* Live with the Privacy Laws." Some of his interesting comments are reprinted below.[10]

> To say these privacy laws have made life more difficult for law enforcement investigators is not to say they have made life impossible ... statutes contain exemptions for "law enforcement purposes," "national security," and "emergencies." ... not a single law prohibits ... gaining access to personal information with proper legal process—a subpoena, warrant, or summons. [Exception: Census law.] ... personal information is available with the consent of the individual.
>
> In permitting individuals to see and correct information ... these new laws have surely resulted in more accurate ... recordkeeping. In all of these respects the new privacy laws have strengthened law enforcement, not weakened it.
>
> ... [P]rivacy laws provide much needed assurance to citizens. ... A law that prevents a security investigator from getting access to student records at a nearby college also protects financial information he himself may have submitted to another college so that his child could qualify for financial aid. A state law against polygraphing may prevent security personnel from testing an employee suspected of dishonesty, but it also prevents the officer's supervisor from plugging him or her into the machine.

The acquisition of information about job applicants can prevent future losses; however, carelessness in gathering the information may result in litigation losses beyond ordinary employee theft. The Fair Credit Reporting Act of 1971 stipulates guidelines for gathering information about individuals applying for employment, credit, or insurance. (State statutes may also specify guidelines.) Those involved in gathering information must comply with various stipulations when appropriate, for example, notifying the individual being investigated and obtaining his/her written permission to acquire information. Furthermore, some federal courts have ruled, upon interpretation of the privacy laws, that a person's rights were violated by the amount and type of personal information sought by a public employer.

Besides following guidelines when information is collected, other privacy requirements and court decisions relate to the release of information. Release of personal information concerning an organization's employees or customers to outsiders can result in litigation. Loss prevention personnel or other employees should not disclose personal information (orally or in writing) unless it is with the individual's consent. The consent should be documented and should specify the type of information and who will receive it. Proper guidelines should be sought from government agencies and attorneys. By following the law, criminal charges, civil liability, punitive damages, and other losses can be avoided.

The Applicant Screening Process

In most organizations, a personnel administrator is involved in equal employment opportunity, recruitment, applicant screening, selection, placement, compensation, benefits, and management-union affairs. Depending on the particular organization, additional responsibilities may include employee discipline, complaints and grievances, and health and safety. Where does the loss prevention manager fit into personnel administration? To begin with, thorough applicant screening can prevent losses. If intensive loss prevention strategies are necessary for an industry, then the loss prevention department can play a prominent role in applicant screening. This role varies depending on such factors as the cost of the manufactured product (e.g., diamonds as opposed to nails), the type of product (e.g., missile components as opposed to sewing kits), and the history of losses (e.g., a television manufacturing plant that has lost ten percent of its inventory in the last year). Another area of joint authority and responsibility is employee discipline. If an employee is caught stealing company property, then administrative action, or prosecution, may occur. Depending on the organization's setup, other areas of joint participation may be necessary (e.g., health and safety).

An additional area where the loss prevention and personnel managers must coordinate their efforts is in screening loss prevention applicants as opposed to regular applicants. Here again, this depends on the organization's needs. Many firms scrutinize loss prevention applicants more thoroughly than other applicants.

An important aid to joint activities, by the loss prevention and personnel departments, is a clear policy with procedural guidelines. In addition, an administrator who is above both departments can smooth relations, settle differences, and reduce infighting.

Today, it is clear that the EEOC regards all screening strategies or techniques as tests in that they have the potential to discriminate against the groups stipulated in the Civil Rights Act of 1964. In recent years, due to several court decisions, the screening and hiring process—including background investigations, formal interviews, tests, and the like—must be job-related and nondiscriminatory.

The primary objective of a screening process is to find the best person for a particular job. Within this process there is a need to increase the chances that the selected applicant is not a risk and will not create losses. Without a thorough screening process, strategies such as locks, alarms, access control, and so forth become almost worthless. Today, especially, there is stiff business competition and industrial espionage which adds skillful spies to the list of dishonest, dangerous, lazy, and other undesirable employees who must be weeded out.

Effective screening also reduces the expensive turnover problem. Organizations with high turnover encounter lost productivity, excessive personnel administration expenses, and lower morale. Many times, these losses and turnover can be reduced by hiring the most appropriate people to fill specific jobs. This can be facilitated by careful screening, even though good compensation, job evaluations, and merit increases are important in reducing turnover.

Because employers are often liable for their employees, the infrequent but costly problem of liability can be lessened through a good screening process. For instance, central station alarm company employees have access to sensitive information about their clients' points of protection, opening and closing times, keys, and code words. If an error is made in the hiring of an employee, and dishonesty and then litigation follow, the employer can be portrayed in a more favorable light if he/she proves that reasonable steps were taken to screen the employee.

Aids in Selecting the Best Person

First, the goals and objectives of the firm should be listed. Next, the job must be clearly defined (i.e., job description) in terms of skills, authority, responsibility, environment, and superior–subordinate relationships. Qualifications comprise the third list. This would include the required education, experience, training, and personality. With these three lists, a clearer picture of the job is evident.

A question of concern for many loss prevention positions is whether there is a greater need for a law enforcement background as opposed to a business background. Here is where reference to the above lists is helpful. The following questions also aid this topic of concern: Does the job description involve dealing with frequent criminal activity? Does the job entail the observation of accounting records and other business-related activity? Or, does the job require both a law enforcement and business background? A careful job analysis can improve selection criteria.

Steps in the Screening Process

The screening process should be carefully planned and cost effective. The planning phase consists of well-thought-out and written step-by-step procedures for screening. A key point for cost-effective screening is to conduct the most expensive part of the process last. It costs much less when a medical examination is placed at the end of the screening steps because unsuitable applicants have been eliminated. As pointed out earlier, the screening process depends on the needs of the organization. More thorough screening procedures are usually used for loss prevention personnel, management, money handlers, and employees exposed to sensitive information. In comparison, retail salespeople and blue-collar workers frequently, although unfortunately, receive superficial screening. A thorough screening process is as follows:

1. Review of application
2. Initial interview
3. Testing (e.g., knowledge, IQ, psychological)
4. Primary interview
5. Background investigation
6. Polygraph examination
7. Medical examination

The above process is only a suggestion. Depending on a multitude of viewpoints, the order and intensity of each step may vary. Furthermore, the law of particular states may eliminate polygraph usage. Antidiscrimination laws and court decisions may also affect the process, as previously stated.

Questions During the Screening Process that Have Been Ruled Discriminatory

The following information relates to questions prohibited during the entire screening process. Court rulings under EEO legislation have stressed repeatedly that questions (and tests) must be job-related.

Questions pertaining to arrest records are unlawful. An arrest does not signify guilt. The courts have stated that minorities have suffered disproportionately more arrests than others. A conviction may be solicited in a question. It is not an absolute bar to employment. Here again, minorities have disproportionately more convictions. Certain offenses can cause an employer to exclude an applicant, depending on the particular job. Thus, the question of convictions must be job-related (e.g., loss prevention) and carefully considered.

Unless a business necessity can be shown, questions concerning credit records, charge accounts, and owning your own home are discriminatory because minority groups are often poorer than others. Unless absolutely necessary for a particular job, height, weight, and other physical requirements are discriminatory against certain minority groups (e.g., Spanish, Asian, and women applicants are often physically smaller than other applicants).

Other unlawful questions, unless job-related, include asking age, sex, color or race, maiden name of applicant's wife or mother, and membership in organizations that reveal race, religion, or national origin.

The questions that can be asked of an applicant are, among others, name, address, telephone number, social security number, past experience and salary, reasons for leaving past jobs, education, convictions, U.S. citizenship, military experience in U.S. forces, and hobbies.

Testing Applicants for "High Integrity" Positions

The testing of applicants varies considerably. Sometimes applicants are not tested. For loss prevention or other positions requiring high integrity, numerous tests are used. Tests measuring knowledge or intelligence play a role in screening applicants, as long as they are job-related. These tests vary, however.

The Minnesota Multiphasic Personality Inventory (MMPI) and honesty tests are emphasized here. These two types are used frequently in the loss prevention field.

The MMPI is basically a widely used personality test. It consists of 550 statements to which a person can respond "true," "false," or "cannot say." Test statements include "I am an agent of God"; "I like tall women." The test reveals habits, fears, sexual attitudes, and symptoms of mental problems. Statements related to particular personality characteristics, like depression, are planted throughout the test. The responses to each personality characteristic are taken out of the test and organized into a scale for each characteristic. Although many MMPI statements seem like nonsense, they help to classify people's personalities. If a test taker falsifies answers, a built-in lie scale reveals this ploy. The lie scale is aided by the rephrasing of similar questions throughout the test. Many loss prevention departments and service businesses use this test, especially to determine fitness for armed duty. The test can be administered by nonprofessionals, but results must be analyzed by a psychologist.

Honesty tests give employers an opportunity to test applicant honesty. These tests are paper-and-pencil tests that measure trustworthiness and attitudes toward honesty. Many companies cannot justify the cost of a polygraph or PSE examination and use paper-and-pencil tests as a cost-effective alternative. These tests also help avoid legal problems associated with the polygraph and PSE. Various testing companies provide such service to the business community. The *Report of the Task Force on Private Security* states[11]:

> Several paper-and-pencil honesty tests were reviewed for this report; this independent evaluation determined that the tests appear to have high face validity. Several validity and reliability studies supporting such tests have been published in scholarly journals.

Legal Limitations and Areas of Inquiry for a Background Investigation

Background investigations are a vital loss prevention strategy. If an employer avoids hiring an undesirable applicant, losses likewise are avoided. Before an individual investigates an applicant's background, a thorough knowledge of applicable laws is necessary. The Federal Consumer Credit Protection Act stipulates background investigation guidelines. For instance, investigators are prohibited from requiring information concerning arrests, indictments, or convictions, where the disposition, release, or parole occurred more than seven years prior to the date of the application.

There are many areas of inquiry that can be part of a background investigation. The intensity of the investigation depends on organizational needs and the type of job. Here criminal history, past employment, personal references, education, and financial condition are emphasized.

An applicant's criminal history, if any, is a prime concern of employers, especially when the applicant is applying for a loss prevention position. As pointed out before, asking about an applicant's arrest record is unlawful. Conviction records are legally obtainable in most jurisdictions; they are usually public records on file at court offices. If an applicant appears to have no convictions, this does not necessarily mean that he/she has never committed a criminal offense. It is possible that the background investigator did not search court records in other jurisdictions where the applicant has lived. This is a prime reason why previous addresses should be solicited from the applicant. Another point is that many employers to not favor prosecution of employee thieves; this results in no conviction records. Thus, the investigator is wise to check with the applicant's past employers.

Past employment is a crucial area of inquiry because it reveals past job performance. An investigator should do this in person or by telephone. Personnel offices may be reluctant to supply information beyond a person's address. A personal visit or telephone call to the applicant's previous supervisor is often worthwhile. When possible, statements by the applicant on the application form should be verified. For instance: name, address, social security number, dates of employment, duties, name of supervisor, and conditions of departure. The investigator must look for inconsistencies.

The personal references supplied by the applicant are usually people who will make favorable comments about the applicant. If an investigator can obtain additional references from contacting listed references, more will be learned about the applicant. Again, inconsistencies must be noted.

Education can be checked out by mail as long as a copy of the applicant's authorization is enclosed. This conforms to privacy legislation. When transcripts and other educational records are received, the investigator should study characteristics (e.g., name, social security number) and look for inconsistencies.

The financial condition of the applicant may reveal the potential for theft. The Fair Credit Reporting Act of 1971 requires the applicant's written consent for a credit investigation. Remember that unless a business necessity can be shown, questions and employment decisions related to credit can be discriminatory, since certain minority groups are often poorer than others.

Since the passage of the Freedom of Information, Privacy, and Fair Credit Reporting acts, and the advancement of equal employment opportunity concepts, many employers have misunderstood these laws and curbed background investigations. In reference to background investigations, these laws stress that background checks must be based on a valid need to know. In addition, applicants being checked should be informed of the purpose of the check, what will be checked, and how the results will be used. They should know that they have the right to review the files and enter corrective information. Consent forms are necessary, signed by the applicant, permitting the investigator to collect information from individuals, government agencies, past and present employers, schools, doctors, and so on.

Background investigations can be expensive, especially in terms of personnel resources. Decisions must be made to use less expensive methods such as written inquiries, telephone calls, and credit checks, as opposed to the more expensive personal visits to the applicant's past employers, neighborhood, schools, and so forth.

Sometimes, when businesses do not have the resources or personnel to conduct adequate background investigations, they can utilize the services of a company that specializes in background investigations. For a fee, these service companies provide background information similar to that collected by a proprietary investigation.

Information Sources

An investigator can become handicapped and frustrated unless he/she knows where and how to obtain information from the billions of records on file. George O'Toole, in his book *The Private Sector*, describes a CIA employee, who investigated prospective employees, as extremely skilled in his job. Before retirement this employee was given a fellowship to write a book titled *Where's What*. O'Toole states that the guidebook lists "6,723 different record systems maintained by the federal government, which contain a total of 3.9 billion files." Also included was a listing of "countless dossiers compiled by state, county, and municipal governments."[12] In O'Toole's book, it is pointed out that there are more records maintained by the private sector than by the federal government. *Where's What* also contains information sources in the private sector.

In *Confidential Information Sources: Public and Private*, John M. Carroll lists four questions that exemplify the essence of his book and describe what is important to know before information is sought[13]:

1. What information is on record?
2. Who has custody of those records?
3. How can I or my surrogate [substitute] get it?
4. How much faith can I place in what I get?

The answers to these questions vary depending on the type of information requested. For instance, federal law and many states' laws prohibit employers from obtaining information on a job applicant's arrest records from police departments. This and other privacy legislation has hindered information acquisition by investigators while protecting individual privacy.

There are basically three types of information available to investigators: public, semipublic, and nonpublic. Public information is available to anybody. Newspapers and libraries are some of the sources of public information. Semipublic information is defined by Fuqua and Wilson (authors of *Security Investigator's Handbook*) as follows: "To be blunt, this means that they are available if you know someone who has access to them. It is not illegal to see them, but it would be illegal to trespass or force your way to them."[14] The third type, nonpublic records, are closed to outsiders; it would be illegal for the private sector investigator to possess the information.

The reality of information acquisition must be explained. I believe that no information is totally secure from unauthorized acquisition. Although difficult to measure, information is often obtained in an unethical or illegal manner; this is the reality of information acquisition. An example can be seen when a private sector investigator pays money to an employee of a bank to secure nonpublic information about an individual. This activity should not be condoned, because individual rights and laws are being violated. Organizations and employees must comply with privacy laws and administrative directives that limit information acquisition by outsiders.

The following list presents some common sources of information for the investigator. A case-by-case approach to each source of information is advised because availability varies.

Government Agencies (Police depts., motor vehicle depts., courts)
Libraries
Newspapers and periodicals
Credit bureaus
Professional and trade associations
Educational institutions
Insurance companies
Hospitals
Utility companies

Computerized information sources are another aid to the investigator. The National Crime Information Center (operated by the FBI) has a computerized databank linking federal, state, and local police. Millions of records contain information on wanted persons, stolen vehicles, license plates, guns, securities (e.g., stocks, bonds, travelers' checks, and money orders), boats, and other articles. Criminal histories are also maintained.

Another source of information is the Law Enforcement Intelligence Unit. It was created in 1956 by police agencies to combat organized crime and because of an inability to obtain enough information from the FBI. Computerized files contain information on thousands of people. Hundreds of member police departments can obtain valuable information from this unit. The unit has received some criticism because of misleading files on citizens.

In the private sector, the Equipment Theft Information Program, Inc. (Addison, IL) maintains a computerized databank of stolen industrial, construction, and farm equipment. This firm is concerned with identifying and recovering equipment.

References

1. U.S. Department of Justice, *Validity and Reliability of Detection of Deception* (Washington, DC: U.S. Government Printing Office, 1978), p. 8.
2. Chris Gugas, "Polygraph or PSE: On the Trail to the Truth," *Security World*, 17, no. 5 (May 1980):48–50.
3. John A. Belt and Peter B. Holden, "Polygraph Usage Among Major U.S. Corporations" *Security Management*, 23, no. 7 (July 1979):27.
4. Ibid., p. 28.
5. James K. Murphy, "The Polygraph Technique," *FBI Law Enforcement Bulletin*, 49, no. 6 (June 1980):4.
6. "Polygraph or PSE," *Security World*, p. 50.
7. "Two Sides to the Truth," *Security World*, 17, no. 1 (January 1980):55.
8. Lester Mood, "Coping with Anti-discrimination Laws," *Administrative Management*, XLI, no. 7 (July 1980):31.
9. Robert E. Smith, "You Can Live with Privacy Laws," *Security Management*, 23, no. 7 (July 1979):45.
10. Ibid., p. 48.

11. U.S. Department of Justice, *Report of the Task Force on Private Security* (Washington, DC: U.S. Government Printing Office, 1976), p. 72.
12. George O'Toole, The Private Sector.
13. John M. Carroll. *Confidential Information Sources: Public & Private* (Stoneham, MA: Butterworth Publishers, 1976), p. 27.
14. Paul Fuqua and Jerry V. Wilson, *Security Investigator's Handbook* (Houston: Gulf Publishing Co., 1979), p. 28.

Chapter 15
Internal Theft Controls*

ROBERT J. FISCHER

It is sad, but true, that virtually every company will suffer losses from internal theft—and these losses can be enormous. "A well-informed security superintendent of a nationwide chain of retail stores recently estimated that it takes between forty and fifty shoplifting incidents to equal the annual loss caused by one dishonest individual inside the organization."[1]

Based on past projects undertaken to evaluate the cost of economic crime, *The Hallcrest Report* estimated employee theft to be at least $100 billion in 1980. The report also estimates that the 1983 figure could well be over $200 billion.[2]

What Is Honesty?

Before considering the issue of dishonest employees, it is helpful to understand the concept of honesty. Honesty is difficult to define. Webster says that honesty is "fairness and straightforwardness of conduct, speech, etc.; integrity; truthfulness; freedom; freedom from fraud." In simple terms, honesty is respect for others and their property. However, the concept is relative. According to Charles Carson, "Security must be based on a controlled degree of relative honesty," since no one fulfills the ideal of total honesty. Figures on bonded employees indicate the following: 15% are relatively honest, 40% are very dishonest and seek opportunities to steal, 45% are generally honest, but can be swayed to either greater honesty or dishonesty. Carson explores relative honesty by asking the following questions:

1. If an error is made in your favor in computing the price of something you buy, do you report it?

*From *Introduction to Security*, 4th ed, by Gion Green, revised by Robert J. Fischer (Stoneham, MA: Butterworths, 1987).

2. If a cashier gives you too much change, do you return it?
3. If you found a purse containing money and the owner's identification, would you return the money to the owner if the amount was $1, $10, $100, $1,000?[3]

Honesty is a controllable variable, and how much control is necessary depends on the degree of honesty of each individual. The individual's honesty can be evaluated by assessing the degree of two types of honesty—moral and conditioned. Moral honesty is a feeling of responsibility and respect that develops during an individual's formative years; this type of honesty is subconscious. Conditioned honesty results from fearing the consequences of being caught; it is a product of reasoning. If an honest act is made without a conscious decision, it is because of moral honesty; but if the act is based on the conscious consideration of consequences, the act results from conditioned honesty.

It is vital to understand these principles because the role of security is to hire employees who have good moral honesty and to condition employees to greater honesty. The major concern is that the job should not tempt an employee into dishonesty.

Unfortunately, there is no sure way by which potentially dishonest employees can be recognized. Proper screening procedures can eliminate applicants with an unsavory past or those who seem unstable and, therefore, possibly untrustworthy. There are even tests that purport to measure an applicant's honesty index. But, tests and employee screening can only indicate potential difficulties. They can screen out the most obvious risks, but they can never truly vouch for the performance of any prospective employee under the circumstances of new employment or under changes that may come about in life away from the job.

The Dishonest Employee

Since there is no fail-safe technique for recognizing the potentially dishonest employee on sight, it is important to try to gain some insight into the reasons employees may steal. If some rule of thumb can be developed which would help to identify the patterns of the potential thief, it would provide some warning for an alert manager.

There is no simple answer to the question of why heretofore honest men and women suddenly start to steal from their employers. The mental and emotional process that leads to this is complex, and motivation may come from any number of sources.

Some employees steal because of resentment of real or imagined injustice, which they blame on management indifference or malevolence. Some feel that they must maintain status and steal to augment their income after financial problems. Some may steal simply to tide themselves over in a genuine emergency. They rationalize the theft by assuring themselves that they will return the money after the current problem is solved. Some simply want to indulge themselves, and many, strangely enough, steal to help others. Or, employees may steal because no one cares, because no one is looking, or because absent or inadequate theft controls eliminate the fear of being caught. Still others may steal simply for excitement.

The Theft Triangle

A simplified answer to the question of why employees steal is the theft triangle. According to this concept, theft—much like fire—occurs when three elements are present: (1) motive, (2) desire, and (3) opportunity.

In simple terms, motive is a reason to steal. Motives might be the resentment of an employee who feels underpaid or the vengefulness of an employee who has been passed over for a promotion. Desire builds on motive by imagining the satisfaction or gratification that would come from a potential action. "Taking a stereo system would make me feel good, because I always wanted a good stereo system." Opportunity is the absence of barriers that prevent someone from taking an item. Desire and motive are beyond the scope of the loss-prevention manager; however, opportunity is the responsibility of security.

A high percentage of employee thefts begin with opportunities that are regularly presented to them. If security systems are lax, or supervision is indifferent, the temptation to steal items that are improperly secured or unaccountable may be too much to resist by any but the most resolute employees.

Many experts agree that the fear of discovery is the most important deterrent to internal theft. When the potential for discovery is eliminated, theft is bound to follow. Threats of dismissal or prosecution of any employee found stealing are never as effective as the belief that any theft will be discovered by management supervision.

Danger Signs

The root causes of theft are many and varied, but certain signs can indicate that a hazard exists. The conspicuous consumer presents perhaps the most easily identified risk. An employee who habitually or suddenly acquires expensive cars and clothes, and who generally seems to live beyond his or her means, should be watched. Such a person is visibly extravagant, and appears indifferent to the value of money. Even though the employee may not be stealing to support such expensive tastes, he or she is likely to run into financial difficulties through reckless spending. The employee may then be tempted to look beyond his or her salary check for ways to support this lifestyle.

Employees who show a pattern of financial irresponsibility are also a potential risk. Many people are incapable of handling their money. They may do their jobs with great skill and efficiency, but they are in constant difficulty in their private lives. These people are not necessarily compulsive spenders, nor do they necessarily have expensive tastes. (They probably live quite modestly, since they have never been able to manage their affairs effectively enough to live otherwise.) They are simply people unable to come to grips with their own economic realities.

Garnishments or inquiries by creditors may identify such an employee. If there seems a reason to make one, a credit check might reveal the tangled state of affairs.

Employees caught in a genuine financial squeeze are also possible problems. If they have been hit with financial demands from illnesses in the family or possibly heavy tax liens, they may find the pressures too great to bear. If such a situation comes to the attention of management, counseling is in order. Many companies maintain funds which are designated to make low-interest loans in such cases. Alternatively, some arrangement might be worked out through a credit union. In any event, an employee in such extremities needs help fast. He or she should get it both as a humane response to the needs and as a means of protecting company assets.

In addition to these general categories, there are specific danger signals which should be noted:

- Gambling on or off premises
- Excessive drinking or signs of alcoholism
- Obvious extravagance
- Persistent borrowing
- Requests for advances
- Bouncing personal checks or checks postdated

What Employees Steal

The employee thief will take anything that may be useful or that has a resale value. The thief can get at the company funds in many ways—directly or indirectly—through collusion with vendors, collusion with outside thieves or hijackers, fake invoices, receipting for goods never received, falsifying inventories, payroll padding, false certification of overtime, padded expense accounts, cash register manipulation, overcharging, undercharging, or simply by gaining access to a cashbox.

This is only a sample of the kinds of attack that are made on company assets using the systems set up for the operation of the business. It is in these areas that the greatest losses can occur, since they are frequently based on a systematic looting of the goods and services in which the company deals, and the attendant operational cash flow.

Significant losses do occur, however, in other, sometimes unexpected, areas. Furnishings frequently disappear. In some firms with indifferent traffic control procedures, this kind of theft can be a very real problem. Desks, chairs, and other office equipment, paintings, rugs—all can be carried away by the enterprising employee thief.

Office supplies can be another problem if they are not properly supervised. Beyond the anticipated attrition in pencils, paper clips, note pads, and rubber bands, these materials are often stolen in case lots. Many firms which buy their supplies at discount are, in fact, receiving stolen property. The market in stolen office supplies is a brisk one and becoming more so as the prices for this merchandise soar.

The office equipment market is another active one, and the inside thief is quick to respond to its needs. Typewriters always bring a good price, and with miniaturization, calculators and minicomputer units make tempting and easy targets.

Personal property is also vulnerable. Office thieves do not make fine distinctions between company property and that of their fellow workers. The company has a very real stake in this kind of theft, since personal tragedy and decline in morale follow in its wake.

Although security personnel cannot assume responsibility for losses of this nature, since they are not in a position to know about the property involved or to control its handling (and should so inform all employees), they should make every effort to apprise all employees of the threat. They should further note, from time to time, the degree of carelessness the staff displays in handling of personal property and send out reminders of the potential dangers of loss.

Methods of Theft

In 1983, 43% of all American businesses failed. Since various studies estimate that 7% to 30% of business failures are the result of some form of employee dishonesty,[4] there is a very real need to examine the shapes it frequently takes. There is no way to describe every kind of theft, but some examples here may serve to give some idea of the dimensions of the problem:

1. Payroll and personnel employees collaborating to falsify records by the use of nonexistent employees or by retaining terminated employees on the payroll
2. Padding overtime reports, part of which extra unearned pay is kicked back to the authorizing supervisor
3. Pocketing unclaimed wages
4. Splitting increased payroll that has been raised on checks signed in blank for use in the authorized signer's absence
5. Maintenance contract service and personnel in collusion to steal and sell office equipment
6. Receiving clerks and truck drivers in collusion on falsification of merchandise count. Extra unaccounted merchandise is fenced.
7. Purchasing agents in collusion with vendors to falsify purchase and payment documents. The purchasing agent issues authorization for payment on goods never shipped after forging receipt of shipment
8. Purchasing agent in collusion with vendor to pay inflated price
9. Mailroom and supply personnel packing and mailing merchandise to themselves for resale
10. Accounts payable personnel paying fictitious

bills to an account set up for their own use
11. Taking incoming cash without crediting the customer's account
12. Paying creditors twice and pocketing the second check
13. Appropriating checks made out to cash
14. Raising the amount of checks after voucher approval or raising the amount of vouchers after their approval
15. Pocketing small amounts from incoming payments and applying later payments on other accounts to cover shortages
16. Removal of equipment or merchandise with trash
17. Invoicing goods below regular price and getting a kickback from the purchaser
18. Underringing on a cash register
19. Issuing (and cashing) checks on returned merchandise not actually returned
20. Forging checks, destroying them when returned with a statement from the bank, and changing cash books accordingly

Morale

In any organization which exists by the cooperative efforts of all its members, it is important that each one of those members feels like an important part of the operation, both as a contributor and as an individual.

This dual role is important to recognize. Each employee must feel like a significant, integrated part of the whole and identify with it, while still maintaining and protecting a personal identity as an individual apart from the structure. If an employee is denied reinforcement in either of these views, his or her sense of personal worth suffers. Anger, anxiety, frustration, or feelings of inadequacy may follow. Any of these attitudes are damaging both to the individual and to the organization as a whole, and they must be headed off.

The best way to do that is to recognize that each employee is, in fact, an important member of the organization, or else he or she wouldn't be there. A function must be performed as a part of the larger function, and the employee is the one performing it. He or she is unique and unduplicatable. His or her reinforcement comes from management, supervisors, and peers. He or she is a vital, irreplaceable organ of the greater body. Of such recognition and respect is high morale created and maintained, and everyone benefits.

Every supervisor must be indoctrinated in the importance of the morale factor in every business. Just as no business can operate without employees, none can operate efficiently with a work force whose morale has been damaged and who respond with an attitude of listlessness and disinterest. Threats and coercion will not restore the optimum level of morale any more than directives and public notices can. This is a condition that comes from human nature. It must be dealt with in those terms.

Additional factors are also important in this regard. Physical surroundings and appropriate rewards are part of the message that tells employees what the organization thinks of them.

In an atmosphere of concern, fairness, and mutual respect, few people would be led to steal. It would be like stealing from themselves. It is important that such an atmosphere be developed and maintained for the good of all.

A program aimed at improving or maintaining employee morale might contain some or all of the following elements:

1. Clear statements of company policy which is consistently and fairly administered.
2. Regular review of wages and wage policy, updated to assure equitable wage levels.
3. House organ or newsletter and bulletin boards kept current.
4. Open, two-way avenues of communication between management and all employees.
5. Clear procedures, formal and informal, for airing grievances and personal problems with supervisors.
6. Vigorous training programs to improve job skills and pave the way for advancement.
7. Physical surroundings—decor, cleanliness, sound and temperature control, and general housekeeping, at a high level.

Procedural Controls

Auditing Assets

Periodic personal audits by outside auditors are essential to any well-run security program. Such an examination will discover theft only after the fact, but it will presumably discover any regular scheme of embezzlement in time to prevent serious damage. If these audits, which are normally conducted once a year, are augmented by one or more surprise audits, even the most reckless criminal would hesitate to try to set up even a short-term scheme of theft.

These audits will normally cover an examination of inventory schedules, prices, footings, and extensions. They should also verify current company assets by physical inventory sampling, accounts receivable, accounts payable (including payroll), deposits, plant, and outstanding liabilities through an ongoing financial audit. In all these cases a spot check beyond the books themselves can help to establish the existence of legitimate assets and liabilities, not empty entries created by a clever embezzler.

Cash

Any business handling relatively few cash payments in and out is fortunate, indeed. Such a business is able to avoid much of the difficulty created by this security sensitive area, since cash handling is certainly the operation most vulnerable and the most sought after by the larcenous among the staff.

Cash by Mail

If cash is received by mail—a practice which is almost unheard of in most businesses—its receipt and handling must be undertaken by a responsible, bonded supervisor or supervisors. This administrator should be responsible for no other cash handling or bookkeeping functions. This official should personally see to it that all cash received is recorded by listing the amount, the payer, and such other pertinent information as procedures have indicated. This list should be made, in duplicate, on sequentially numbered forms, with both copies signed by the manager in charge of opening the mail as well as by the cashier who receipts for the money. The cashier will then keep one copy of the record for the file; the other will be forwarded to accounting. Both will verify the numbering on the cash receiving list.

There is clearly a danger here at the very outset. If cash is diverted before it is entered on any receipt, there is no record of its existence. Until it is channeled into company ledgers in some way and begins its life as a company asset, there is no guarantee that it will not serve some more private interest. This requires supervision. In the case of a firm doing a large catalogue business which receives large amounts of cash in spite of pleas for checks or money orders, it has sometimes been felt that the operation should be conducted in a special room reserved for the purpose.

Daily Receipts

All cash book entries must be checked against cash on hand at the end of each day. Spot checks on an irregular basis should be conducted.

Cash receipts should be deposited in the bank and each day's receipts balanced with the daily deposit. Petty cash, as needed, should be reimbursed by check.

All bank deposits should be accompanied by three deposit slips. One is receipted by the bank and returned by the cashier to the person making the deposit; the second is mailed to the office accounting department; and the third is the bank's copy.

Each day deposit slips should be balanced with the day's receipts.

Bank Statements

Bank statements should be received and reconciled by someone who is not authorized to deposit or withdraw funds, or to make a final accounting of receipts or disbursements. When bank statements are reconciled, cancelled checks should be checked against vouchers for any possible alterations and for proper endorsement by the payee. Any irregularities in the endorsements should be promptly investigated. If the statement itself seems in any way out of order by way of erasure or possible alteration, the bank should be asked to submit a new statement to the reconciling official's special personal attention.

Petty Cash

A petty cash fund, set aside for that purpose only, should be established. The amount to be carried in such a fund will be based upon past experience. These funds must never be commingled with other funds of any kind and should be drawn from the bank by check only. They should never be drawn from cash receipts. No disbursements of any kind should be made from petty cash without an authorized voucher signed by the employee receiving the cash and countersigned by an authorized employee. No voucher should be cashed that shows signs of erasure or alteration. All such vouchers should be drawn up in ink or typed. In cases of typographical error, new vouchers should be prepared rather than correcting the error. If there is any reason for using a voucher on which an erasure or correction has been made, the authorizing official should initial the change or place of erasure.

Receipts substantiating the voucher should accompany it and should, where possible, be stapled or otherwise attached to it.

The petty cash fund should be brought up to the specified amount required by a check to the amount of its depletion. The vouchers upon which disbursements were made should always be verified by an employee other than the one in charge of the fund. All vouchers submitted and paid should be cancelled in order to avoid reuse.

Petty cash should be balanced occasionally by management, at which time vouchers should be examined for irregularities.

Separation of Responsibility

The principle of separation of responsibility and authority in matters concerning the company's finances is of prime importance in management. This situation must always be sought out in the survey of every department. It is not always easy to locate. Sometimes even the employee who has such power is unaware of the dual role. But the security specialist must be knowledgeable about its existence and suggest an immediate change or correction in such operational procedures whenever they appear.

An employee who is in the position of both ordering and receiving merchandise, or a cashier who authorizes and disburses expenditures, are examples of this double-ended function in operation. All situations of this nature are potentially damaging and should be eliminated. Such procedures are manifestly unfair to company and employee alike. To the company because of the loss that might incur; to the employee because of the temptation and ready opportunity they present. Good business practice demands that such invitations to embezzlement be studiously avoided.

It is equally important that cash handling be separated from the recordkeeping function. A cashier who becomes his or her own auditor and bookkeeper has a free rein with that part of the company funds. The chances are that the cashier will not steal, but he or she could—and might. He or she might also make mathematical mistakes without someone else double checking the arithmetic.

In some smaller companies this division of function is not always practical. In such concerns it is common for the bookkeeper to act also as cashier. If this is the case, a system of countersignatures, approvals, and management audits should be set up to help divide the responsibility of handling company funds as well as accounting for them.

Promotion and Rotation

Most embezzlement is the product of a scheme operating over an extended period of time. Many embezzlers prefer to divert small sums on a systematic basis, feeling that the individual thefts will not be noticed and, therefore, the total loss is unlikely to come to management's attention.

These schemes are sometimes frustrated either by some accident that uncovers the system or by the greed of the embezzler, who is so carried away with success that he or she steps up the ante. But while the theft is working, it usually is difficult to detect. Frequently, the thief is in a position to alter or manipulate records in such a way that the theft escapes the attention of both internal and outside auditors. This sometimes can be countered by upward or lateral movement of employees.

Promotion from within, wherever possible, is always good business practice, and lateral transfers can be effective in countering possible boredom or the danger of reducing a function to rote, and thus diminishing its effectiveness.

Such movement also frustrates embezzlers. When they lose control of the books governing some aspect of the operation, they lose the opportunity to cover their thefts. Discovery would inevitably follow careful audits of books they could no longer manipulate. If regular transfers were a matter of company policy, no rational embezzler would set up a long-term plan of embezzlement unless he or she found a scheme that was audit-proof, and such an outcome is highly unlikely.

To be effective as a security measure, such transfers need not involve all personnel, since every change in operating personnel brings with it changes in operation. In some cases, even subtle changes may be enough to alter the situation sufficiently to reduce the totality of control an embezzler has over the books. If such is the case, the swindle is over. He or she may avoid discovery of the previous looting but cannot continue without danger of being unmasked.

In the same sense, embezzlers dislike vacations. They are aware of the danger if someone else should handle their accounts, if only for the two or three weeks of vacation. So they make every effort to pass up the holiday.

Any manager who has a reluctant vacationer should recognize that this is a potential problem. Vacations are designed to refresh the outlook of everyone. No matter how tired they may be when they return to work, vacationers have been refreshed emotionally and intellectually. Their effectiveness in their job has probably improved and they are, generally speaking, better employees for it. The company benefits from vacations as much as employees do. No one should be permitted to pass up authorized vacation—especially one whose position involves control over company assets.

Access to Records

Many papers, documents, and records are proprietary, or at least are available to only a limited number of people who need such papers in order to function. All other persons are deemed off-limits. They have no apparent need for the information. Such papers should be secured under lock and key—and, depending on their value or reconstructability, in a fire-resistant container.

Forms

Certain company forms often are extremely valuable to the inside as well as the outside thief. They should be secured and accounted for at all times. They should be sequentially numbered and recorded regularly so that any loss can be detected at a glance.

Blank checks, order forms, payment authorizations, vouchers, receipt forms, and all others which authorize or verify transactions, are prime targets for thieves and should, therefore, be accounted for.

Since there are many effective operational systems in use for the ordering, shipping, or receipting of goods, as well as the means by which all manner of payments from petty cash or regular debt discharge are authorized, no one security system to protect against illegal manipulation within such systems would apply universally. It can be said, however, that since every business has some means to authorize transactions of goods or money, the means by which such authorizations are made must be considered in the security program. Security of such means must be considered an important element in any company's defense against theft.

Generally speaking, all forms should be prenumbered and, where possible, used in numerical order. Any voided or damaged forms should be filed and recorded, and forms reported lost must be accounted for and explained. All such numbered forms of every kind should be inventoried and accounted for periodically.

In cases where purchase orders are issued in blocks to various people who have the need for such authority, such issuance must be recorded, and disposition of their use should be audited regularly. In such cases, it is customary for one copy of the numbered purchase order to be sent to the vendor, who will use that number in all further dealings on that particular order; another copy will be sent to accounting for purpose of payment authorization and accrual if necessary; and one copy will be retained by the issuing authority. Each block issued should be used sequentially, although, since some areas may have more purchasing activity than others, purchasing order copies as they are forwarded to accounting may not be in overall sequence.

Purchasing

Centralized Responsibility

Where purchasing is centralized in one department, controls will always be more effective. Localizing responsibility as well as authority reduces the opportunity for fraud, accordingly. This is not always possible or practical, but in areas where purchasing is permitted by departments needing certain materials and supplies, there can be confusion occasioned by somewhat different purchasing procedures. Cases have been reported where different departments pay different prices for the same goods and services and thus bid up the price the company is paying. Centralization of purchasing would overcome this problem.

Purchasing should not, however, be involved in any aspect of accounts payable or the receipt of merchandise other than informationally.

Competitive Bids

Competitive bids should be sought wherever possible. This, however, raises an interesting point that must be dealt with as a matter of company policy. Seeking competitive bids is always good practice, both to get a view of the market and to provide alternatives in the ordering of goods and materials, but it does not follow that the lowest bidder is always the vendor to do business with.

Such a bidder may lack adequate experience in providing the services bid for, or may have a reputation of supplying goods of questionable quality, even though they may meet the technical standard prescribed in the order. A firm may also underbid the competition in a desperate effort to get the business, but then find it cannot deliver materials at that price, no matter what it has agreed to in its contract.

In order to function wisely and to be able to exercise good judgment in its area of expertise, purchasing must be permitted some flexibility in its selection of vendors. This means that it will not always be the low bidder who wins the contract.

Since competitive bidding provides some security control in reducing favoritism, collusion, and kickbacks between the purchasing agent and the vendor, these controls would appear to be weakened or compromised in situations where the purchasing department is permitted to select the vendor on considerations other than cost. This can be true to some degree, but this is a situation in which business or operational needs may be in some conflict with tight security standards—and in which security should revise its position to accommodate the larger demands or efficiency and ultimate economy. After all, cheap is not necessarily economical.

Controls in this case could be applied by requiring that in all cases where the lowest bid was not accepted, a brief explanation in outline form be attached to the file along with all bids submitted. Periodic audits of such files could establish if any pattern of fraud seems likely. Investigation of the analysis or assumptions made by purchasing in assigning contracts might be indicated in some situations to check the validity of its stated reasoning in the matter.

Other Controls

Copies of orders containing the amount of merchandise purchased should not be sent to receiving clerks. These clerks should simply state the quantity actually received with no preconception of the amount accepted. Payment should be authorized only for that amount actually received.

Vendor invoices and receipts supporting such vouchers should be cancelled to avoid the possibility of their resubmission in collusion with the vendor.

Purchasing should be audited periodically, and documents should be examined for any irregularities.

Payroll

It is important that the payroll be prepared by persons who will not be involved in its distribution. This is consistent with the effort to separate the various elements of a particular function into its component parts, and then distribute the responsibility for those parts to two or more persons or departments.

Every effort should be made to distribute the payroll in the form of checks rather than cash, and such checks should be of a color different from those used in any other aspect of the business. They also should be drawn on an account set aside exclusively for payroll purposes. It is important that this account be maintained in an orderly fashion. Avoid using current cash receipts for payroll purposes.

Personnel Records

Initial payroll information should be prepared from personnel records which, in turn, have come from personnel as each employee is hired. Such a record should contain basic data such as name, address, attached W-2 form, title, salary, and any other information that the payroll department may need. The record will be countersigned by a responsible executive verifying the accuracy of the information forwarded.

This same procedure should be followed when an employee terminates employment with the company. All such notifications should be consolidated into a master payroll list, which should be checked frequently to make sure that payroll's list corresponds to the employment records of personnel.

Unclaimed Payroll Checks

Unclaimed paychecks should be returned to the treasurer or controller after a reasonable period of time for redeposit in the payroll account. Certainly, such cases should be investigated to determine why the checks were returned or why they were issued in the first place. All checks so returned should be cancelled to prevent any reuse and filed for reference. Since payrolls reflect straight time, overtime, and other payments, such payments should be supported by time sheets authorized by supervisors or department heads. Time sheets of this nature should be verified periodically to prevent overtime padding and kickback.

Time cards should be marked to prevent reuse.

Payroll Audits

The payroll should be audited periodically by external auditors for any irregularities, especially if there has been an abnormal increase in personnel or net labor cost.

To further guard against the fraudulent introduction of names into the payroll, distribution of paychecks should periodically be undertaken by the internal auditor, the treasurer, or other responsible official. In large firms this can be done on a percentage basis, thus providing at least a spot check of the validity of the rolls.

Accounts Payable

As in the case of purchasing, accounts payable should be centralized to handle all disbursements upon adequate verification of receipt and proper authorization for payment.

These disbursements should always be by checks that are consecutively numbered and used in that order. Checks that are damaged, incorrectly drawn, or for any reason unusable must be marked as cancelled and filed for audit. All checks issued for payment should be accompanied by appropriate supporting data, including payment authorizations, before they are signed by the signing authority. It is advisable to draw the checks on a check-writing machine which uses permanent ink and is as identifiable to an expert as handwrtiting or a particular typewriter. Checks should be on safety paper which will show almost any attempted alteration.

Here, as in order departments, periodic audits must be conducted to examine the records for any sign of nonexistent vendors, irregularities in receipts or payment authorizations, forgeries, frauds, or unbusinesslike procedures that could lead to embezzlement.

General Merchandise

Merchandise is always subject to pilferage, particularly when it is in a transfer stage, as when it is being shipped or received. The dangers of loss at these stages are increased in operations where controls over inventory are lax or improperly supervised.

Separation of Functions

To control sensitive aspects of any operation involving the handling of merchandise, it is desirable to separate three functions. Receiving, warehousing, and shipping should be the responsibility of three different areas. Movement of merchandise from one mode to another should be accompanied by appropriate documents which clearly establish the responsibility for specific amounts of merchandise passing from one sphere of authority to another.

Receipting for a shipment places responsibility for a correct count and the security of the shipment on the receiving dock. This responsibility remains there until it is transferred and receives a proper receipt from the warehouse supervisor. The warehouse supervisor must verify and store the shipment, which is his or her responsibility until it is called for (by the sales department, for example), or directed to be shipped by an authorized document. The warehouse supervisor ensures that the goods are assembled and passed along as ordered, and receives a receipt for those goods delivered.

In this process, responsibility is fixed from point to point. Various departments or functions take on and are relieved of responsibility by the receipts. In this way a perpetual inventory is maintained as well as a record of responsibility for the merchandise.

Requisitions must be numbered to avoid the destruction of records or the introduction or unauthorized transfers into the system. Additionally, stock numbers of merchandise should accompany all of its movement to describe the goods and thus aid in maintaining perpetual inventory records.

In small firms where this separation of duties is impractical, and receiving, shipping, and warehousing are combined in one person, the perpetual inventory is essential for security, but it must be maintained by someone other than the person actually handling the merchandise. The shipper-receiver-warehouser should not have access to these inventory records at any time.

Inventories

Inventories will always be an important aspect of merchandise control, no matter what operations are in effect. Such inventories must be conducted by someone other than the person in charge of that particular stock. In the case of department stores, for purposes of inventory, personnel should be moved to a department other than their regularly assigned one.

In firms where a perpetual inventory record is kept, physical counts on a selective basis can be undertaken monthly or even weekly. In this procedure a limited number of certain items randomly selected can be counted and the count compared with current inventory record cards. Any discrepancy can be traced back to attempt to determine the cause of the loss.

Physical Security

It is important to remember that personnel charged with the responsibility of goods, materials, and merchandise must be provided the means to properly discharge that responsibility. Warehouses and other storage space must be equipped with adequate physical protection to secure the goods stored within. Authorizations to enter such storage areas must be strictly limited, and the responsible employees must have means to further restrict access in situations where they may feel that the security of goods is endangered.

Receiving clerks must have adequate facilities for storage or supervision of goods until they can be passed on for storage or for other use. Shipping clerks must also have the ability to secure goods in dock areas until they are received and loaded by truckers. Without the proper means of securing merchandise during every phase of its handling, assigned personnel cannot be held responsible for merchandise intended for their control, and the entire system will break down. Unreasonable demands, such as requiring a shipping clerk to handle the movement of merchandise in such a way that he or she is required to leave unprotected goods on the dock while filling out the rest of the order, lead to the very reasonable refusal of personnel to assume responsibility for such merchandise. And when responsibility cannot be fixed, theft can result.

The Mailroom

The mailroom can be a rich field for a company thief to mine. Not only can it be used to mail out company property to an ally or to a set-up address, but it deals in stamps—and stamps are money. Any office with a heavy mailing operation must conduct regular audits of the mailroom.

Some firms have taken the view that the mailroom represents such a small exposure that close supervision is unnecessary. Yet the head of the mailroom in a fair-sized Eastern firm got away with over $100,000 in less than three years by his manipulation of the postal meter. Only a firm that can afford to lose $100,000 in less than three years should think of its mailroom as inconsequential in its security plan.

Trash Removal

Trash removal presents many problems. Employees have hidden office equipment or merchandise in trash cans and have then picked up the loot far from the premises in cooperation with the driver of the trash pick-up vehicle. Some firms have had a problem when they put out trash on the loading dock to facilitate pick-up. Trash collectors made their calls during the day and often picked up unattended merchandise along with the trash. On-premises trash compaction is one way to end the use of trash containers as a safe and convenient vehicle for removing loot from the premises.

Every firm has areas which are vulnerable to attack—what and where they are can be determined only by thorough surveys and regular reevaluation of the entire operation. There are no shortcuts. The important thing is to locate the areas of risk and set up procedures to reduce or eliminate them.

When Controls Fail

There are occasions when a company is so beset by internal theft that problems seem to have gotten totally out of hand. In such cases, it is often difficult to localize the problem sufficiently to set up specific countermeasures in those areas affected. The company seems simply to come up short. Management is at a loss to identify the weak link in its security, much less how theft is accomplished after security has been compromised.

Undercover Investigation

In such cases, many firms similarly at a loss, in every sense of the word, have found it advisable to engage the services of a security firm which can provide undercover agents to infiltrate the organization and observe the operation from within.

Such an agent may be asked to get into the organization on his or her own initiative. The fewer people who know of the agent's presence, the greater the protection, and the more likely he or she is to succeed in the investigation. It is also true that when large-scale thefts take place over a period of time, almost anyone in the company could be involved. Even one or more top executives could be involved in serious operations of this kind. Therefore secrecy is of great importance. Since several agents may be used in a single investigation, and since they may be required to find employment in the company at various levels, they must have, or very convincingly seem to have, proper qualifications for the level of employment they are seeking. Over- or under-qualification in pursuit of a specific area of employment can be a problem, so they must plan their entry carefully. Several agents may have to apply for the same job before one is accepted.

Having gotten into the firm's employ, the agent must work alone. The agent must conduct the investigation and make reports with the greatest discretion to avoid discovery. But he or she is in the best possible position to get to the center of the problem, and such agents have been successful in a number of cases of internal theft in the past.

These investigators are not inexpensive, but they

earn their fee many times over in breaking up a clever ring of thieves.

It is important to remember, however, that such agents are trained professionals. Most of them have had years of experience in undercover work of this type. Under no circumstances should a manager think of saving money by using employees or well-meaning amateurs for this work. Such a practice could be dangerous to the inexperienced investigator and would almost certainly warn the thieves, who would simply withdraw from their illegal operation temporarily until things had cooled down, after which they could return to the business of theft.

*Interviewing and Interrogation**

Interviewing and interrogation involve gathering information from people. During interviewing, the investigator obtains information willingly. But during an interrogation, the interviewee is often unwilling to supply information. The investigator needs to know the techniques associated with each type of situation.

There is no one correct method of conducting an interview or interrogation. The circumstances of each particular situation dictate the characteristics of these investigative functions.

Why are interviews or interrogations conducted? A primary reason is to learn the truth. Other possible reasons: obtain evidence (e.g., a confession to aid in prosecution), eliminate suspects, recover property, and obtain information that results in corrective action.

An emphasis is placed on investigations in the private sector even though many of the ideas presented here are used in public sector investigations. The reader is reminded that interviews and interrogations may be carried out to assist in a variety of loss-prevention efforts that are not necessarily criminal in nature.

The following lists contain practical ideas for interviews and interrogations.

Preliminaries

1. Maintain records.
2. Plan the questioning.

---

*The section on interviewing and interrogation is reproduced from Philip P. Purpura, *Security and Loss Prevention* (Stoneham, Mass.: Butterworths, 1984), pp. 228–230, with permission from the publisher.

3. Make an appointment, if necessary.
4. If a procedure or law question arises, consult with a superior or attorney.
5. Question in privacy, if possible.
6. Make sure someone of the same sex as the interviewee is present.
7. Identify yourself to the interviewee.
8. Openly tape record the questioning, if possible.

Interviewee

1. Consider the interviewee (e.g., background, intelligence, education, biases, emotional state).
2. Communicate on the same level.
3. Watch for nervousness, perspiration, and fidgeting.
4. A person may be reluctant to talk in order to protect him/herself or others.
5. A person may talk to relieve guilt or to cause problems for an enemy not involved in the loss.

Objectives of the Interviewer

1. Establish good rapport (e.g., ask: "How are you?").
2. Maintain good public relations.
3. Maintain eye contact.
4. Do not jump to conclusions.
5. Maintain an open mind.
6. Listen attentively.
7. Be perceptive to every comment and any slips of the tongue.
8. Maintain perseverance.
9. Control the interview.
10. Carefully analyze hearsay (statements by a person as to what another person told him or her; unverified information).

Strategies by the Interviewer

1. Ask open-ended questions: those questions that require lengthy answers. Example: What happened at the plant before the accident? Close-ended questions require short yes or no answers that limit responses. Example: Were you close to the accident?
2. Silence makes many interviewees feel uncomfortable. Silence by an investigator, after an interviewee answers an open-ended question, may cause the interviewee to begin talking again.
3. Build up interviewee memory by having the interviewee begin the story of an incident at the very beginning.

4. To test honesty, ask questions to which you know the answers.

The reader is probably familiar with many movies and television programs that portray the interrogation process as the *third degree*. The usual setup is one bright light hanging over the seated suspect in a dark room. The investigators standing around are constantly asking questions and use violence when they try to break the suspect. The *Miranda warnings* and court action involving public and private police have virtually eliminated this abuse.

During Interrogation (An Extension of the Interview)

1. Discuss the seriousness of the incident.
2. Request the story several times. Some investigators request the story backwards to catch inconsistencies.
3. Appeal to emotions. Examples: Everybody makes mistakes. You are not the first person who has been in trouble. Don't you want to clear your conscience?
4. Point out inconsistencies in statements.
5. Confront person with some of the evidence.

The above lists are only a part of an enormous amount of interviewing and interrogation strategies. Many other approaches are possible. For instance, hypnosis has been gaining popularity in refreshing witness memory. Special training programs are available for competent investigators.

Prosecution

Every firm has been faced with the problem of establishing policy regarding the disposal of a case involving proved or admitted employee theft. They are faced with three alternatives: to prosecute, to discharge, or to retain the thief as an employee. The policy they have established has always been difficult to arrive at, because there is no ready answer. There are many proponents of each alternative as the solution to problems of internal theft.

However difficult it may be, every firm must establish a policy governing matters of this kind. And the decision as to that policy must be arrived at with a view to the greatest benefits to the employees, the company, and to society as a whole. An enlightened management would also consider the position of the as-yet-to-be-discovered thief in establishing such policy.

Discharging the Thief

Most firms have found that discharge of the offender is the simplest solution. Experts estimate that 90% of those employees discovered stealing are simply dismissed. Most of those are carried in the company records as having been discharged for "inefficiency" or "failure to perform duties adequately."

This policy is defended on many grounds, but the most common are:

1. Discharge is a severe punishment and the offender will learn from the punishment.
2. Prosecution is expensive.
3. Prosecution would create an unfavorable public relations atmosphere for the company.
4. Reinstating the offender in the company—no matter what conditions are placed on the reinstatement—will appear to be condoning theft.
5. If the offender is prosecuted and found not guilty, the company will be open to civil action for false arrest, slander, libel, defamation of character, and other damages.

There is some validity in all of these views, but each one bears some scrutiny.

As to learning (and presumably reforming) as a result of discharge, experience does not bear out this contention. A security organization found that 80% of the known employee thieves they questioned with polygraph substantiation admitted to thefts from previous employers. Now, it might well be argued that, since they had not been caught and discharged as a result of these prior thefts, the proposition that discharge can be therapeutic still holds, or at least has not been refuted. That may be true and it should be considered.

Prosecution is unquestionably expensive. Personnel called as witnesses may spend days appearing in court. Additional funds may be expended investigating and establishing a case against the accused. Legal fees may be involved. But can a company afford to appear so indifferent to significant theft that it refuses to take strong action when it occurs?

As to public relations, many experienced managers have found that they have not suffered and decline in esteem. On the contrary, in cases where they have taken strong, positive action, they have been applauded by employees and public alike. This is not always the case, but apparently a positive reaction is usually the result of vigorous prosecution in the wake of substantial theft.

Reinstatement is sometimes justified by the circumstances. There is always, of course, a real danger of adverse reaction by the employees, but if reinstatement is to a position not vulnerable to theft, the message may get across. This is a most delicate matter that can be determined only on the scene.

As far as civil action is concerned, that possibility must be discussed with counsel. In any event, it is to be hoped that no responsible business person would decide to prosecute unless the case was a very strong one.

Borderline Cases

Even beyond the difficulty of arriving at a satisfactory policy governing the disposition of cases involving employee theft, there are the cases which are particularly hard to adjudicate. Most of these involve the pilferer, the long-time employee, or the obviously upright employee in financial difficulty who steals out of desperation. In each case the offender freely admits guilt and pleads that he or she was overcome by the temptation.

What should be done in such cases? Many companies continue to employ such employees, provided they make restitution. They are often grateful, and they continue to be effective in their jobs.

In the last analysis, each individual manager must make the determination of policy in these matters.

Only he or she can determine the mix of toughness and compassion that will guide the application of policy throughout.

Hopefully, every manager will decide to avoid the decision by making employee theft so difficult—so unthinkable—that it will never occur. That goal may never be reached, but it's a goal to strive for.

References

1. *U.S. Government Organization Manual* (Washington, D.C.: U.S. Government Printing Office, 1966), p. 60.
2. William C. Cunningham and Todd H. Taylor, *Private Security and Police in America: The Hallcrest Report* (Portland, OR: Chancellor Press, 1985), p. 243.
3. Charles P. Carson, *Managing Employee Honesty* (Stoneham, MA: Butterworth Publishers, 1977).
4. Richard C. Hollinger and John P. Clark, *Theft by Employees* (Lexington, MA: Lexington Books, 1983), p. 4.
5. *The Use of Polygraphs and Similar Devices by Federal Agencies, Hearing Before A Subcommittee on Government Operations, House of Representatives* (Washington, D.C.: U.S. Government Printing Office, 1974), p. 29.
6. *The Use of Polygraphs*, p. 29.

Appendix 15a
Loss Prevention Checklists*

VICTOR HAROLD

Every theft occurs for some reason. Each person has an individual reason for stealing. Stealing from the company is often rationalized by the thief saying, "They could afford it." The real reason for theft on any level, whether it be the taking of a paper clip, postage, petty cash, or an outright major misappropriation of funds, is a subject all its own and beyond the scope of this.

Psychologist Robert Katz explains that:

> Motivation to thievery is in all of us. Those who believe theft can be successfully accomplished, attempt it. The guilt or stress a person feels is not directly proportional to the value of the theft. In other words, the greater the dollar amount of the crime, the less guilt felt by the criminal. Subsequent similar criminal acts place less stress on the thief than the prior ones.

---

*Reprinted with permission of Victor Harold from *How to Stop Theft in Your Business*.

Whatever the reason for corporate theft, it is increasing at an alarming rate. Experts tell us losses due to internal crime are far greater than all other crimes put together—in excess of $20 billion per annum in known losses since 1984. Known losses are those which have been publicly established. Unknown losses are those circumstances in which employers refuse to prosecute, insured situations, thieves that got away with it, and honest errors which have led to real losses. A management group specializing in security safeguards estimates that each and every vulnerable business can lose a minimum of 10% of their gross volume primarily to internal theft.

To enable the corporate owner to defensively react to the potential of theft, and to prevent losses of such a magnitude that shutdown may be the only solution, a basic security survey is needed. These checklists will show weak security areas and aid the corporate owner in securing the facility from intrusion, theft, and loss.

Why Have a Checklist

There is no easy way to check out company security. Endless hours must be spent by executives to determine not only the specific area of loss, but also pinpoint the individuals responsible for the shortages. Often, management will retain outside security consultants to work undercover. They can help ascertain the problem areas, uncover the thieves, and advise on remedial measures to avoid future problems.

What if the company is small—too small to have the financial resources necessary to retain outside security consultants? How can the small company owner accomplish an adequate check of her or his own business? In many cases the owner cannot but can, nevertheless, create circumstances which make it difficult for fraud and theft to occur. Sometimes, in the process of setting up a secure facility, a business owner may uncover or recognize a theft-prone situation. When such a circumstance presents itself, he or she can take appropriate action.

Businesspeople need an adequate security checklist detailing the more important, most asked questions. Following is such a list of questions designed as an aid and guide for experienced as well as novice personnel. Maximum use of it will enable the company to get answers which will aid the businessperson in setting up a well secured facility.

Pertinent questions posed in the checklist should be answered affirmatively. If they cannot be answered affirmatively, management is warned that a potential trouble spot exists and should be given immediate attention.

A Few Caveats

1. A security checklist, no matter the depth, can never be as good as the people to whom the checkout task is assigned.
2. This checklist is only a guide. You may obtain answers which will lead to more questions. Ask them. Keep asking the questions until you are thoroughly satisfied with the answers.
3. When and if necessary, get a security professional to assist you. The local policy force is free, very professional, and can be a genuine aid.
4. You may uncover a small prior or current theft situation. Be alert to the fact that a small matter may cover a major problem. Keep asking penetrating questions, or have your professional take over.
5. If you have a theft sensitive product, or if stealing can occur in various departments, make sure your employees are consistently kept aware of your intent to discharge upon discovery and to prosecute successfully.
6. Minor thefts should not take up major time. Even major thefts do not require a great deal of disproportionate administrative involvement. Do not neglect your company's executive mission.
7. Finding out that theft has occurred in your business may evoke disgust and may trigger thoughtless behavior. You must avoid an overreaction which, if it emerges, may expose you to legal liability. Consult the police or your security professional.

The Theft Fraud Embezzlement Syndrome

Employee Theft Signals

There are warning signs that generally indicate that further investigation is warranted. Management is advised to check two ways. One, determine why the signal is showing itself so clearly. Two, conduct an indepth investigation of the employee responsible for the signal's origination. Some fraud/theft signals are:

1. The employee comes in too early.
2. The employee stays late.
3. The employee wears too loose clothing.

4. Ladders or boxes near a window
5. The employee goes outside a lot, citing coffee breaks, a walk, personal phone calls, etc.
6. Cautions others that a supervisor is approaching
7. Documents contain changes or erasures
8. Talks furtively to strangers
9. Records are sloppy
10. Refuses an assistant
11. Refuses to go on vacation
12. Changes vacation schedules several times
13. Appears to have many personal problems
14. Resents the income of others
15. Is a very "aggressive" collector
16. Has recently purchased things not in line with income
17. Doesn't pay heed to office procedures
18. Establishes own rules
19. Frequent change of suppliers
20. Is usually in debt to other employees
21. Appears to be a "workhorse" type
22. Is lenient with petty cash
23. Too many personal phone calls
24. Personal calls are muffled and never overheard by others
25. Customers' billing complaints
26. Shows sensitivity to routine questions regarding job
27. Protective of records
28. Debts easily written off with little or no explanation
29. Raw material purchases up—production doesn't follow
30. Receivable collection declines
31. More than normal employee hours show up
32. Creditors deal only with a specific employee
33. Visits to a specific employee from individuals reluctant to explain their business

Some advice: If an employee exhibits embezzlement behavior, it does not mean that fraud has appeared/occurred or is inevitable. Some situations may require a work pattern where furtive behavior is a method of doing business. If it isn't, and there may be more reasons to suspect the employee, obtain immediate professional help. J.J. Gilbert, a security expert, recently told a seminar group, "If a manager's suspicion of fraud has been aroused, there's an 80% chance that his feelings are accurate."

Assets

1. Do you have an inventory control system?
2. Is an actual inventory taken periodically?
3. Do you separate office supplies from noninventory related equipment?
4. Do you have periodic reviews of the insurance policies to determine adequate coverage against fire, theft, embezzlement, and other types of losses?
5. If inventories are to be used in the business, do you use special requisition forms indicating inventory withdrawal for company use?
6. Is inventory responsibility assigned to a reliable employee?
7. Is the inventory kept in a secure area?
8. Is access to the inventory controlled?
9. Is the inventory area separate from other departments?
10. Is the inventory area safe from physical damage such as breakage, fire, weather, spoilage, etc.?
11. Have you retained an outside consultant to ascertain further steps to be taken to prevent loss?
12. Are scrap materials promptly sold?
13. Are strict controls and accounting records kept for the sale of scrap material?
14. Is the sale of scrap or salvageable material a separate department and not handled by employees involved in purchasing, production, or sales?
15. If the inventory is composed of easily sold consumer goods, is access to the inventory area strictly controlled?
16. If some of the assets consist of securities, are they kept in a safe area?
17. Have you at least two corporate officers maintaining joint custody of the securities?
18. Does the accounting department have a detailed record of the notes, securities, and other valuables held in custody?
19. Are the securities records kept up-to-date?
20. If the firm holds collateral, is a meticulous record of it maintained by responsible executives?
21. Are securities and collateral balanced regularly?
22. Do you confirm the collateral periodically?
23. Are the securities on hand in the name of the company and properly endorsed?
24. Are important business documents such as leases, mortgages, and contracts kept safe and secure?
25. Are the important business documents kept in the custody of a responsible executive?
26. Do you avoid making it known that some pilferage is permissible and will be tolerated?
27. Even if losses diminish, do you consistently mention shortages will not be acceptable?

28. If a markdown policy on damaged or outdated goods exists, are the markdowns made by a top management person?
29. If merchandise is sold "as is," is the item secretly marked to avoid a return and refund at the regular price?
30. Is return merchandise accepted only with a valid sales slip?
31. Is your policy on returns and refunds clearly posted and also mentioned by the salespeople?
32. Are cash refunds handled by a supervisor?
33. Is returned merchandise reticketed and put back into the sales area almost immediately?
34. Can you have a policy of check refunds by mail?
35. Can you have a policy of crediting the customer's account?
36. To find out if a mystery name was used, are refunds spot-checked by telephoning the customer to ascertain if the service was satisfactory?

Petty Cash

1. Periodically, do you analyze the methods of safekeeping your petty cash funds?
2. Are the funds kept in an office safe?
3. Do you avoid disbursing the funds from an envelope?
4. Do you avoid storing the funds in a desk drawer?
5. If an office safe is not available, are the funds kept in a locked metal box which is bolted down to prevent unauthorized removal of the entire box?
6. Is access to petty cash prohibited to all but responsible parties?
7. Do you frequently determine ease of removal of all or part of the petty cash such as during a lunch hour?
8. Have you adopted a policy for the safekeeping of the petty cash?
9. Are periodic checks being made to determine strict policy adherence?
10. Are periodic analyses made to determine if the amount of petty cash on hand is in excess of requirements?
11. Is an effort being made to cut down the use of petty cash to a point where only a small amount of funds are on hand?
12. Does the company keep accurate records of the amounts of capital it advances to the petty cash fund?
13. Do you advocate the use of preprinted petty cash vouchers?
14. Are the vouchers of the type on which the amount of petty cash can be handwritten as well as numerically indicated?
15. Are the vouchers signed by the petty cash user?
16. Are uses of petty cash questioned?
17. Are the reasons for the use of petty cash filed along with the vouchers?
18. If a reimbursing check is issued, does a senior executive void the petty cash voucher?
19. When the petty cash fund is reimbursed, are you certain the reimbursement check is drawn to the individual in charge of the fund?
20. Is it company policy to avoid cashing personal checks out of the petty cash fund?
21. If checks must be cashed, are they properly approved?
22. If checks must be cashed, are matching funds returned to petty cash as soon as possible?
23. Is it company policy to avoid accepting casually written IOUs for advances made to personnel?
24. If advances must be made, does the company obtain proper receipts or documentation?
25. When petty cash vouchers for the advanced amount are returned, are they combined with the initial receipt?
26. Are the petty cash funds counted and balanced daily?
27. Is the counting and balancing done by someone not involved in disbursement?
28. Do you avoid using petty cash funds to buy postage?
29. Are you determining the feasibility of using checks for such purchases?
30. Are the stamp funds kept separate from petty cash?
31. Are the stamp funds kept under the control of an employee other than the petty cash manager?
32. If cash is used to purchase stamps, do you require the postmaster to issue receipts for the amount of postage required?
33. Do you have the petty cash funds audited frequently?
34. Are the audits made on a surprise basis?

Cash Expenditures

1. Is it company policy to have each and every payment made by check?
2. Are all checks numbered?
    a. Do you account for spoiled and voided checks?
    b. Are spoiled and voided checks kept in a file along with those returned with the monthly statements?

3. Is there a security measure used for safekeeping blank checks?
   a. Are current checkbooks kept in a secure vault?
   b. Do you periodically look through the checkbooks, especially at the back, to make sure checks are not missing?
   c. Are surplus checkbooks kept apart and secured from those in use?
4. Do you have a check signing procedure?
   a. Are two signatures required for sums over a stipulated amount?
   b. Is it company policy to prevent signing blank checks?
   c. Are all checks made out with a typewriter and checkwriter?
   d. Before any check is signed and mailed to a vendor, does a responsible employee certify that the items being paid for have been received?
   e. Does the company avoid the use of facsimile signatures?
5. Is there a check mailout procedure?
   a. Do you avoid having the same people who made out the checks or verified the inventory mail the checks?
   b. Do you enclose a memo with each check requesting notification when the vendor receives the check?
   c. Do you avoid "hand" delivery?
6. Is it company policy to avoid advance payments?
7. Is it company policy to avoid issuing checks made out to "cash"?
8. Is a clause printed on your checks voiding them if not deposited within a reasonable time?
9. Are the checks made of a special paper which is warranted against alterations?
10. Do you have a method to handle bank reconciliations?
    a. Is the reconciliation handled by an employee who had nothing to do with originating the checks?
    b. Or one who prepares the deposits?
11. Is there a procedure used to follow-up outstanding checks not presented for payment?
    a. Is it handled by a responsible executive?
    b. Have you a policy which stipulates the elapsed time before follow-up is initiated?
    c. Do you ascertain from the vendor why the check has not been deposited?
12. Is there a periodic returned check inspection system?
    a. Do you check for unfamiliar vendors?
    b. Do you check for alterations of dates, amounts, etc.?
    c. Do you check for illegible and improper signatures?
    d. Do you check for proper bank cancellations?
13. Is every voucher approved for payment?
    a. Is the function separate from the officer drawing the check?
    b. Is the voucher carefully checked for the service or product rendered to the company?
14. Are the items or invoice numbers for which the check is drawn written upon the check?
    a. Are the checks marked "Paid in Full" or "Partial Payment"?
    b. If the check is a partial payment, is the balance clearly stated on the check?

Cash Items

1. Do you periodically identify the individuals handling cash receipts?
   a. Has an indepth background check by an outside firm been accomplished?
   b. Have they been bonded?
   c. Have you established a maximum limit on the amount under their supervision?
2. Do you periodically chart the flow of cash from the place it is initially received?
   a. Are checks and cash delivered directly to the office?
   b. If delivered to the post office, does a responsible employee pick up?
   c. Is other incoming mail handled at the post office?
   d. Does the same responsible individual pick up the mail daily?
   e. Has the ratio of incoming cash to checks stayed the same?
3. Are large amounts of cash and negotiable instruments handled in the office only?
   a. Have you determined the adequacy of your office security?
   b. Do you periodically question ease of office entry and exit?
   c. Is the cashier's office enclosed and separated from other facilities?
   d. Is office entry forbidden to unauthorized personnel?
4. Does a responsible employee open the mail, especially that mail which contains cash and other negotiable items?
   a. Have you double checked and determined if that employee has been bonded and has had

a complete background check?
b. Are the incoming receipts counted by more than one person?
c. After the incoming receipts have been counted, are they listed on a special form?
d. Does another responsible employee check for accuracy?
e. Is final responsibility for the incoming receipts delegated to a senior executive?
f. Do you rotate the cash counting responsibility?
g. Are all cash payments given an immediate receipt?
5. Are remittances brought to the main office?
a. Do you avoid having remittances handled by satellite offices?
b. Do you avoid using field representatives for collections?
c. Do you vary the collection routine?
d. Are the collection routes varied?
e. Do you use flexible collection times?
6. Have you a list of all individuals having access to customer accounts?
a. Is access prohibited to all who receive incoming remittances?
7. If cash registers are used, do you have reliable and bonded people check out and clear the machines at shift's end?
a. Is the person who retains the clearing key a responsible employee?
b. Is the register cleared by a responsible employee other than the cashier?
c. Does your cashier account for overrings and for bad sale ringups?
d. Is a manager called to note the cancelled sale?
e. Do your registers print their own receipts from a self-contained tape roll?
f. Are separate receipts written up by sales people?
g. Are the receipts validated by the registers?
h. Does the customer always receive a validated receipt?
i. Are you immediately checking on those cashiers which turn in an excess of overrings?
j. Are the periodic register checkouts by supervisory personnel accomplished on a surprise basis?
k. Do you avoid having an expected routine, such as having a certain day on which registers are checked?
8. Are you certain the cash registers are of the type which must be completely closed before another new sale is recorded?

9. If register closeouts are handled by the cashier, are the totals verified by management, and the cashier receipted for each closeout?
10. Are shortages logged and signed by the cashier?
11. Do you place limits on register shortages?
12. Do you have a policy regarding shortages and overages?
13. Are the register receipts clearly marked with the store's name and date of sale?
14. Have you removed obstacles or merchandise surrounding the register or cashier which prohibit plain view?
15. Are you alert to signals between cashiers and customers?
16. Have you used a professional shopper to purchase an odd amount item and pay with the exact amount and note if that odd figure was rung up?
17. Do you supply complete uniforms without pockets to the cash-handling personnel?
18. Are you aware that with the new UCC laser reading registers, items can be moved or faced away from the reading device and not recorded?
19. Under the guise of checking a register's accuracy, do you periodically recalculate a shopper's goods?
20. Have you a policy for keeping only a small amount of cash in the registers?
21. If your security system transmits to an agency, do you have a holdup transmit device in a register cash drawer?
22. Do you reconcile the cash register totals with the deposit slips?
23. If your business has a large cash flow, do you make several daily deposits?
24. Are your deposit slips made out in triplicate?
25. Do you retain one and send two to the bank?
26. Does the teller sign and certify the duplicate?
27. Is the duplicate deposit slip mailed back to the firm?
28. Is it received by another responsible employee, and not by the one who prepared the deposit?
29. Are all checks immediately stamped "For Deposit Only"?
30. If you have a return policy, is it marked on the same sales receipt?
31. If an exception to a sales policy is being made, has that exception been authorized by management, and noted on the receipt?

Payroll

1. Are prospective employees thoroughly investigated?

2. Are you aware that the prospective employee's references may not provide sufficient background data?
3. Are gaps in the employment history investigated?
4. Do you use a method or device to check employees in or out?
5. Is a senior supervisor used to prepare the work reports?
6. Do you use job descriptions which indicate the exact nature of the work for which the employee was hired?
7. Does the job description indicate the amount of time required to perform the particular function?
8. Are payrolls prepared by employees not connected with supervision, or who have contact with a multitude of fellow employees?
9. Are the payrolls prepared from timesheets or labor reports?
10. Do you separate the payroll calculation function from the check preparation job?
11. Is the payroll distributed to the employees by an individual who does not have any other payroll or labor supervision function?
12. Are records of hirings, layoffs, firings, vacations, sick time, transfers, etc., kept up to date, duplicated, and verified by department heads?
13. Is undistributed payroll kept secure and made the responsibility of an executive not connected with payroll?
14. Is an executive memo issued validating salary increases?
15. If cash payrolls are used, do the employees sign a receipt?
16. Periodically, are the payrolls distributed on a surprise basis by an executive acquainted with the employees?
17. If the surprise distribution is used, does a responsible executive observe the disbursement?
18. If an employee is not present to accept payroll, is it company policy to avoid having a coworker accept the payroll on behalf of the absentee?

Shipping and Incoming

1. Is the shipping area sealed off from the street?
2. Does the shipping department have a responsible manager?
3. Are all shipping personnel bonded?
4. When merchandise is brought to the firm, is off-loading carefully supervised?
5. Is the merchandise counted and verified?
6. Are key items immediately checked for presence?
7. From time to time, do you determine that packaging on outgoing merchandise is secure and adequate?
8. Are the labels securely attached?
9. Have the labels been prepared by a department not associated with shipping?
10. Are the labels matched to the orders and verified?
11. Do you use a signature method to assign accountability to responsible employees?
12. Are the shipping documents time stamped and signed after the shipper accepts the cargo?
13. If computer data is used, is the waybill information put on line immediately?
14. If your facility originates the shipment, is your copy of the waybill separated from other copies?
15. If your shipments are large, do you attempt to use sealed cargo containers?
16. If you use sealed containers, is the seal number recorded on the waybill?
17. Are loose cartons kept in a secure area prior to shipping?
18. Do the shipping documents and cargo travel together?
19. Are the cargo vehicles sealed, and the seal numbers documented?
20. Are the vehicle drivers instructed that deviation from the prescribed route is prohibited?
21. Are the drivers instructed to avoid picking up unauthorized passengers?
22. Are your vehicles prominently marked with your identification, and numbers on the sides, front, back, and top?
23. When checking in items, is the checker unaware of the amount or count of goods?
24. Are damaged goods immediately removed to a secure area?
25. If incoming goods are stored for any length of time, is the location of the storage noted on the documents or some other appropriate record?
26. Are your inbound and outbound areas physically separated?
27. Are the cargo doors to the shipping area closed and locked when not in use?
28. Are employee vehicles prohibited from parking near the shipping area?
29. Are you certain the consignee's agent or driver has been properly identified?
30. If there is any doubt regarding the consignee's driver, do you instruct your shipping personnel to verify her or his identification with the consignee?

31. Do you use equipment which can photograph the driver and the delivery receipts?
32. Do you use at least two employees to release the shipment, one to pull the items from storage, the other to release shipment after obtaining signed delivery receipts from the driver?
33. Have you instructed the shipping supervisor to monitor the procedures, provide assistance, and spend most of the time on the floor of the department?
34. Periodically, is a physical inventory of the storage area taken, preferably on a surprise basis?

Sales

1. Are all sales invoices serially numbered?
2. Are blank invoices distributed to authorized sales personnel only?
3. Do you record which invoice numbers are assigned to a particular individual?
4. Does the individual sign a receipt for the invoices?
5. Are undistributed invoices kept in a safe area?
6. Are periodic surprise audits made of the blank invoices to determine if any are missing?
7. Does a responsible employee of the firm check the invoice pricing against the company's price lists?
8. Are sales invoices checked against shipping documents?
9. After the goods have been shipped, are the shipping documents routinely sent to the accounting department?
10. If over-the-counter cash sales are a practice, does the customer receive a serially numbered sales slip?
11. Are the blank invoices for these transactions kept in a safe place?
12. Are similar secure distribution procedures in effect for cash sales slips as with corporate invoices?
13. If cash registers are used in the business, are secure procedures outlined in the "Cash Items" section being followed?
14. If credit extension is a prime method of doing business, is it company policy to run a complete credit check of each new client?
15. Is it company policy to recheck the client's credit at least three times yearly?
16. Does your firm rely on its own credit data sources, as opposed to relying on credit information supplied by the new client?
17. Are credits, rebates, and discounts sent to the customer for their acknowledgement?
18. Are periodic audits of the credit made and verified with the client?
19. Are the credit invoices serially numbered?
20. Does the accounting department check all pricings and extensions and indicate that it has done so?
21. Does the accounting department verify the shipping order with the sales invoice?
22. Does accounting originate a monthly summary of the accounts receivable trial balance?
23. Is a copy of the accounts receivable trial balance routinely given to the credit department?
24. Do you balance monthly the general ledger control accounts with the trade accounts receivable ledger?
25. When the company has miscellaneous receivables such as sale of assets, scrap sale, dues from officers, etc., are the receivables separated from the trade accounts receivables?
26. Are accounts receivable aged regularly?
27. Is the aging done by an employee who does not handle the other ledgers?
28. Is a copy of the aging given to the credit department?
29. Are the bad debt write-offs handled by an employee who does not have contact with other accounting records?
30. Does the employee contact the customer to ascertain if the customer has knowledge of the write-off?
31. When bad debt recoveries are made, does a different employee enter the recovery?
32. Is a separate record maintained of items written off by the employee handling the bad debt write-offs?
33. Are invoices sent to the customer after shipment is made, and not before?
34. Do you have monthly statements sent to customers having a balance outstanding?
35. Is the amount owed by the customer checked against the accounts receivable trial balance?
36. Is the statement mailed to the customer by an employee other than the one who prepared it?
37. Is incoming mail pertaining to the receivables directed to an employee other than the one who prepared or mailed the statement?
38. Do you have independent confirmation of the receivables?
39. Is confirmation made by outside auditors?
40. Are the confirmations made on a surprise basis?
41. Is it company policy to prevent outside sales-

people from accepting payments from the customers?
42. If goods are delivered by the salespeople, are they prohibited from accepting payment on delivery?
43. Have you determined the popularity of a salesperson and why the customer would rather wait to be served by that individual?

Purchasing

1. Are the purchase orders serially numbered?
2. Are the requisition forms serially numbered?
3. Are the purchase and requisition forms kept in a safe place?
4. Are the blank forms checked and counted periodically?
5. Are the blank forms checked and counted upon receipt from the printer?
6. Are your purchases priced competitively or subject to competitive pricing?
7. Are the purchases, receiving, and warehousing functions separated and administered by different personnel?
8. Are requisition approvals made by an executive other than the employee who originated the order?
9. Are received goods checked for quality and quantity?
10. Are the quality and quantity checks performed by an executive not associated with the purchasing department?
11. Are the purchase orders checked against merchandise received?
12. When the seller's invoice arrives, is it checked against the purchase order and the received report?
13. After all reports are verified, are they given to another employee for check preparation?
14. Are the checks mailed by the last employee, and not given back to a purchasing department employee?
15. Are the invoices, orders, and other supporting documents marked "Paid" as soon as the check is drawn?
16. Are duplicate payments avoided by careful periodic scrutiny of all ledgers, records, and statements?
17. Are the open account balances periodically confirmed for the sake of the vendor?
18. Are the balances confirmed by outside auditors?
19. Are you certain returned merchandise is properly credited?
20. If an employee of the firm purchases goods from the company, are such accounts kept separate and strictly supervised?
21. If an employee of the firm purchases goods from a vendor to your company, are such purchases kept separate and strictly supervised?
22. Is the function of ordering supplies separated from the inventory check-in procedure and further separated from the payment process?

Chapter 16
Executive Protection Planning*

CHARLES J. DIECIDUE

Major corporations are confronted with serious problems today as a result of possible terrorist activities. Situations such as kidnappings of business executives and their families, along with threats against employees and/or corporate properties to extort sums of money, are causing increasing alarm and concern. Present indications show that this type of fanatical activity is on the upsurge. In order to achieve a realistic view of the current situation, it may be useful to examine the recent worldwide growth of terrorism.

In the past few years, individuals or small groups with unlimited capacity for violence through such tactics as bombing, assassination, extortion, hijacking, and kidnapping, have had a disproportionately large impact on the world. They have succeeded in gaining worldwide attention for themselves and their causes. They have compelled governments and large corporations to negotiate with them on their terms, often winning mammoth concessions.

Rise of Terrorism in U.S.

Most Americans once regarded political terrorism as something that only happened abroad. Unfortunately, it has been introduced into the United States. In the 1960s we experienced senseless explosions, concerted attacks on the police, skyjackings of commercial airliners, and other forms of harassment and sabotage against institutions both public and private. The 1970s initiated an era of kidnapping and hostage-taking of corporate and political figures. In this decade, even more brazen than ever, terrorists have continued their destruction and kidnapping. Hijacking of ships, kidnapping of journalists and negotiators, and blatant executions of hostages have put major powers into a no-win position, forcing them to acquiesce to terrorists' demands.

In the past the crime of kidnapping was primarily a method of extorting money. However, kidnapping has also become one of the most effective political weapons of the terrorist. Terrorists have abducted government officials, diplomats, and corporate executives. Corporate executives, in fact, have become their favorite targets. Placing lives in the balance, terrorists in turn have received release for imprisoned colleagues, military aid, and money and safe passage for themselves to nations offering asylum. Quite ironically, political kidnapping works.

In today's society, the prime threat of kidnapping still lies in the potential for extortion of large sums of money from corporations, financial institutions, and wealthy individuals. In fact, kidnappings for purposes of ransom are increasing at a faster rate than those stemming from political motives.

While kidnapping in the United States is still a rare crime, it has increased in recent years. Although political kidnapping has been heretofore almost nonexistent in this country, target lists of corporate executives have been discovered in the headquarters of extremist groups. Political terrorists in the U.S. could choose targets for kidnappings as they did in the past for bombing attacks. Therefore, executives who own or work for major corporations (especially those with overseas operations), utilities and financial institutions, can be potential targets of a political kidnapping in America.

---

*Reprinted in revised form with permission of Cahners Publishing Co., from transcripts of a seminar presented at the 1978 International Security Conference, Chicago.

Formulating a Protection Plan

The purpose of this chapter is to increase awareness of the nature of the terrorists threat, and to provide a set of guidelines for the formulation of procedures that will reduce the probability of falling victim to a criminal attack. The security suggestions here are by no means all-inclusive. They are intended to provide only a basic foundation in the preparation of a kidnapping program.

In the last analysis, the safety of corporate executives depends to a great degree on the company's own protection efforts. Law enforcement agencies can only offer assistance.

Executive protection programs usually meet the most resistance from the executives themselves. Security is a tough item to sell. Convincing top executives to accept certain precautions which are intermingled with their personal activities is a painstaking task. Executives worry about their image, the cost of protection, and the burden that such procedures place on themselves and their families. More and more today, however, executives are beginning to face the issue of their own protection.

Threat Assessment

The security director or manager must be concerned with every conceivable probability regarding criminal activity against the company and its personnel. If the probability of kidnapping exists, it is the manger's responsibility to prevent or forestall the likelihood. Statements such as, "Kidnapping—that will never happen here," or "You can't do anything about it," signify a sorrowful attitude of fatalism and defeatism which is quite common today, even among security directors. It must be recognized that security is a relative concept. Only degrees of security can be obtained, not absolute protection. Therefore, no security program can be effective unless it is based upon a clear understanding of the actual risks it is designed to control. Until that threat is assessed accurately, precautions and countermeasures—even those of the highest quality, reliability, and reputation—cannot be chosen except by guesswork.

Planning Essential

The value of a security program depends not only upon the excellence of the resources utilized, but upon their relevance. There is nothing more vital to the long-term success of a security program than planning. Planning is essential to develop and implement policies and programs. Consequently, security policy must be sufficiently broad and flexible, as well as dynamic, to insure its effectiveness over time.

Establish Objectives

Prior to the initiation of the planning effort, security objectives should be established, and the type of input that must be furnished to the overall kidnapping program should be determined. Once these objectives are formulated and agreed upon by the executive board members, the planner (security director) can begin to become familiar with both the evolving hardware and the environment within which the kidnapping program is to operate. Precautions should be taken to avoid overreaction, such as creating an Orwellian atmosphere, or, conversely, establishing minimum standards by the creation of a program of ineffectual trivialities.

Home and Office Survey

A security survey of the home and office should be conducted by the security director or another competent individual or firm. The survey concerns three types of entry: surreptitious entry, entry by trick, and openly forced entry.

If an intrusion is necessary to carry out a kidnapping, the abductor will proceed where the least amount of labor and risk is required. Decrease entry potential and increase the probability of apprehension, and the criminal will more than likely pass by for an easier mark. The methods of decreasing the potential for an intrusion follow common sense.

Time is an intruder's ally. We can make it the enemy. There are no burglar-*proof* locks or alarm systems—given enough time, a skilled professional burglar can crack Fort Knox. But every minute it takes to gain access increases the probability of apprehension and an involuntary stay in the penitentiary. Burglary-resistant locks, doors, and windows are stumbling blocks. A well-lighted home area that is visible from the street increases the intruder's exposure and reduces the time she or he is willing to spend there. An electronic alarm system is another time restricting device, and transmitting a silent signal directly to the proper authorities increases the risk of apprehension.

Risk Management Analysis

Weighing budgetary considerations, every company should strive for maximum security. By analyzing risk management elements in the following three stages, the security planner will have a head start in gathering some of the information needed to identify and measure entry vulnerabilities.

Preliminary Risk Analysis

- Who are the top executives in the organization?
- Who are the individuals who have access to large amounts of money?
- Who are the individuals who generate publicity?

Current reports state that 50 percent of kidnapping victims are ordinary people, but many are linked to a source of money, a source of power, or a source of publicity. Kidnappers want all three: money, power, and publicity.

Procedural Vulnerability

- Are door cylinders changed each time any servant or employee with key access severs with the company or home?
- Is there key control?
- Has responsibility been fixed to make sure that all doors and windows are locked and alarm systems activated at proper hours or close of business day?
- Is a check made of all areas such as workshops, closets, and storerooms to make sure there are no stay-behinds?
- Are alarm systems and emergency power batteries tested on a scheduled basis to see that they are operable?
- Is responsibility for security in the home fixed?

Physical Vulnerability

List all means of access into the home or office, such as doors, windows, skylights, elevator shafts, ventilators—any opening more than 18″ square. Against these lists, describe any burglary-resistant devices and alarm systems that are installed. For example: if a central station system is utilized, some questions to be asked are:

- Is there line supervision?
- Are there tamper-proof contacts on all doors and windows?
- What type of alarm system is being used (sonic, photoelectric beam, microwave, etc.)?
- Is the telephone pair box protected against a possible bypass?
- Are the telephone pair lines marked with red flags denoting the alarm company?
- What was the date of the last test of the alarm system?
- What type of test was conducted?
- How far is the central station from the premises?
- What is the average response time by police or guards when an alarm is transmitted?
- Do guards have keys?
- What is the procedure in checking the premises if an alarm has been transmitted?

Other considerations in evaluating vulnerability to intrusion include:

- In proprietary alarm systems, what type of audible bell and/or sirens are in use?
- Are highly pick-resistant locks utilized?
- Is there proper illumination around the home?
- Are telephone numbers of the police, FBI, fire department, and security director easily accessible?
- Check shrubbery that could provide cover for a potential intruder.
- Closed circuit television, aesthetically placed to view visitors at the front or rear doors, can be integrated with a voice intercom.
- As a minimum requirement, a simple peephole should be provided in entrance doors.
- Glass in doors should be replaced with burglary-resistant polycarbonate material to preclude easy entry.
- Explore the possible use of a *buddy* alarm system with neighbors.

Protection in Transit

Kidnapping is a sophisticated crime that usually requires elaborate plans for success. Development of the scheme depends upon observation of the victim's routine. Individuals are most vulnerable to kidnapping attempts when moving from one place to another. Seventy percent of all successful executive kidnappings have taken place while the victim was in transit—in most cases while driving.

The Vehicle

- The vehicle should be in excellent condition, serviced by a reliable organization.
- It should be equipped with an audible alarm system to announce tampering, especially at the hood and door areas. Alarm systems should be extended into the interior of the auto, allowing activation by driver or passengers.
- All vehicles should be equipped with a key-operated locking gas cap. (An individual was apprehended by federal authorities in New York City who manufactured capsules designed to be thrown into a gas tank, causing it to explode in ten seconds. He had sold 150 of these destructive devices, known as *fireflies*, composed of metallic sodium and calcium carbide in a gelatin capsule. A simple locked gas cap would defeat or at least deter such an attack.)
- If possible, two-way radio should be employed in the vehicle.
- The gas tank should be kept at least half-full at all times.
- For greater anonymity, the vehicle should be registered in the name of the corporation rather than the individual executive.
- A company car pool may be established so that each executive does not use the same vehicle every day.
- Company logos on vehicles should be avoided.
- Executives' names should not be posted in company parking areas.
- Executives should avoid using the same taxi stand repeatedly.
- Chauffeurs and relief drivers should be screened carefully before being hired and should be trained in evasive driving maneuvers and the proper procedures to be used in the event of a kidnapping attempt.
- The chauffeur should be instructed to give a prearranged trouble signal before the executive enters the vehicle if the chauffeur is under duress.
- If possible, the chauffeur should not be given advance notice of the executive's destination or schedule.

Executives and their drivers should practice the following rules of the road to thwart potential kidnappers:

- Before entering the vehicle, check the interior. If you observe anything suspicious, such as an unrecognized package, do not enter the vehicle or touch the suspected object.
- After entering the vehicle, lock all doors.
- Do not take the same routes at predictable times.
- Be alert for possible set-up distress situations and carefully evaluate each one.
- Travel with others when feasible.
- Be alert to an auto following you. If you know that you are being followed, phone or radio ahead immediately. Note location, direction you are traveling and description of the auto that is following you and its license number, if possible. Do not stop or force a confrontation. Seek police assistance. If unable to locate the police, activate the alarm or blow the horn to attract attention. The best time to prevent a successful kidnapping is during the attempt. Surprise is the key element in an abduction. Abductors do not wish attention and do not want their victims killed at this time.

If the vehicle is attacked, consider the following. Try to evade the kidnapper's vehicle by veering left and right so that it cannot come broadside or cut you off. Stay in the vehicle. Escape attempts on foot are not advised. Again, attract attention by horn or alarm.

If the auto is about to be boxed in or blocked in, do not stop. Professional training in evasive driving becomes very important in such situations. Large corporations are now spending a great deal of money training their chauffeurs to drive evasively. Attempt to accelerate to move around the obstruction. If this is not feasible, ram the blocking vehicle at the end opposite the engine, causing it to swing or rotate around the engine weight. Outracing an abductor's car is not an end in itself. Speed is only a means to achieve a temporary evasion. Speed entails great risks and should be used only to get to a specific point quickly.

If the abductors are close enough and attempt to fire weapons at the driver, surrender. Do not try to be a hero.

On Foot

- Do not walk close to the building line.
- Walk in populated areas, preferably with others.
- If the walk is not too far, carry a distress alarm switch.
- If you are suspicious of being followed, go to a store, restaurant, bar or other public place, and call the police.
- Always carry change in case you need to make an emergency telephone call.

Traveling by Plane

- If company planes are used, they should be properly secured; pilots should be screened and cleared before employment.
- It is preferable, if possible, to use commercial air carriers.
- The greatest emphasis regarding air travel should be on the auto trip to and from the airport. Accordingly, executives who fly should keep time of arrival and departure confidential.
- If feasible, tickets should be purchased at the airport several hours before departure.

Traveling Abroad

- Executives traveling abroad should familiarize themselves with certain foreign phrases such as "I need the police."
- Establish signals with secretaries, family, and business associates to alert them to trouble.
- Executives associated with overseas operations should be thoroughly briefed about the host country and its political or social climate—as well as the type of kidnappings that have occurred there, who the abductors were, and what methods they employed.
- Make contacts in various police departments, armed forces, or with highly placed foreign government officials. Such contacts can facilitiate collection of local intelligence regarding the safety of company personnel in that country. It is always helpful to know someone in a high position.

Everyday Security Awareness

Family Education

- Security awareness must permeate through every member of the family, including children. The idea is not to create anxiety but to generate security-minded attitudes and habits, whereby every family member will benefit.
- The spouse should be oriented and trained in proper procedures to be taken, not only in kidnapping, but in all emergency situations.
- All servants should be screened before employment. When hired, they should be trained in proper security procedures.

Home Security Suggestions

- The identity of all unknown visitors, such as repair personnel, telephone and utility service personnel, should be verified before they are allowed access to the home. Ask them to show identification, and call their employer to verify their authenticity. Utility companies usually have a telephone number for this purpose.
- Servants should be aware of all new delivery drivers who have possible access to the home.
- Check home (and office) windows regarding easy observance from outside, especially with binoculars.
- Avoid social publicity as to where and when vacations will be taken, or other social functions that pinpoint the executive's location.
- All cleaning personnel should be supervised.
- Be alert to any suspicious individuals loitering on foot; in vehicles, or on bicycles near the home.
- Have an unlisted telephone number.
- Give little information to strangers on the telephone.
- Treat unexpected packages with suspicion.
- Do not become routine in shopping, hair appointments, etc.
- Educate children in the use of the telephone, conversing with strangers, answering the door, etc.
- Children should be accompanied to the school bus stop, and they should use private conveyance when feasible.
- Executives with children in school should insist that school authorities never release the children from school without parental consent. Many school authorities can be easily convinced by a well-dressed individual pulling up in a chauffeured limousine, announcing that he must take the children immediately because the parents have been injured, or some other similar pretext. If a caller states that he is the child's parent, this fact should be confirmed by school authorities. If any doubt exists, the child should not be released. If it is impossible for either parent to be contacted, a trusted individual should be designated to give authorization for release.
- Advise children to be with friends at all times.
- Needless to say, children should be instructed never to enter the autos of strangers, even if the stranger claims to "know Mommy and Daddy."
- Parents should always know where their child is going, and with whom.

If an Abduction Occurs

Telephone Threats

- Telephones of top executives should be equipped with coil pick-ups attached to tape recorders, enabling a possible threat to be recorded. If this is not possible, establish a signal that would alert an associate or secretary to listen in.
- In speaking to the kidnapper over the phone, express willingness to cooperate and ask to speak to the hostage. Beware of hoax kidnappings. Always ask the kidnapper to describe the hostage: the clothes worn, color of eyes and hair, etc.
- Immediately after the first confrontation, hoax or not, notify local law enforcement authorities. Have law enforcement clear all statements given to the press.

Mail Threats

- If a threatening letter is received, avoid having numerous individuals touch it.
- Keep a record of all who handled the letter.

Hostage Behavior

In the event kidnapping does occur, how should the hostage behave? The chances for the hostage's survival depends in part on his or her conduct while in captivity. Although each abduction represents a special set of circumstances, law enforcement experts suggest several common-sense rules for hostages:

- Remain as calm and businesslike as possible.
- Be alert, do not panic, and do not struggle. Even the most rational of kidnappers may react violently when a hostage tries to resist or escape.
- Keep a precise mental diary of all incidents, no matter how trivial, that occur during the period of abduction.
- While being transported, make mental notes of direction traveled, length of time involved, and how fast the vehicle was traveling.
- Make note of landmarks, physical and acoustic.
- Be cognizant of any exceptional sensations, odors, vibrations, that may aid in determining the destination.
- Touch different areas of the auto glass and the auto in order to leave fingerprints.
- Take special note of the abductors—their physical appearance, height, weight, mannerisms and habits, their use of slang or nicknames, their speech patterns and any other conversations which reflect their motives and attitudes.
- Do not become argumentative or provoke your abductors. Remain impartial in any philosophical or social discussions which are initiated by the captors. Conversely, do not become overly subservient, siding with their ideas and philosophy.
- Do not discuss rescue efforts, ransom terms, amnesty offers, or anything else that may obstruct the plans that law enforcement or family are making concerning your rescue.
- Do not attempt to escape unless you have carefully assessed the probability for success. Do so only with the odds completely in your favor.
- Always bear in mind that every possible effort will be expended by your family, company, and law enforcement authorities for your immediate and safe release.

The Role of Law Enforcement

Without a doubt, law enforcement officials will make every possible effort to effect the safe return of the kidnapped hostage. Accordingly, at the crucial point of the kidnapping, the initial notification, it is most important that law enforcement officials are notified and apprised of the total situation.

People who think they can handle a kidnapping by themselves, without police assistance, are very foolish. Law enforcement officers have many informers who can find out things very rapidly that a private party cannot find out. They have a pulse on that is going on in the street. It is very important that they be notified of all details of the incident immediately.

Law enforcement's function throughout the investigation will be to pursue all active leads, check possible individuals involved through records and informants, coordinate activities with the telephone company, postal authorities, etc., and set up highly sophisticated electronic equipment.

The family will, of course, be advised of all appropriate courses of action that may result in return of the hostage. Law enforcement personnel in any kidnapping will always deal with the individual(s) coordinating response for the family. They will not make any move without concurrence from the family.

Law enforcement should be notified of every offer and communication from the kidnappers. Family or friends of hostages should be particularly wary of working out secret or private deals with the kidnappers.

It should be remembered that the law enforcement official's only aim is the safe return of the hostage. The emotional entanglements caused by kidnapping make it very difficult for law enforcement to function smoothly and fluidly. To aid in the efficient operation of law enforcement investigation, wholehearted assistance should be given by the family and friends of the victim. The ultimate benefit gained by this assistance will be the safe return of the hostage.

Chapter 17
Operations of a Guard Force

JOSEPH G. WYLLIE

In the five years from 1982 to 1987, the Uniformed Security Guard industry in the United States flourished. Various industries throughout the United States have engaged the services of private security firms. Others replaced their in-house security guard forces with full-time contract security guards.

A close look at the existing situation in the United States today readily indicates the vital need for security guards. The most recent statistics on record with most large metro police departments and the FBI show a definite increase in crime of all types throughout this country. Due to inadequate police protection in large as well as small cities, private concerns are turning to security guards as a deterrent to major crime.

It is a known fact that private security guards, both contract and in-house, outnumber police as much as two to one. In 1973, security guards and investigative services in the United States amounted to $2.5 billion in sales. The New York Times estimated that in the year 1986 this amount had risen to over $20 billion in sales.

In 1973, *in-house guard forces* amounted to approximately 65 percent of the total force of security guards. In 1986, however, the growth of contract security guards had increased to 65 percent, while in-house forces decreased to approximately 35 percent.

Companies presently using in-house security guards are taking a close look at existing costs with the future possibility of converting to contract guards. The salary increases and the fringe benefit packages negotiated at union contract renewals benefit in-house security guards by increasing their hourly rate costs. In many firms, it has become prohibitive to continue with an in-house force that has become so expensive.

Due to the confidentiality of security programs in various large corporations employing in-house guards, there is little information available. Several large in-house guard forces are in actuality police departments. One well regarded in-house guard force with a 150- to 200-person unit is Grumman Aerospace, Bethpage, L.I., New York.

A wealth of information is available regarding private security contract guards. All available records show that the big three in the United States are Pinkerton's Inc., founded in 1850, the oldest and largest with 40,000 employees. Next in size is Burn's International Security Services, Inc., founded in 1914, with 38,000 employees and then the Wackenhut Corporation, founded in 1954, with 25,000 employees. The balance of the contract guard firms is made up of several substantial firms such as Guardsmark, Globe Security Systems, and Wells Fargo. Then there are some medium sized firms as well as thousands of local guard companies. It has been estimated that there are between 5,000 and 6,000 local guard companies throughout the United States.

The basic mission of a guard force is to protect all property within the limits of the client's facility boundaries and protect employees and other persons on the client's property. This type of service offered by a guard agency must start with the basic requirements. The firm's concept of service must be one of integrity and professionalism, implemented by people with years of experience and expertise in their particular field of security. A guard agency may have an assignment that could be protection of a nuclear power plant to the actual guarding of the Alaska Pipeline. The agency may supply a single guard for a small business or provide total security for NASA's Kennedy Space Center. No matter what the size, the commitment to the concept of service remains the same. Total dedication to the job at hand is the paramount issue.

Today's demands for facility security are complex and diverse. Meeting such demands can require an expertise in the guard agency that defies description. The varied requirements of highly sophisticated clients, with far-flung operating facilities, has called for a highly professional approach to begin to meet some of the client's security problems. Guard clients today could be anywhere from the government of an emerging nation concerned with the development of its internal security capability to a giant petrochemical complex or a nuclear power generating facility.

The watchman service of thirty years ago has evolved into the modern *system approach* to total security. The system approach requires, first, an in-depth analysis of the client's situation and requirements. This assessment of the client's needs is vital to the development of a plan of action for a total security concept.

In the case of a small guard job, such analysis and planning will be brief yet thorough. As larger clients present more detailed requirements, the procedure grows accordingly. And, as occurs frequently in the large agencies, due to worldwide activities, the resolution of extremely complicated and diverse security situations for giant business, industrial, or government facilities at home or abroad results in major analytical and planning procedures for the guard agency management.

Security Guards of the Future

The security guard of the 1980s is a far cry from the night watchman of the 1950s, 1960s, and in most cases even the 1970s. To be considered as an effective guard force in today's competitive market requires a training program that can guarantee a guard of the caliber to make a judgement that could save a multibillion dollar facility from total destruction.

In the past the security guard was principally concerned with protection of the site or facility. The post orders mainly were concerned with calling the fire department in case of fire or explosion. If there was a break-in or intrusion by an individual or group of individuals, then the local police were to be called. For any other problem during the tour of duty, an up-to-date alert list was available for phone calls to the client for decisions not to be handled by the guard force.

With the advent of the terrorism problems throughout the world, the security guard now has an awareness of security that never was considered until the early 1970s. It is a known fact that a successful act of sabotage against a nuclear power plant could result in serious and disastrous consequences to the health and safety of the public.

Security personnel, who are responsible for the protection of special nuclear material on-site and in-transit and for the protection of the facility or shipment vehicles against industrial sabotage, must be required to meet minimum criteria to assure that they will effectively perform their assigned security related job duties.

The new security awareness for guards in our nuclear oriented society has made a radical change in their basic mission. The guard's duties will shift from patrolling to operating of sophisticated equipment. At the time, the guard will remain ready for immediate deployment as part of a coordinated armed response unit.

Capable, confident guards will be no less important in the future, but they will have increasingly greater responsibility and awareness and will need professional training in electronics and other new security requirements. Obviously, it takes more than a snappy uniform and a shiny badge to make a security guard. The guard of the 1990s will be a unique individual, highly trained, specializing in a type of physical security that was unknown as much as ten years ago.

To be competitive in the security guard market of the future, the security firm must be ready to meet the new challenges with new concepts, bold innovations, and unrelenting insistance on high standards.

Guards at a Nuclear Power Plant

The Federal Government has mandated in article 10 CFR 73.55 the basis for security of nuclear power plants. The article is titled *Requirements for Physical Protection of Licensed Activities in Nuclear Power Reactors Against Industrial Sabotage*. In this article, under the physical security organization, four paragraphs stipulate the type of organization that is required:

1. The licensee shall establish a security organization, including guards, to protect the facility against radiological sabotage.
2. At least one full-time member of the security organization who has authority to direct the physical protection activities of the security organization shall be on-site at all times.

3. The licensee shall have a management system to provide for the development, revision, implementation and enforcement of security procedures.
4. The licensee shall not permit an individual to act as a guard, watchman, or armed response person, or other member of security organization unless such individual has been trained, equipped, and qualified to perform each assigned security job duty.

Once the security force has been organized, then the licensee shall demonstrate the ability of the physical security personnel to carry out their assigned duties and responsibilities. Each guard, watchman, armed response person, and other members of the security organization shall requalify at least every 12 months. Such requalification must be documented. Each licensee shall submit a training and qualification plan outlining the processes by which guards, watchmen, armed response persons, and other members of the security organization will be selected, trained, equipped, tested, and qualified to assure that these individuals meet the requirements of this paragraph.

The training and qualification plan shall include a schedule to show how all security personnel will be qualified, within two years after the submitted plan is approved.

Executive Protection

Executive protection is a service designed to guard wealthy persons and top executives around the world —especially in Latin America and Europe—from kidnappers and assassins. This offshoot of the security guard service has grown tremendously since the 1960s; in the 1980s it became a very essential service for most large corporations, both in the United States and overseas.

Most security firms offering executive protection start with a survey that will identify vulnerabilities within the corporate and residence environment, plus review of the various executives' social, recreational, and travel activities. The normal transition from the original survey would be the preparation and implementation of a crisis management program. This program will assist senior executives in developing a corporate response during a crisis situation while maintaining continuity of operations. Within this program will be the development of plans, organizations, and procedures to reduce vulnerabilities to potential threats prior to a crisis, while minimizing loss of assets and reducing corporate liability. Crisis management programs also focus analysis and decisionmaking and demonstrate corporate awareness and preparedness. The security firm's executive protection division provides real-world training, using scenerio formats to exercise the crisis management team's functional areas, which include: legal, personnel, finance, public relations and the negotiator and security. There is also an area where the security firm can provide hostage situation assistance to corporations at the time of an actual hostage situation or extortion demand. The executive protection division also can provide advice and assistance in:

(a) Identifying and obtaining trained hostage negotiators for those corporations desiring such assistance
(b) Developing terms for negotiation, when so asked by the corporation
(c) Establishing methods of ransom payment usually in third country locations
(d) Helping set up operational security matters at the affected location

An additional service offered to the corporation by the security firm is threat analysis. This service includes the following:

(a) evaluation of the threat in specific areas of interest
(b) summary of terrorist/criminal activity and propaganda and assessment of their activity
(c) updated reports of internal situations in specific areas of interest or as requested by the corporation

The executive protection service is responsible for the design and implementation of long-term bodyguard operations, tailored to the corporation's specific situation, usually but not always in response to a stated or perceived threat situation. The international bodyguard operations are coordinated through the security firm's international department utilizing resources from their additional subsidiaries and affiliates worldwide. The domestic bodyguard operations are coordinated through the security firm's network of offices throughout the United States.

The executive protection division of the security firm also can coordinate counterterrorist driver's training through recognized United States and overseas schools specializing in such training. The division negotiates the best price available with the school that, in their professional opinion, presents the best course for the driver. The savings are passed on to the corporate client. On many occasions, the personal chauffeurs of top executives are enrolled in these schools.

Another service that the executive protection division provides is awareness and survival training. This service is offered to corporate executives who travel frequently to high-risk areas. It is also offered to individuals who, by reason of earned or inherited wealth, are targets of criminal or terrorist elements. This training is usually given as an integral part of crisis management but applies to all levels of executives and managers, male or female.

The security firm in the 1980s must now be able to supply executive protection programs which apply experience—validated flexible procedures which bring about tangible improvement in the protected person's security posture. With the terrorism situations in the United States and overseas, the professional security guard firms are now in a position to offer this additional sophisticated service.

Security Services for Air Travel

In 1973, the FAA instituted an antihijack program that is in effect in all major airports throughout the United States. The service is known in the security industry as Predeparture Screening. The security employees at the airports are called PDS screeners. The service supplied has a number of benefits to the air travellers and the airlines which include:

- Prevention of kidnapping and extortion (hijacking and bomb threats)
- Controlled passenger entry
- Acts as a deterrent
- Ensures boarding pass validity
- Detection of weapons and hazardous materials
- Gives a positive public image

The security person(s) must conduct a thorough inspection of ALL passengers and ALL carry-on baggage before anyone or anything is allowed aboard the plane.

Explanation of definitions:

1. Carry-on baggage—Any item the passenger is carrying. This includes all baggage, valet bags, coats, gifts, purses, musical instruments, recording devices, and other items.
2. Dangerous item—Any item that may threaten the safety and security of airline passengers and aircraft. This includes incapacitating gasses, knives, stilettos, fountain pens that might fire a projectile, and cigarette lighters with hidden weapons.
3. Weapons—The word *weapon* is used to denote handguns, rifles, shotguns, and knives.
4. Explosives—Any item that can be triggered to explode. This includes dynamite, black powder, hand grenades, and shells.

The PDS screener's responsibility is to search all passengers and baggage and ensure that none of the above items are carried aboard the aircraft. The PDS screener is to prevent hijacking and criminal acts from occurring.

PDS inspectors use three devices in their screening: (1) X-ray machine, (2) magnetometer, and (3) hand wand.

All carry-on baggage, including pocketbooks, must go through the X-ray machine. All passengers must go through the magnetometer. If the passenger triggers off the magnetometer the second time, then the screener must use the hand wand to find the metal involved.

In 1981, the FAA changed their regulations which required an armed law enforcement officer present at each concourse predeparture screening check point. The FAA changed the requirement to let the airlines use specially trained security personnel from the private security industry to provide the required coverage. The individual involved is now called a check-point security supervisor or CSS and this person is now operating in place of the armed police officer. The CSS is an unarmed, specially trained security officer.

All major airports in the United States are using specially trained security guards for their predeparture screeners (PDS) and their check point security supervisors (CSS). These services have become an integral part of the major airlines security programs.

Liabilities Connected with Guard Force

Various legal aspects of industrial security and plant protection must be fully understood by the security guard.

A guard force is not engaged in law enforcement as such; therefore, the guard is not a law enforcement officer, like a police officer or sheriff. Guards are engaged in the protection of goods and services. The plant management makes the rules regarding the conduct of persons engaged in production. The final end is a smooth flow of production—not law enforcement.

Rules and regulations do not have the same force as law. An employee cannot be deprived of freedom because of breaking a rule or regulation to help production. The most that can be done is to dismiss the employee. Violation of law by someone working in the plant brings the same repercussions as breaking the law elsewhere—the case is under the jurisdiction of law enforcement agencies local, state, or federal. The work performed by a security guard is not related to police work. Execution of the job and training are different. The security guard must leave law enforcement to the responsible agency.

In special situations a security guard may make arrests. A security guard, peace officer, or any other person may arrest an offender without a warrant if the offense is a felony or an offense against public peace. A felony is ordinarily an offense punishable by confinement in a penitentiary for a period of more than one year. Arrests such as these should be made only with the consent of a superior, except in an emergency situation, and only on company property. False arrests and searches can result in civil and criminal suits. A security guard has no authority in a civil case and if required to testify in any civil case, the security guard should report the facts to the supervisor of the force and in turn demand a subpoena in order to testify.

Before making the arrest, the security guard should know that the law has actually been violated, that the violation is a crime, that information proves beyond a reasonable doubt that the person committed the crime. No arrest is legal until after the actual violation of the law. No person may be arrested on a charge of suspicion. The arrest is made by actual restraint of the person or by the guard saying, "You are under arrest." Actual touching of the person is unnecessary—it is enough if the person submits to your custody. The guard has no authority beyond the company property line other than that of a *private citizen*. No person is to be transported as a prisoner off company property by a security guard. The guard must notify the local law enforcement agency and turn the prisoner over to them on the company property. Crimes that may occur on company premises: murder, arson, assault, burglary, larceny, intoxication, violation of sabotage and espionage laws.

When a crime is committed on company property, the guard on duty must take prompt measures to afford protection of the crime scene. In the event of a serious crime, the security guard will not investigate the area. The guard should refrain from touching any evidence in the crime scene area and should prevent unauthorized persons from handling such evidence. The nature of the crime and the type of evidence in the area require that the security guard be extremely careful in moving about so as not to obliterate or otherwise destroy crime evidence.

The security guards will rope off or isolate the area and avenue of entry or escape believed to be used. No one should be allowed to enter or leave the area pending the arrival of representatives of the law enforcement agency having primary investigative jurisdiction. The guard should then obtain the names and addresses of any possible witnesses to be furnished to the law enforcement agency.

Power and Authority of the Security Guards

The accentuation of professionalism in the ranks of law enforcement in the United States has filtered down to the ranks of the contract security guard. Although some of the duties of the security guard are similiar to the duties of the police officer, their overall powers are entirely different.

Recent court decisions have found the security guard is not encumbered by the so-called Miranda warnings of rights. The security guard is *not* a law enforcement officer. Some recent state of Missouri Supreme Court decisions have made the arrest powers of a security guard much easier to understand in this current wave of lawlessness.

As you can see from the following information, the security guard in today's society must of necessity receive a basic training to learn the rules and regulations governing the guard's power and authority.

Private Security Guards Don't Have to Tell Suspects about Their Rights

Between movies and T.V. everybody has heard the expression, "Read them their rights." It refers to the warnings the suspect in custody is supposed to receive before interrogation. Otherwise, a confession can't be introduced at trial. That's why cops chant "You have the right to remain silent. ...."

Recently, some people have claimed that private security guards must also precede their questioning with a recitation of these rights. *Here's the most recent of a series of decisions that indicate private guards need not give these so-called Miranda warnings**:

> The assistant security manager of the K-mart store in Willowbrook, Illinois, one day saw a shopper take a scarf from a rack, tear off the price tag and put the scarf in her purse. Outside the store, the security officer showed the shopper his badge and asked her to come

---
*Excerpted from *You & the Law*, February 2, 1981.
Prepared by: The Research Institute of America, Incorporated.

back to the security office. When asked for the receipt for the scarf, the shopper said, "Oh, I must have forgotten to pay for it." The guard made some comment about the shopper driving 50 miles "to steal at K-mart". Reportedly, the shopper said, "Sure, why not?" Another store employee was present in the security office and later corroborated the guard's testimony.

When the shopper was brought to trial for the theft under an Illinois shop-lifting law, the first thing her lawyer did was attempt to suppress the confession, because the guard had not read her rights. But he failed at the beginning and at the end of the trial. His client's conviction sent him to the Appellate Court in Illinois—but he did no better there.

The higher court pointed out that the U.S. Supreme Court in the Miranda case had defined "custodial interrogation" as "questioning initiated by law enforcement officers after a person had been taken into custody or otherwise deprived of his freedom of action in any significant way". This prompted the court to agree with all others that questioning by private security guards is not a "custodial interrogation" because the private guards are not "law enforcement officers." This is so even when they are acting pursuant to a specific shoplifting statute such as the Illinois Retail Theft Act. (*People v. Rattano*, 401 N.E.2d 278)

Don't let the freedom from Miranda restraints give your security people the impression that anything goes. Whether obtained by the police or by private guards, a suspect's confession must be voluntary to stand up in court.

Missouri

Arrest Power of Private Person (Security Guard)

The Missouri Supreme Court in *State v Fritz*, 490 S.W. 2nd 30, 32 (1973), explained: "in Missouri a private citizen may make an arrest on a *showing* of commission of a *felony* and *reasonable grounds* to suspect the arrested party."

In *State v Parker*, 378 S.W. 2nd 274, 282 (1964), the Court held that a private citizen does have the right "to arrest for certain crimes, such as the commission of a *felony* or the commission of *petit larceny* in *his presence*. But he should be sure both of the crime and of the person. ... a private citizen has the right ... to arrest in order to prevent a *breach of peace* or an affray."

The more recent Missouri cases which hold that a private person possessed of knowledge that a recent felony has been committed may arrest anyone he has reasonable grounds to believe has committed the offense, are: *Helming v Adams*, 509 S.W. 2nd 159 (1974); *State v Goodman*, 449 S.W. 2nd 656 (1970); *State v Keeny*, 431 S.W.95 (1968).

Thus, it is clear that in Missouri a security guard may arrest someone who had committed a *felony*, even though not committed in the presence of the guard, as long as the guard knows that a felony has in fact occurred, and she or he has reasonable grounds to believe the person to be arrested is the perpetrator of the crime.

Regarding *misdemeanors*, according to a 1970 opinion by the Missouri Attorney General: "A private citizen may only arrest for those *misdemeanors* which involve *breaches of the peace, petit larceny committed in his presence*, or pursuant to those powers granted him by virtue of (Section 573.125, Missouri Statutes)."

Section 573.125, Missouri Statutes, pertains to *shoplifting*. It provides that a *merchant*, *agent*, or *employee*, who has *reasonable grounds* or *probable cause* to believe that a person has taken money or merchandise, may detain the person in a reasonable manner and for a reasonable length of time in order to make an investigation. The detention does not constitute an unlawful arrest, and it does not render the merchant, agent, or employee either criminally or civilly liable.

The entire shoplifting statute is set forth:

537.125. Shoplifting: Detention of suspect by merchant—liability presumption:

1. As used in this section:
   (a) "Merchant" means any corporation, partnership, association or person who is engaged in the business of selling goods, wares and merchandise in a mercantile establishment;
   (b) "Mercantile establishment" means any mercantile place of business in, at or from which goods, wares and merchandise are sold, offered for sale or delivered from and sold at retail or wholesale;
   (c) "Merchandise", means all goods, wares and merchandise offered for sale or displayed by a merchant;
   (d) "Wrongful taking" includes stealing of merchandise or money and any other wrongful appropriation of merchandise or money.
2. Any merchant, agent, or employee, who has reasonable grounds or probable cause to believe

that a person has committed or is committing a wrongful taking of merchandise or money from a mercantile establishment, may detain such person in a reasonable manner and for a reasonable length of time for the purpose of investigating whether there has been a wrongful taking of such merchandise or money. Any such reasonable detention shall not constitute an unlawful arrest or detention, nor shall it render the merchant, agent, or employee, criminally or civilly liable to the person so detained.

3. Any person willfully concealing unpurchased merchandise of any mercantile establishment, either on the premises or outside the premises of such establishment, shall be presumed to have so concealed such merchandise with the intention of committing a wrongful taking of such merchandise within the meaning of subsection 1, and the finding of such unpurchased merchandise concealed upon the person or among the belongings of such person shall be evidence of reasonable grounds and probable cause for the detention in a reasonable manner and for a reasonable length of time, of such a person by a merchant, agent, or employee, in order that recovery of such merchandise may be effected, and any such reasonable detention shall not be deemed to be unlawful, nor render such merchant, agent, or employee criminally or civilly liable.

Use of Force by Private Person

Missouri has a specific statute covering a private person's use of force in making an arrest. The Statute, as well as the comment interpreting the statute, are reproduced below:

563.051 Private person's use of force in making an arrest

1. A private person who has been directed by a person reasonably believed to be a law enforcement officer to assist such officer to effect an arrest or to prevent escape from custody may, subject to the limitations of subsection 3, use physical force when and to the extent reasonably necessary to carry out such officer's direction unless she or he knows or believes that the arrest or prospective arrest is not or was not authorized.

2. *A private person acting on his own account* may, subject to the limitations of subsection 3, use physical force to effect arrest or prevent escape only when and to the extend such is *immediately necessary* to effect the *arrest*, or to *prevent escape from custody*, of a person whom he reasonably believes to have committed a crime and who in fact has committed such crime.

3. A private person in effecting an arrest or in preventing escape from custody is justified in using *deadly force only*,
   (a) When such is *authorized under other sections* of this chapter; or
   (b) When he *reasonably believes such to be authorized under the circumstances and he is directed or authorized by law enforcement officer to use deadly force*; or
   (c) When he reasonably *believes such use of deadly force is immediately necessary to effect the arrest of a person who at that time and in his presence*:
      (a) Committed or attempted to commit a *class A felony or murder*; or
      (b) *Is attempting to escape by use of a deadly weapon*.
   (d) The defendant shall have the burden of injecting the issue of justification under this section.
   (L.1977, S.B. No. 60, P. 662, & 1, eff. Jan. 1, 1979.)

Comment to 1973 Proposed Code

Based on Model Penal Code 3.07; Illinois Criminal Code, Ch. 38, 7–5; New York Revised Penal Law 35.30.

In *State v Parker*, 378 S.W. 2d 274, 282 (Mo. 1964), the Missouri Supreme Court stated:

> The private citizen is limited in the power of arrest; but he does have the right, without warrant or other process, to arrest for certain crimes, such as the commission of a felony or the commission of petit larceny in the presence. But he should be sure of the crime and the person. ... All authorities seems to agree that a private person has the right (where not abrogated by statute) to arrest in order to prevent a breach of peace or an affray. We know of no statute which abrogates this right of the citizen in this state.

Authorities cited included *Pandjiris v Hartman*, 196 Mo. 539, 94 S.W. 270 (1906) and *Wehmeyer v Melvihill*, 150 Mo. App. 197, 130 S.W. 681 (1910). *This section deals with the private person acting on his own, or with other private persons, in making arrests, subsection 2; and when he is summoned or directed to assist a law enforcement officer, subsection:*

1. *The section distinguishes the occasions when deadly force can be used. Subsection 1 prescribes the amount of non-deadly physical force that a person can use if*

summoned by a law enforcement officer. As with other sections of this Chapter, the section allows a person to act on appearances provided he does so reasonably. To be justified under subsection 1, the private person must, first, be summoned by one he reasonably believes to be a law enforcement officer; second, use only that amount of force which he reasonably believes necessary to carry out the orders of the officer; and lastly, believe the arrest lawful.

Subsection 2 prescribes the amount of nondeadly physical force a private person may use when acting on his own account, which impliedly includes acting in conjunction with other private persons. The applicability of Subsection 2 is contingent on the private person having a *reasonable belief* that the person to be arrested has committed a crime and that such person in fact has committed such crime. Again the defense is dependent on *using physical force only as a final means of effecting an arrest.*

The section makes a slight modification in Missouri law. *The section authorizes the use of physical force even when the crime was committed out of the presence of the private person.* However, the in presence requirement announced in *State v Parker,* supra has not been strictly adhered to by Missouri courts. For example, in *State v Keeney,* 431 S.W. 2d 95 (Mo. 1968), the Missouri Supreme Court held that where a private person had been advised by the victim of a crime as to the description of the robber's automobile and 16 minutes later such person observed the automobile fitting the description in another state, he had the authority to arrest the occupants of the automobile and search the same. The safeguards that a private person must reasonably believe the person did in fact commit the crime removes the need for the "in presence" requirement as to the use of non-deadly physical force.

Under *subsection 3* the use of *deadly force* by a private person effecting an arrest is *authorized only if it is allowed under another section of this Chapter, as for example in self-defense under 563.031; or when he is directed by a law enforcement officer to use deadly force and he reasonably believes such to be authorized; or when it is necessary in the arrest of a person who has committed a Class A Felony or murder or who is attempting to escape by using a deadly weapon. Subsection 3 (2) authorized the use of deadly force when the private person is directed to use deadly force by the officer he has been summoned to assist.* The private person must, however, reasonably believe the use of deadly force to be authorized under the circumstances. Mistakes will not vitiate the applicability of the justification unless such mistakes were unreasonable.

Subsection 3(3) is similar to the corresponding paragraph in the preceding section, 563.046, subsection 3(2). However, there are two significant differences. First, *as to the private person, the situations giving rise to the use of deadly force must occur "at that time and in his presence." Thus, the private person must personally detect the crime and immediately thereafter attempt to effect the arrest. Secondly, the situations in which the private person is justified in using deadly force are more limited than those in which a law enforcement officer may use deadly force. For the private person, it must involve a Class A Felony or murder or attempted escape by use of a deadly weapon.*

(A Class A Felony is a crime for which the penalty is death, life imprisonment, or imprisonment for a term of twenty years or more. Examples in addition to murder are: robbery, kidnapping, causing catastrophe, rape, and first degree assault with a deadly weapon.)

Definitions for Misdeameanors and Felonies

The Missouri Statute defining these terms is as follows:
556.016 classes of crimes:

1. An offense defined by this code or by any other statute of this state, for which a sentence of death or imprisonment is authorized, constitutes a "crime". Crimes are classified as felonies and misdemeanors.
2. A crime is a *felony* if it is so designated or if persons convicted thereof may be sentenced to death or imprisonment for a term which is *in excess of one year.*
3. A crime is a *misdemeanor* if it is so designated or if persons convicted thereof may be sentenced to imprisonment for a term of which the *maximum is one year or less.*

(L.1977, S.B. No.60, p. 662, 1, eff. Jan. 1, 1979).

Crimes are classified for the purpose of sentencing into Class A through Class D felonies, and Class A through Class C misdemeanors.

4. *Larceny* is included in the Missouri Statute entitled *Stealing* Section 570.030, Missouri Statutes. It provides that stealing is a Class C *felony* if the value of the property or services appropriated is *one hundred fifty dollars ($150.00) or more.*
5. *Good Samaritan Act*: The Missouri Good Samaritan Act which is reproduced below applies only to those persons trained to provide first aid. The

Statute would therefore apply to guards who have had legitimate first aid training.

190.195 Personal liability for civil damages removed in certain emergency care situations.

Any person who has been trained to provide first aid in a standard, recognized training program may render emergency care or assistance to the level for which he or she has been trained, at the scene of an emergency or accident, and shall not be liable for civil damages for acts or omissions other than damages occasioned by gross negligence or by willful or wanton acts or omissions by such person in rendering such emergency care.

Laws 1973, p. 306, 20, effective July 1, 1974.

Training

In view of the demands of industry for fully trained security guards, a new phase of the guard industry has come into view. To give the necessary training required for the basic guard who could be working on a one-guard site up to the basic guard working at a nuclear power plant, a new look has been given to guard training.

Training today must be organized so as to provide the initial or basic training as well as the follow-up programs necessary to maintain quality standards for the personnel. Most professional security agencies offer at least a basic security officer's program. These programs can run as long as twenty-four hours and cover subjects ranging from laws of arrest to weapon safety. The present system attempts to package the training in a practical delivery system and to keep quality high in terms of testing.

Many of the basic training courses are tailored to individual client needs. In recent years a number of states have mandated requirements for security officers and most states have mandated requirements for weapon training.

Another offshoot of the training for guards has been the predeparture screening services required at all airports across the United States. In screening carry-on baggage with an X-ray machine, the ability to detect the outline, shape or form of a weapon or an explosive device is critical. The FAA has mandated a training program that must be uniformly given to every screener at every airport in the country. It has been necessary to develop audio-visual training programs prepared in cooperation with the Federal Aviation Agency. Besides instruction in the techniques of discerning the outlines of X-ray screens, the course offers new pointers in the use of magnetometers, both the walk-through models and the hand-held wands.

As an illustration of the completeness of the Basic Guard Training Program, you will see in the chapter appendixes an outline of an existing program. The importance of training cannot be emphasized enough. The modern day security officer has a great many responsibilities and is often required to make important decisions under sometimes trying conditions. Only thorough training will provide assurance of an effective and competent guard operation. The primary duties of guards and supervisors are treated below.

Report Writing

Very few people like paperwork, yet it seems that the occupation does not exist where paperwork is not required. For the security officer, the paperwork is in the form of reports. There are four basic reasons for completing so many reports.

1. *To inform.* Written communications reduce the chances of misunderstandings or errors. Verbal communications, however, are highly prone to misunderstandings, errors in reproduction, and can be easily ignored.
2. *To record.* Never trust memory. No memory is perfect. Exact amounts, costs, dates, times, and similar data are easily forgotten unless recorded.
3. *To demonstrate alertness.* By recording incidents, the security officer makes both supervisor and client aware of the job being done. It is very easy for people to get the impression that security officers do little but stand around. One way of avoiding this type of image is to conscientiously document all incidents.
4. *To protect yourself.* There may come a time when it becomes necessary for a security officer to prove to have witnessed an event, accomplished a certain action, or notified the proper authorities of an incident. The reports will accomplish all four of these goals.

The report should be clear and concise. A good report answers five basic questions:

1. *What?* The report must state what happened as accurately as possible.
2. *Where?* The exact location of an occurrence can have great bearing in establishing guilt, innocence, or liability.
3. *Who?* When writing a report, the officer should

answer as many whos as possible; for example, who did it and who was notified?
4. *When?* The when of an incident may establish an alibi, or help to prevent damage, theft, or injury.
5. *Why?* The why involves judgment and opinion and may not be easily proven, but it may be very important in judgment of guilt or liability.

In addition to answering these questions, there are simple guidelines to follow when preparing a report to assure that the final result is clearly written and well organized.

1. Use simple language which anyone can understand. When using technical words and phrases, be sure the meaning is clear. Avoid using slang terms or words that have multiple meanings.
2. Be sure that you use the proper spellings and addresses of the individuals involved in the report.
3. Prepare the report in such a manner that the happenings are in logical sequence and, when possible, show the approximate time of the occurrence.
4. Do not ramble. It is preferable to use short paragraphs, with each covering one particular point.
5. Do not use vague descriptions. Write only specific observations.
6. When descriptions of individuals are obtained, list all the usual manners of description such as height, weight, color of hair, etc., but also include unusual details such as presence of a mustache, sideburns, eyeglasses, and any peculiarities of walk of speech. Notice and report all information possible on types and color of dress.
7. Avoid contradictory statements that would tend to discredit the overall information.
8. Facts, not fiction, are important. If you include your opinion, label it as your opinion, not as a fact.

Any problem, from a missing light bulb to a major safety hazard, should be reported. The security officer should continue to provide written reports on any incident until appropriate action is taken to correct the situation. In this way, you can demonstrate your importance to the client.

Weapons Safety

No part of the training of a security officer is more critical than firearms training. Your life, as well as the lives of others, depends upon your skill with a revolver and knowledge of its proper and safe use.

Safety is the basic reason for the existence of security personnel. They are employed to assure the safety of persons and property and should always reflect this concept.

Weapons safety, unlike any other aspect of a security officer's job, places a great demand upon skills, knowledge, and the judgment necessary to best use both. Judgment can be exercised only when the factual basis for making such judgment is present. In this case, the principles of firearms safety must be well understood by security officers before any judgment can be made.

The first principle of weapons safety is control. The officer must control the firearm when wearing, storing, and firing it.

Wearing a Firearm

When on duty, the officer's weapon must be readily available for immediate use. It should be worn in a manner that permits swift access while also offering maximum safety. To satisfy this requirement, the weapon should be worn at the belt line and on the same side as the strong shooting hand.

The weapon should always be carried in its holster. Any other method, such as tucked in the belt, is hazardous and has contributed to self-inflicted gunshot wounds. The holster strap or flap should be kept securely snapped over the gun. This prevents the weapon from accidentally falling or being jarred out of the holster. It will also prevent someone from grabbing the revolver.

The weapon must always be loaded when worn on duty. An unloaded revolver is a hazard to the wearer. Drawing an unloaded gun in an invitation to be shot.

When the shifts change and the revolver must be transferred from one officer to his relief, the weapon should be empty. Never transfer a loaded weapon. More accidents occur at this time than at any other time of duty.

When transferring a weapon, unload the gun and hand it to the person receiving it with the breach open. The cartridges should be transferred separately. An additional benefit is derived from this procedure. The relief officer must check and load the weapon prior to assuming the duties of the post.

Storing a Weapon

Common sense demands that all firearms be kept out of the reach of children and irresponsible adults.

Unloaded weapons should be locked up at all times and cartridges should be secured separately from the weapon. Never store a loaded weapon.

Firing a Weapon

The security officer must keep the weapon under control while firing it. This statement may seem obvious, but it is often misunderstood. Control, in this case, refers to the mental discipline required to know when not to fire as well as the physical control necessary to hit the target. Consider these situations:

1. An armed intruder is firing at you. There is a crowd of bystanders behind. Do you return fire?
2. A saboteur is on a four-story rooftop in a crowded facility, well silhouetted against the sky. Do you shoot?
3. An arsonist is standing in front of a light frame building. You do not know if anyone is inside. Do you shoot?

The answer to all three questions is "no." In the first situation, returning fire would most assuredly endanger the bystanders. In the second case, the path of the bullet, after passing the target, could injure or kill a person several blocks away. In the third instance, the bullet could penetrate the frame building and kill an occupant, even after passing through the target. Never underestimate the penetrating power of a gun.

Control in firing also means having the mental discipline to never draw a weapon unless there is the intention to kill the target to protect life itself.

The Guard and the Revolver

No publication can describe all the cases in which a guard should and should not use a firearm. It is possible, however, to present some general guidelines and some specific examples. The guard who considers these carefully, and discusses them with the supervisor and fellow guards, should be able to develop good judgment in the use of a firearm.

The first thing an armed security officer should keep in mind is the fact that an error in the use of a firearm will probably have a long lasting, perhaps permanent, effect. It is necessary therefore to give long and careful consideration of the answers to the questions: "Why do I have a firearm?" "When should I use it?" "When should I not use it?"

While a private security officer, like a police officer, is armed, you should not confuse your rights and responsibilities concerning firearms with those of your public counterpart. There are specific and definite laws governing the police officer and the use of a firearm. There are laws, just as specific and just as definite regarding a private officer's use of a weapon.

A firearm is a symbol of a guard's authority and duty to carry out specific tasks as ordered by your employer. The police officer's duties and responsibilities are obviously much broader. The police officer can arrest suspects, a security guard cannot. A police officer can use a gun to stop a speeding automobile; a guard cannot. A police officer can use a weapon to protect property and, again, a guard cannot.

To simplify matters a bit, the security officer may use a firearm to protect a life, and only to protect a life. That life may be your own or that of a bystander. In any case, your use the gun only to protect a life.

When *not* to use a firearm? Fortunately, there are many more of these instances. Do not use a weapon:

- to prevent a theft
- to stop a fleeing suspect
- to stop a speeding automobile
- to stop someone from bothering or harassing.
- on someone who would like to harm you but cannot; for example, a knife wielder or club wielder who is restrained by a fence or gate, or by other people
- to fire warning shots at a fleeing criminal
- to attempt to frighten people

Safety

Accident prevention is said to be everybody's job, but, as everybody's job, no one does too much about it. It does, however, fall well within the domain of security personnel. It is the security officer's responsibility to observe all unsafe conditions and to warn people of potential hazards. It is also your responsibility to report any violations of safety rules and to set a good example by your own behavior.

Far too many accidents happen due to unsafe conditions which were not noted, reported, or corrected. After finding an unsafe condition, the officer must do one of two things: correct the condition or report it to someone who can make the correction. If a storm blows a power line down, the security officer should report it. If, on the other hand, you find a bag of oil rags in the corner, you would simply place them in a metal covered container and report it later. Safety is purely a matter of common sense.

Corrective action should be taken when possible, or the proper authority should be called to handle the situation.

It is important that the security officer undertake the sometimes thankless task of safety. It is important both to the client and to the people he is protecting from injuries due to careless safety practices.

Safety Checklist

1. Are the floors kept clean and free of dirt and debris?
2. Are rough, splintered, uneven or other floor defects repaired or the hazards suitably marked?
3. Are nonskid waxes used to polish floors?
4. During bad weather, are storm mats placed near entrances and floors mopped frequently?
5. Are stairways equipped with handrails?
6. Are steps equipped with handrails?
7. Are stairways well lighted?
8. Are electric fan or heater extension cords tripping hazards?
9. Are cords of electric fans or heaters disconnected from the power source when not in use and at the end of each working day?
10. Are electric fans or heaters adequately grounded?
11. Are cigarette or cigar stubs placed in suitable ashtrays or containers?
12. Are grounds free of debris, etc.?
13. Are sufficient containers provided for trash, ashes, etc.?
14. Are floors free of oil spills, grease or other substances which create a slipping hazard?
15. Are windows clean?
16. Is broken glass in evidence?
17. Are the aisles clearly defined and free of obstruction?
18. Is material neatly stacked and readily reached?
19. Does piled material project into aisles or passageways?
20. Are tools lefts on overhead ledges or platforms?
21. Is the lighting adequate?
22. Are materials stored under or piled against buildings, doors, exits or stairways?
23. Are walks kept clear of obstructions, slipping and tripping hazards, broken glass, snow and ice?

Bomb Threats

Bomb threats are a serious concern to all security personnel. Fortunately, most bomb threats turn out to be false alarms, but the next encounter with such a threat may turn out to be real, so none should be taken lightly. All bomb threats should be treated with quick, calm, steady professional action.

Normally, local police authorities will be notified by client management when a bomb threat occurs. Upon receiving a bomb threat, a security officer's first duty is to notify the client immediately and to take the action ordered. If ordered to call the police, you should do so and then evacuate anyone in or near the facility. The handling of bombs and bomb disposal are police duties. The security force's job is to assist the police in finding the bomb and in evacuation proceedings.

The security officer should *NOT* attempt to examine a bomb, regardless of any previous experience you may have had in the world of explosives. Many bombs are extremely complicated and designed to explode when any attempt is made for deactivation. Only trained demolition experts are qualified to safely handle a bomb.

Bomb Search

The number of locations where a bomb may be hidden are innumerable, and only the most obvious places can be searched in a reasonable amount of time. However, most facilities have areas which are generally more vulnerable than others and should be checked first. The following thoughts should be kept in mind when searching for a bomb:

1. Do not touch anything that does not have to be disturbed. If lights are off, do not turn them on. If fuse panels are turned off, do not activate them. These may be wired to detonate explosives.
2. Most bombs which have actually been found were of the time-mechanism variety. The timing devices are usually cheap alarm clocks which can be heard ticking at surprising distances. Be on the alert for ticking sounds.
3. Bombs found in searches were usually found near an exit. Look closely in areas near doorways.
4. Be alert for objects which look out of place, or are of unusual size or shape.
5. Thoroughly check any areas which are accessible to the public. Rest rooms and janitors' closets are frequently used as hiding places.
6. A bomb search should be conducted for a period of twenty to thirty minutes. This should provide ample time for a reasonable search, without creating unnecessary danger to the searchers.
7. A methodical search technique is necessary to ensure that no areas are overlooked. An orderly

investigation of all rooms within the facility is mandatory. It is wise to prepare a checklist of places to be searched in advance so that a thorough search can be conducted.

8. As you search, be alert to:

- freshly plastered or painted places
- disturbed dirt in potted plants
- pictures or other hanging objects not straight
- ceiling tiles that have been disturbed
- torn furniture coverings
- broken cabinets or objects recently moved
- trash cans, air conditioning ducts, water fountains
- elevator shafts, phone booths

Precautions

A security officer can assist police by observing the following precautions:

Don't:

- Touch a bomb.
- Smoke in the immediate vicinity of a suspected bomb.
- Expose the bomb to sun. Direct rays of the sun or light of any kind may cause detonation.
- Accept identification makeup as legitimate. Don't take for granted the identification markings on packages and boxes as they may have been forged. Keep in mind that bombs are usually camouflaged in order to throw the recipient off guard. Don't take for granted that the package is bona fide because of its having been sent through the mail. Many bombs are forwarded in this manner. Others are sent through express agencies, while some are delivered by individual messengers.
- Take for granted that it is a high explosive bomb. Be prepared in the event that it is of the incendiary type. Have sand and extinguishers on hand.
- Use two-way radios as transmitting could detonate a bomb.
- Have unnecessary personnel in the immediate area of the suspected bomb or explosive.

Do:

- Evacuate the building or area around the suspected bomb, only if the client orders it. In large cities, this function is usually performed by the fire department. Only vital and necessary personnel should be allowed within 100 yards of the package.
- Remove all valuable equipment, important files, computer tapes, etc. at least 100 yards away from the package.
- Open all windows and doors in the immediate vicinity of the suspected devices. This allows the blast to escape, thereby reducing pressure on the walls and interiors. It will also reduce window breakage and the hazards caused by flying glass and debris.
- Shut off all power services to the area *immediately*. This reduces the possibility of gas explosion or electrical fires.

Types of Explosives

Blasting caps or detonators are:

- Metallic cylinders approximately 2 inches long, 3/16 inches in diameter closed at one end (may be larger or smaller).
- Partially filled with a small amount of relatively easily fired or detonated compound.
- When fired, the resultant shock or blow is sufficient to detonate explosives.
- *Very* dangerous to handle, as they can be detonated by heat, friction, or a relatively slight blow.

Nitroglycerin is:

- Colorless to yellow liquid with a heavy, oily consistency.
- Highly dangerous—extremely sensitive to heat, flame, shock or friction.

Dynamite is:

- High explosive, usually cylinderical in shape, size: 1¼ inches diameter and approximately 8 inches long, (may be up to 12 inches diameter, and 30 inches long).
- Outer wrapper often covered in parafin and *usually* marked "DANGEROUS—HIGH EXPLOSIVE."
- Shock sensitive—needs a blasting cap for detonation.

Fire Protection

Of the many jobs a security officer performs, one of the most important is that of fire protection. To do the job effectively, you must be familiar with fire fighting equipment and know how and when to use it.

Fire is comprised of three elements: heat, fuel and oxygen. Remove any one of these three and the fire will go out.

If a fire should break out the following directions will most effectively safeguard persons and property against harm and damage:

1. Call the fire department first.
2. Direct all employees out of the burning building and keep them out after evacuation.
3. Notify and enlist the help of the company fire brigade if one exists.
4. Check and close fire doors.
5. Shut off machinery, power and gas.
6. Check to see if gate valves are in working condition, if a sprinkler system exists.
7. Now and only now, attempt to control the fire by means of an extinguisher.
8. Post someone to direct the firefighters to the fire.
9. Remove motor vehicles from the area.
10. Once the fire has been contained, keep a close watch on the area to see that the fire does not start again.
11. Be sure all extinguishers used are immediately recharged.
12. Complete a written report covering all of the information about the fire.

Fire Prevention

The best way to fight a fire is to prevent a fire from starting. Following is a list of things that you should be alert for while on patrol to eliminate sources of fire and obstructions that might lead to fire spreading:

1. Look for violations of no-smoking regulations.
2. Investigate any unusual odors, especially smoke and gas. Don't be satisfied until you have found the cause and action has been taken.
3. Check for obstructed passageways and fire doors.
4. Look for obstructions in front of fire-alarm boxes, extinguishers and fire hydrants.
5. On every patrol, check all gas or electric heaters, coal and kerosene stoves to see that they do not overheat.
6. Check to see that boxes, rubbish or hazardous materials are not left close to stoves, boilers, steam or smoke pipes.
7. Check to see that all gas or electric appliances not in use are disconnected.
8. Check to see that all discarded and disposable materials have been placed in their proper containers.

Emergency Medical Assistance

It is possible that a security officer will be present when someone needs medical assistance. The first reaction should be to summon help. If this is not possible, the officer should be prepared to assist the victim. Guards should be trained in emergency medical assistance (EMA) procedures in the event a severe accident occurs. Someone's life may depend on your knowledge of EMA.

At the Scene

People at the scene of an accident will be excited. A security officer must remain calm, dealing with the most serious injury or condition first. The most urgent medical emergencies which require prompt action to save a life are: severe bleeding, stoppage of breathing, and poisoning. Shock may accompany any of these, depressing the body functions and keeping the heart, lungs and other organs from functioning normally.

What to Do First

1. Don't move the injured person, unless it is absolutely necessary to save the victim from danger. If the victim has been injured internally, or if the spine is broken, unnecessary movement may kill or cripple him.
2. Act fast if the victim is bleeding severely, has swallowed poison or has stopped breathing because of drowning, gas poisoning or electric shock. Every second counts. A person may, for example, die within three minutes of the time breathing stops, unless given artificial respiration.
3. Because life-and-death emergencies are rare, in most cases a guard can start EMA with these steps: Keep the patient lying down quietly. If she or he has vomited and there is no danger that the neck is broken, turn the head to one side to prevent choking. Keep the victim warm with blankets or coats, but don't overheat or apply external heat.
4. Summon medical help. The doctor should be told the nature of the emergency, and asked what should be done.
5. Examine the patient gently. Cut clothing, if

necessary, to avoid movement or added pain. Don't pull clothing away from burns.
6. Reassure the patient, and try to remain calm. Calmness will convince the patient that everything is under control.
7. Always be prepared to treat shock.
8. Do not force fluids on an unconscious or semiconscious person. Fluids may enter the windpipe and cause strangulation. Do not try to arouse an unconscious person by slapping, shaking or shouting. Do not give alcohol to any victim.
9. Following any incident where EMA would be rendered, a detailed written report should be made covering all of the circumstances. Be sure to include the treatment given.

Controlling Bleeding

The adult human body contains approximately six quarts of blood. Although an adult can readily withstand the loss of a pint, the amount usually taken for transfusion purposes, that same loss by a child may have disastrous results. In an adult, lack of consciousness may occur from the rapid loss of as little as a quart of blood. Because a victim can bleed to death in a very short period of time, immediate stoppage of any large, rapid loss of blood is necessary.

Direct Pressure

The preferred method for control of severe bleeding is direct pressure by pressing a hand over a dressing. This method prevents loss of blood from the body without interfering with normal circulation.

Apply direct pressure by placing the palm of the hand on a dressing directly over the entire area of an open wound on any surface part of the body. In the absence of compresses, the fingers or bare hand may be used, but only until a compress can be obtained and applied.

Do not disturb blood clots after they have formed within the cloth. If blood soaks through the entire compress without clotting, do not remove, but add additional layers of padding and continue direct hand pressure, even more firmly.

On most parts of the body, a pressure bandage can be placed to hold pads of cloth over a wound. Properly applied, the bandage will free the hands for another EMA.

To apply the bandage, place and hold the center directly over the pad on the wound. Maintain a steady pull on the bandage to keep the pad firmly in place while wrapping the ends around the body part. Finish by tying a knot over the pad.

Elevation

If there is no evidence of a fracture, a severely bleeding hand, arm or leg should be elevated above the level of the victim's heart. Once elevated, the force of gravity will reduce blood pressure at the site of the wound and slow the loss of blood. Elevation is used in addition to direct pressure.

The combination of pressure and elevation will stop severe bleeding in most cases; however, there are times when additional techniques are required. One additional technique is pressure on the supplying artery.

Pressure on the Supplying Artery

If severe bleeding from an open wound of the arm or leg does not stop after the application of direct pressure plus elevation, the use of pressure points may be required. Use of the pressure point technique temporarily compresses the main artery which supplies blood to the affected limb against the underlying bone and tissues.

If the use of a pressure point is necessary, do not substitute its use for direct pressure and elevation, but use the pressure point in addition to those techniques. Do not use a pressure point in conjunction with direct pressure any longer than necessary to stop the bleeding. However, if bleeding recurs, reapply pressure at a pressure point.

Pressure Point: Open Arm Wound

Apply pressure over the brachial artery, forcing it against the arm bone. The pressure point is located on the inside of the arm in the groove between the biceps and the triceps, about midway between the armpit and the elbow.

To apply pressure on the brachial artery, grasp the middle of the victim's upper arm, your thumb on the outside of the victim's arm and your other fingers on the inside. Press your fingers toward your thumb to create an inward force from opposite sides of the arm. The inward pressure holds and closes the artery by compressing it against the arm bone.

Pressure Point: Open Leg Wound

Apply pressure on the femoral artery by forcing the artery against the pelvic bone: The pressure point is located on the front center part of the diagonally slanted *hinge* of the leg, in the crease of the groin area, where the artery crosses the pelvic bone on its way to the leg.

To apply pressure to the femoral artery, position the victim flat on the back, if possible, and place the heel of your hand directly over the pressure point. Then lean forward over your straightened arm to apply the small amount of pressure needed to close the artery. To prevent arm tension and muscular strain, keep your arm straight while applying the technique.

Call for Assistance

Whenever possible, get medical assistance as soon as you have made the victim comfortable and are sure the person's life is not in immediate danger. Often you can do more harm than good if you don't summon proper help immediately.

If in doubt as to a victim's well-being, keep the person quiet, preferably lying down and covered. Sometimes a concussion victim will appear perfectly normal and insist upon returning to work only to collapse later. In any case, do not allow the victim to move around. Remember, your greatest contribution to a victim's well-being may be to restrain efforts to move the person in a mistaken belief that such efforts are helpful. It is usually best to let the victim remain calm and relaxed before transporting to the medical station. Obtain professional help whenever possible.

Reporting a Medical Case

When reporting a medical case, the following information must be given clearly so that the necessary equipment and medical assistance can reach the victim in the shortest possible time:

- Exact location and phone number from which you are reporting
- Type of injury, if evident
- Seriousness of injury
- Number of persons involved
- Visible symptoms, such as heavy bleeding, poison stains, etc.
- Cause of injury, if known, so that adequate personnel may be sent to the area to handle such dangerous conditions as leaking gas, flowing chemicals, etc.

Guard Supervision

In every business organization, different management levels exist that are responsible for various tasks. At the top of the structure are people who must decide the organizational goals and policies. At the opposite end of the operational spectrum are those who are immediately responsible for the accomplishment of established goals. Between top management and these workers are the people who must explain management's objectives to all employees. These people give guidance and leadership. They represent top management to the workers by setting standards, developing work schedules, training employees, and exercising necessary controls to insure quality performance. A guard supervisor is one of these important people.

The Supervisor

A supervisor, the person in the middle, is the key to success. The greater your ability to carry out your responsibilities, the more efficiently the company will operate.

In addition to job skills, a modern supervisor must be familiar with up-to-date personnel practices and the legal requirements that affect the jobs of your personnel. You must also know how to deal with the day-to-day problems of a security department.

One of the most important ways a supervisor can get the best results from the people is to let them know they have your full support. You can reinforce this knowledge by giving the employees the necessary authority to do their jobs, and by seeing that this authority is respected. You should step in to share responsibilities and, if things go wrong, help to clear up the problem without condemnation. As happens on occasion, a good worker may run into controversy. When this occurs, it is comforting to know that the boss will stand by. This does not mean insisting someone is right when clearly he or she is not, but rather it is accepting some of the responsibility for a poor plan and helping someone to carry the blame. All these steps will demonstrate a supervisor's support of the crew, and people support a leader who supports them.

Another important trait of a good supervisor is

willingness to accept suggestions from the workers. In fact, encourage such comments. It is natural for people to offer suggestions. A supervisor who makes it clear that you are not interested in such input cuts off an important flow of communication between yourself and your staff. Once the employees realize their supervisor is not interested in their ideas, maybe even resents them, they will not take the time to devise a better system of doing things.

Making the mistake of ignoring the thoughts and ideas of another person will hinder working relationships within the company. One person cannot think of everything. Those employees most knowledgeable in a specific area could be of assistance and should not be overlooked. The people who handle the day-to-day situations are in the best position to suggest changes in the organization's policies and operations.

The best way to get more suggestions from the staff is to simply ask for them. Whenever a problem arises, the supervisor should discuss the situation with the people involved to further encourage input. By offering them the chance to do some of the thinking, the manager is openly demonstrating interest in their ideas. Most employees would love to do some brainwork.

Keep Communications Open

While not every idea submitted will be a workable one, no suggestion deserves the fifteen-second brush-off. The supervisor must be appreciative of all suggestions, regardless of caliber. Each and every idea merits consideration. The employee should be thanked for the time and interest and encouraged to keep trying, on the premise that the next idea could be a winner.

Leadership

The guard supervisor sets the example of professional quality for the staff. The subordinates are a mirror of the management. If a guard appears sloppy, unshaven, in need of a haircut and a shoeshine, his supervisor probably needs to take a good look at her own appearance. If a guard speaks sharply to the client's customers or employees, it may be a reflection of the woman who is in charge. Perhaps the supervisor should pay careful attention to her own manner. The guard force reflects the company's image and the supervisor should ensure that the proper appearance is being projected.

Techniques for Setting the Example

1. Be physically fit, well-groomed and correctly dressed.
2. Master your emotions. Erratic behavior, ranging from anger to depression, is noneffective.
3. Maintain an optimistic outlook. Excel in difficult situations by learning to capitalize on your own capabilities.
4. Conduct yourself so that your own personal habits are not open to censure.
5. Exercise initiative and promote the spirit of initiative in your subordinates.
6. Be loyal to those with whom you work and those who work with you. Loyalty is a two-way street.
7. Avoid playing favorites.
8. Be morally courageous. Establish principles and stand by them.
9. Share hardships with your people to demonstrate your willingness to assume your share of the difficulties.

The Professional Security Supervisor

Today's security work requires a person with an exceptionally high degree of skill, training, and information. The person who demonstrates these qualities is recognized by others as a professional. You exude the confidence and skill to make it possible for the rest of the community to have faith in your ability to act in their interest. The security officer who meets these standards is a professional in the fullest meaning of the word, and is respected as such.

A professional person has:

Education. By virtue of having completed certain education programs and having passed official examinations, professional people are recognized as possessing distinctive kinds and amounts of knowledge and skill. These are types of knowledge and skill in which the average citizen feels deficient, and therefore, turns to professionally trained people for help, in the form of advice or other services.

Standards of Performance. Professional people are expected to be dedicated to high ideals. They are assumed to operate under a superior code of ethics. To this end, the professional organizations establish standards of ethical performance, as well as standards of competence. Professional people take pride in these standards and expect members of their profession to meet them. Because of the continuous flow of social and economic changes in our world, training and the improvement of standards is a continuing problem for every security authority.

It is the understanding of fundamental principles which distinguishes the competent professional person from the mere technician. This is as true in security work as it is in medicine, law, and other professional fields.

A security supervisor is personally judged by the general public. The client, as well, looks upon you as the contact with the organization and will measure the company by the supervisor. The security personnel, as well, look to the supervisor to set an example. As in other areas, therefore, the leader must maintain a professional code of ethics. Professionalism is vital to any position of authority and this fact is no less true for the security supervisor.

Train Personnel Effectively

The responsibilities of a guard supervisor include providing sound, effective training to the staff. An understanding of every operational requirement of the security officers will give the supervisor more awareness of the difficult facets of their work, areas where you may be able to offer assistance when and where it is needed.

The supervisor can facilitate this aspect of your job by determining the duties of each security officer and establishing a master training plan that will teach the new employees their respective tasks. This plan will also serve as refresher training for other personnel who have been on the force for a long period of time.

Treat Employees Courteously

Mutual respect is essential to an efficient working relationship. Employees should not be treated as natural enemies; nor should they be made to feel inferior. You must in turn report to your bosses; you should treat staff in the same courteous manner you expects from *your* superiors.

Consideration is a key word. A demand should be accompanied by an explanation. Advance notice of any situations that might alter an employee's plans, such as overtime, post reassignments, or special orders, is a simple courtesy that will prevent unnecessary ill will. Reprimands or criticisms made in private, away from the watchful eyes of one's peers, precludes humiliation of a staff member.

Develop Loyalty

An effective supervisor is loyal to the employees, the company, and the client. Constant criticism of the company and management is destructive to employee morale. While criticism is a necessary and unavoidable part of any activity, it must be offered constructively to resolve a problem, improve a system, lower costs, and other worthwhile purposes. Criticism for the sake of criticism has no worth and no place in business.

A responsible supervisor does not indulge or pass on gossip or rumors about other employees. A supervisor who is loyal to the personnel is usually repaid with loyalty from the unit.

When You Must Criticize

"To err is mortal; to forgive, divine." The supervisor is sitting on the semicolon of this statement. Not only must you recognize errors, see that they are corrected, and discourage further mistakes, but you are also expected to maintain composure while doing so.

It is a fact of life that most people resent being told that they have done something wrong, especially if the person who does the criticizing is tactless and forceful. Harsh criticism can hurt a person's morale, damage the ego, and create lasting antagonisms. When faced with the job of criticizing an employee, the supervisor should try to follow these seven simple rules:

1. *Be sure of the facts.* Ask the right people the right questions, and do so objectively. Only when you are sufficiently satisfied that an error has been made should you call in the employee. If being criticized for something you *did* can cause resentment, being criticized for something you *didn't* do will really breed antagonism.
2. If the mistake is important and has upset you, *cool off before you talk to the employee.* When you are angry, you are more likely to say something personal. Avoid personal criticism; address your comments to correcting the mistake, not to punishing the security officer.
3. *Discuss the situation in private.* Nothing embarrasses a person more than being reprimanded before one's peers or, worse yet, one's subordinates. Take time to move away from inquisitive eyes and ears. Your criticism will be better and lasting resentment may be avoided.
4. *Ask questions first—don't accuse.* This fits in neatly with the "Be sure of your facts" rule.

Don't come into the discussion with your mind made up. Ask for the employee's side of the story. Everyone appreciates being heard, especially when a mistake has been made.
5. *Before you criticize*, let your worker know that you appreciate some of the good work produced. Medicine is easier to swallow if you mix it with sugar!
6. When the situation dictates that an oral reprimand be given, *explain to the employee the reasoning behind your actions.* An employee deserves to know why there is criticism and how this will affect the future. For example, if a security officer is being criticized for the first tardiness, the officer should not be made to feel that the job is in jeopardy. However, if the reprimand is for continual absences or latenesses, and the job *is* on the line, the employee should know this as well.
7. If at all possible, *leave a good impression* with your employee at the end of the discussion. This does not mean you should make light of mistakes. Rather, it will remove some of the tension and embarassment if, when the employee returns to work, you pat the person on the back or say something like, "At least we know you're human."

These seven rules will help the supervisor to deal tactfully with the situation when you *must* criticize. You should remember that the goal of criticism is to leave the person with the feeling of having been helped.

Personnel Counseling

Every supervisor must be prepared to discuss an employee's personal problems when asked to do so, but only to the extent that the individual desires, and within limits carefully set by the supervisor.

The biggest problem for the manager, in a counseling situation, is to steer a proper course between practical and constructive advice, and particularly to stay clear of amateur psychiatry. When an employee seeks personal counseling, the supervisor should consider these guidelines:

1. *Watch your general attitude.* Always show a continuing sincere interest in your people as individuals with homes and families and not simply as subordinates. If there is sickness at home, remember to ask about progress. If someone's daughter is graduating from high school, show some interest in that also.
2. *Make yourself available.* If someone indicates a desire to talk to you about a matter that has come up, answer by saying that if it is important to the employee, you'll be glad to take whatever time is necessary. The employee will probably agree to have the interview after hours, when nobody else is around. In any case, it is obvious that you should make it possible to have the employee talk to you in private. Hold the meeting as soon as possible after the request.
3. *Some meetings you will have to initiate.* This can occur, for example, when a usually competent and reliable person shows a marked falling off in interest or quality of work, or is unusually tardy or frequently absent, all indicating that some personal situation is interfering with efficiency. Don't keep putting the meeting off . . . it will never be any easier than at the present moment.
4. *Be as prepared as possible.* If you have initiated the meeting, be sure of your facts with specific examples of the kinds of behavior that are giving you concern. If the employee has asked for the meeting, refresh your memory about any personal situations that may previously have come to light about the employee.
5. *Put the employee at ease.* You will already have achieved part of this by arranging for a private meeting. Maybe a cup of coffee or a soft drink is indicated.
6. *Be a good listener.* Whether the problem is real or imagined, give the employee a chance to explain the situation without interruption.
7. *Be wary of advice on personal matters.* On emotional and personal problems, your best contribution will be to serve as a sounding board. You can, of course, give advice on any company policy that may be involved, avenues of financial assistance available through the company, and other matters where you are sure of your ground. But with a personal problem, your main function as a counselor should be to help the individual recognize what the problem is, and to explore possible alternate solutions, with final decisions left to the individual. Always remember, when you arc dcaling with personal and emotional problems, you will rarely be in possession of enough facts to take the responsibility for recommending specific solutions.
8. *Avoid assuming the psychiatrist's function.* If you have reason to believe that the employee has more than the normal kinds of anxiety, suggest professional counsel.*

---

*Excerpts taken from "How To Communicate Better With Workers" by Carl Heyel.

Chapter 18
Civil Law: The Controller for Private Security*

ROBERT J. FISCHER

Tort Law: Source of Power and Limits

Tort law may be invoked for either an intentional or negligent act. In some cases tort law may be imposed even though an individual is not directly at fault. Such a legal obligation is called strict liability, and does not generally affect the security officer. Strict liability, however, is of concern to enterprises which contract or employ security services.

Negligence

The Restatement of Torts (Second) Section 307, states that "[It] is negligence to use an instrumentality, whether a human being or a thing, which the actor knows, or should know, to be incompetent, inappropriate or defective and that its use involves an unreasonable risk of harm to others." This statement has particular importance to security employers and supervisors in three areas: negligent supervision of employees, negligent training, and supervisory negligence. Security officers have been held liable for negligent use of firearms and force.

In all cases of negligence, the plaintiff (a person who brings an action; the party who complains or sues) must prove the case by a preponderance of the evidence (more than 50%) in all of the following areas:

1. An act or failure to act (an omission) by the defendant
2. A legal duty owed to the plaintiff by the defendant; the person defending or denying; the party against whom relief or recovery is sought
3. A breach of duty by the defendant
4. A foreseeable injury to plaintiff
5. Damages

Intentional Torts

An intentional tort occurs when the person who committed the act was able to foresee that the action would result in certain damages. The actor intended the consequences of his or her actions, or at least intended to commit the action that resulted in damages to the plaintiff. In general the law punishes such acts by punitive measures which exceed those awarded in common negligence cases.

The most common intentional torts are listed in the following sections.

Assault

Intentionally causing fear of harmful or offensive touching, but without touching or physical contact. In most cases, courts have ruled that words alone are not sufficient to place a person in fear of harm.

Battery

Intentionally harmful or otherwise offensive touching of another person. The touching does not have to be a human physical contact, but may be through an instrument such as a cane or rock. From a security point of view, the contact must be nonconsensual and not privileged. Privileged contact is generally granted to merchants who need to recover their merchandise; privilege is generally a defense against

---
*From *Introduction to Security*, 4th ed, by Gion Green, revised by Robert J. Fischer (Stoneham, MA: Butterworths, 1987).

charges of battery if the merchant's actions were reasonable. If, however, the touching were unreasonable the plaintiff would have a case for battery. The same argument holds for searches: If a search is performed after consent has been given, no battery has occurred. However, if consent is not given, the search is illegal, and a battery has probably occurred.

False Imprisonment or False Arrest

Intentionally confining or restricting the movement or freedom of another. The confinement may be the result of physical restraint or intimidation. False imprisonment implies that the confinement is for personal advantage rather than in order to bring the plaintiff to court. This is one of the torts most frequently filed against security personnel.

Defamation

Injuring the reputation of another by publicly making untrue statements. Slander is oral defamation, while libel is defamation through the written word. The classic case of a security officer yelling "Stop thief!" in a crowded store has all the necessary elements for slander if the accused is not a thief. Although it is generally true that truth is an absolute defense in defamation issues, the courts may also look at the motivation. True statements published with malicious intent can be prosecuted in some jurisdictions.

Malicious Prosecution

Groundlessly instituting criminal proceedings against another person. To prove malice, the plaintiff must show that the primary motive in bringing about criminal proceedings was not to bring the defendant to justice. Classic cases include proceedings brought about to extort money, or to force performance on contracts. Although there is no liability for reporting facts to the police or other components of the criminal justice system, if the prosecution resulted from biased statements of fact, incomplete reports, or from the defendant's persuasion (political, sexual, religious, etc.), then liability for malicious prosecution might be proved.

Invasion of Privacy

Intruding upon another person's physical solitude, disclosing private information about another person, or publicly placing someone in a false light. Four distinct actions are in this category: 1) misappropriation of the plaintiff's name or picture for commercial advantage, 2) placing the plaintiff in a false light, 3) public disclosure of private facts, 4) intrusion into the seclusion of another. For security purposes, invasion of privacy generally occurs during a search or an observation of an individual. If signs outside fitting rooms advise customers that they may be observed, some legal observers believe that shoppers should not expect privacy, and thus cannot legitimately complain of invasion of privacy. Concern over liability for invasion of privacy is increasing; this liability may be the result of reference checks, background investigations, or the use of truth detection devices.[1]

Trespass and Conversion

Trespass is the unauthorized physical invasion of property, or remaining on property after permission has been rescinded. Conversion means taking personal property in such a way that the plaintiff's use or right of possession of chattel is restricted. In simpler terms, conversion is depriving someone of the use of personal property.

Intentional Infliction of Mental Distress

Intentionally causing mental or emotional distress to another person. The distress may be either mental or physical, and may result from highly aggravating words or conduct.

Arrest

Arresting a person is a legal step which should not be taken lightly. A citizen's power to arrest another is granted by common law and in many jurisdictions by statutory law. In most cases, it is best to make an arrest only after an arrest warrant has been issued. However, most citizen's arrests occur when the immediacy of a situation requires an arrest without warrant. The exact extent of citizen's arrest power varies, depending upon the type of crime, the jurisdiction (laws), whether the crime was committed in the presence of the arrester, or on the status of the citizen (private citizen or a commissioned officer).

In most states, warrantless arrests by private citizens are allowed when a felony has been committed and reasonable grounds exist for believing the person arrested committed it. Reasonable grounds means that the arrester acted as would any average citizen who, having observed the same

Table 18–1. Statutory Arrest Authority of Private Citizen

	Minor Offense		Major Offense		Certainty of Correct Arrest
	Type of minor offense	Type of knowledge required	Type of major offense	Type of knowledge required	

State	Crime	Misdemeanor amounting to a breach of the peace	Breach of the peace	Public offense	Offense	Offense other than an ordinance	Indictable offense	Presence	Immediate knowledge	View	Upon reasonable grounds that is being committed	Felony	Larceny	Petit Larceny	Crime involving physical injury to another	Crime	Crime involving theft or destruction of property	Committed in presence	Information a felony has been committed	View	Reasonable grounds to believe being committed	That felony has been committed in fact	In escaping or attempting	Summoned by peace officer to assist in arrest	Is in the act of committing	Reasonable grounds to believe person arrested committed	Probable cause
Alabama				•				•				•										•				•	
Alaska	•							•				•										•				•	
Arizona		•						•				•										•				•	
Arkansas												•												•		•	
California				•				•				•										•				•	
Colorado	•							•								•		•									
Georgia			•					•	•			•													•	•	
Hawaii	•							•								•		•							•		
Idaho				•				•				•										•				•	
Illinois							•				•	•								•							
Iowa				•				•				•										•				•	
Kentucky												•										•				•	
Louisiana												•							•			•				•	
Michigan												•										•			•		
Minnesota				•				•				•										•				•	
Mississippi		•			•		•	•				•										•				•	
Montana			•					•				•										•				•	
Nebraska								•			•	•										•				•	
Nevada				•				•				•										•				•	
New York					•			•				•										•					
N. Carolina[a]		•														•		•							•		•
N. Dakota				•				•				•										•				•	
Ohio								•				•												•		•	
Oklahoma				•				•				•										•				•	
Oregon	•							•								•		•									•
S. Carolina												•	•											•	•		
S. Dakota				•				•				•										•				•	
Tennessee				•				•				•										•				•	
Texas			•					•		•		•									•		•				
Utah				•				•				•										•				•	
Wyoming								•		•												•				•	

[a] Statute eliminates use of word *arrest* and replaces with *detention*.

Source: Charles Schnabolk, *Physical Security: Practices and Technology* (Stoneham, MA: Butterworth Publishers, 1983), pp. 64–65.

facts, would draw the same conclusion. In some jurisdictions a private citizen may arrest without reasonable grounds, as long as a felony was committed.

Most states allow citizen's arrest for misdemeanors committed in the arrester's presence. However, a minority of states adhere closely to the common law practice of allowing misdemeanor arrests only for offenses that constitute a breach of the peace and occur in the arrester's presence (see Table 18–1).

Although the power of citizen's arrest is very significant in the private sector, because it allows security officers to protect their employer's property, there is little room for errors of judgment. The public police officer is protected from civil liability for false arrest if the officer has probable cause to believe a crime was committed, but the private officer (citizen) is liable if a crime was not committed, regardless of the reasonableness of the belief. This liability is because citizen's arrests generally can be made only if a crime has definitely been committed.

This distinction arose from the case *Cervantez v. J.C. Penney Company* (156 Cal. Rptr 198 [1978]). In the Cervantez case, an off-duty police officer moonlighting as a store detective for J.C. Penney Company made a warrantless arrest of two individuals for misdemeanor theft. Later they were released because of lack of evidence. The plaintiffs sued the company and the officer for false arrest, imprisonment, malicious prosecution, assault and battery, intentional infliction of emotional distress, and negligence in the selection of its employee. The primary issue in the Cervantez case was whether the officer could rely on the probable cause defense. The court's decision rested on whether the officer acted as a police officer in California or as a private citizen.

The store and officer argued that the probable cause defense was sound because the detective was an off-duty police officer and thus could arrest on the basis of probable cause. The plaintiffs argued that the store detective should be governed by the rules of arrest applied to private citizens, and that the officer was therefore liable for his actions because no crime had been proved to be committed. The plaintiffs contended that the officer was employed as a private security officer, and thus his arrest powers were only those of a private citizen.

The California Supreme Court ruled that the laws governing the type of arrest to be applied depend on the arrester's employer at the time of the arrest. Since the officer was acting as a store detective when he made the arrest, his arrest powers were no greater than those of a private citizen. Thus probable cause could not be used as a defense against false arrest.

Some states have avoided the problem of the Cervantez case by extending the probable cause defense to private citizens. The most common extension involves merchants and shoplifting arrests. Many states have a mercantile privilege rule, which allows the probable cause defense for detentions, but not for arrests. The law permits a private citizen or his employees to detain in a reasonable manner and for a reasonable time a person who is believed to have stolen merchandise, so that the merchant can recover the merchandise or summon a police officer to make an arrest. Some states have extended this merchant clause to cover public employees in libraries, museums, or archival institutions.

The exact extent of the protection afforded to the merchant and his or her employees or agents depends on the individual state's statutes. Some states offer protection against liability for false arrest, false imprisonment, and defamation; others offer only protection against false imprisonment, but not against false arrest. It is interesting to note that very few states allow a merchant to search a detainee. The private citizen's authority to search is unclear, and will be discussed later in this chapter.

Detention

Detention is a concept which has grown up largely in response to the difficulties faced by merchants in protecting their property from shoplifters and the problems and dangers they face when making an arrest. Generally, this privilege differs from arrest in that it permits a merchant to briefly detain a suspected shoplifter without turning the suspect over to the police. An arrest requires that the arrestee be turned over to the authorities as soon as practicable and, in any event, without unreasonable delay.

All the shoplifting statutes refer to *detain*, not to *arrest*. The distinction was based on the fact that an arrest is for the purpose of delivering the suspect to the authorities and exercising strict physical control over that person until the authorities arrive. A detention, or temporary delay, would not be termed an arrest, as defined. The distinction is difficult to defend but the statutes (with the exception of Rhode Island, which permits detention and search only by a peace officer) are clear. In Florida, for example:

> A law enforcement officer, a merchant, a merchant's employee, or a farmer who has probable cause to believe that merchandise or farm produce has been unlawfully

taken by a person and that he can recover it by taking the person into custody may, for the purpose of attempting to effect such a recovery or for prosecution, take the person into custody and detain him in a reasonable manner for a reasonable length of time. In the case of a farmer, taking into custody shall be effectuated only on property owned or leased by the farmer. In the event the merchant, merchant's employee, or farmer takes the person into custody, a law enforcement officer shall be called to the scene immediately after the person has been taken into custody.[2]

California was one of the first states to establish merchant immunity, in a 1936 Supreme Court decision. In *Collyer v. S.H. Kress & Co.* (5 Cal. 2d 175, 54 p. 2d 20 [1936]), the court upheld the right of a department store official to detain a suspected shoplifter for twenty minutes.

Most statutes include the merchant, employee, agent, private police, and peace officer as authorized to detain suspects, but they do not include citizens at large, such as another shopper. Most of the statutes also describe the purpose of the detention and the manner in which it may be conducted. These purposes are: search, interrogate, investigate suspicious behavior, recover goods, and await a peace officer.

The manner in which the detention is to be conducted is generally described as "reasonable" and for "a reasonable period of time." Only five states describe the time of the detention: West Virginia, Maine, and Montana provide for a maximum of thirty minutes; Indiana and Louisiana provide for a maximum of one hour. No state describes the reasonableness of the manner in which the detention is to be performed.

The privilege of detention is, however, subject to some problems. There must be probable cause to believe larceny has or is about to take place before a merchant may detain anyone. Probable cause is an elusive concept, and one which has undergone many different interpretations by the courts. It is frequently difficult to predict how the court will rule on a given set of circumstances which may, at the time, clearly indicate probable cause to detain. Secondly, reasonableness must exist both in time and manner of the detention or the privilege will be lost.

Interrogation

No law prohibits a private person from engaging in conversation with a willing participant. Should the conversation become an interrogation, the information may not be admissible in a court of law. The standard is whether the statements were made voluntarily.

A statement made under duress is not regarded as trustworthy and is therefore inadmissible. This principle applies equally to police officers and private citizens. A confession obtained from an employee by threatening loss of job or physical harm would be inadmissible—and would also make the interrogator liable for civil and criminal prosecution.

The classic cases involving interrogation are *Escobedo v. State of Illinois* (378 U.S. 478, 32 OhioOp. 92d) 31, 84 s.Ct. 1758, 12 L.Ed.(2d)977 (1964) and *Miranda*. Today Miranda has become the leading case recognized by most American citizens in reference to their rights. On March 13, 1963, Ernesto Miranda was arrested at his home and taken to the Phoenix police station. There he was questioned by two police officers, who during Miranda's trial admitted that they had not advised him that he could have a lawyer present. After two hours of interrogation, the officers emerged with a confession. According to the statement, Miranda had made the confession "with full knowledge of my legal right, understanding any statement I make may be used against me." His confession was admitted into evidence over defense objections during his trial. He was convicted of kidnapping and rape. On appeal the Arizona Supreme Court upheld the conviction indicating that Miranda did not specifically request counsel. The U.S. Supreme Court reversed the decision based on the fact that Miranda had not been informed of his right to an attorney, nor was his right not to be compelled to incriminate himself effectively protected.

Although the principle behind the *Miranda v. Arizona* (384 U.S. 436, 86 S.Ct.1602, 16L.Ed.2d. 694) decision was the removal of compulsion from custodial questioning (questioning initiated by law enforcement officers after a person has been taken into custody or otherwise deprived of freedom), it generally only applies to public law enforcement officers. The police officer must show that statements made by the accused were given after the accused was informed of the facts that he or she need not speak, that the statements might be used in court, that an attorney may be present, and that if the accused cannot afford an attorney, one will be appointed for the accused prior to questioning. These Miranda warnings are not necessary unless the person is in custody or is deprived of freedom. Based on this distinction, most courts agree that private persons are not generally required to use Miranda warnings because they are not public law enforcement officers.

In another case *In re Deborah C.* (177 Cal. Rptr 852 [1981]), the California Supreme Court upheld the principle that private citizens are not required to use Miranda warnings and that statements made by the accused in citizen's arrests are admissible in a court of law. The court felt that the Miranda rationale did not apply to the retail store environment, because store detectives lack the psychological edge that police officers have when questioning someone at a police station.

A few states require citizens to use a modified form of Miranda before questioning, and some—a definite minority—prohibit questioning. Wisconsin law states, "The detained person must be promptly informed of the purpose of the detention and be permitted to make phone calls, but shall not be interrogated or searched against his will before the arrival of a peace officer who may conduct a lawful interrogation of the accused person."[3]

Search and Seizure

A search may be defined as an examination of a person and/or property for the purpose of discovering evidence of guilt in relation to some specific offense. The observation of items in plain view is not a search as long as the observer is legally entitled to be in the place where the observation is made. This includes public and private property which is normally open to the public, for example, shopping malls, retail stores, hotel lobbies, and so on.

Common law says little about searches by private persons, and is inconclusive. However, searches by private persons have been upheld by the courts where consent to search was given and where searches were made as part of a legal citizen's arrest. The best practice to follow is to contact police officials, who can then ask for a search warrant or search as part of an arrest. However, since searches often need to be conducted on short notice without the aid of a police officer, several factors are important to understand.

First, in a consent search the searcher must be able to show that the consent was given voluntarily. Second, the search cannot extend beyond the area for which consent to search was given. It is advisable to secure a written agreement of the consent to search. Third, the consent must be given by the person who possesses the item. Possession, not ownership, is the criteria for determining whether a search was valid. Although many firms issue waivers to search lockers and other work areas, an officer must remember that the consent to search may be withdrawn at any time. If the consent is withdrawn, continuing a search might make the officer and the company liable for invasion of privacy. Some companies have solved this problem by retaining control over lockers in work areas. In this situation, workers are told that the lockers are not private, and may be searched at any time.

A search made as a part of an arrest is supported by case law. In general, the principle of searching the arrestee and the immediate surroundings—defined as the area within which one could lunge and reach a weapon or destroy evidence—has been repeatedly held as constitutional. The verdict on searches incident to arrest is still mixed. In *People v. Zelinski* (594 P 2d 1000 [1979]), the California court disapproved of searches made incident to an arrest, but did approve of searches for weapons for protective reasons. New York courts tend to support searches, indicating that private officers, like their public counterparts, have a right to search incident to an arrest. In general, it appears that unless the security officer fears that a weapon may be hidden on the arrestee, the officer should wait until the police arrive to conduct a search—unless permission is given for such a search.

Even in the statutes governing retail shoplifting, the area of search is limited. Only Oklahoma specifically mentions that a search is permitted. Some states neither forbid nor condone searches; rather, they allow security personnel to investigate or make reasonable inquiries as to whether a person possesses unpurchased merchandise. In other states, searches are strictly forbidden, except looking for objects carried by the suspected shoplifter. However, courts generally favor protective searches where the officer fears for his or her own safety.

Exclusionary Rule

In a historic decision, the Supreme Court ruled that any and all evidence uncovered by public law enforcement agents in violation of the Fourth Amendment will be excluded from consideration. That means all evidence, no matter how trustworthy, or indicative of guilt, will be inadmissible if it is illegally obtained. This landmark case (*Mapp v. Ohio*, 367 U.S. 643 [1961]) was the most important case that contributed to the development of the *exclusionary rule*, which states that illegally seized evidence (and its fruits) are inadmissible in any state or federal proceedings. Mapp had, however, been preceded in 1914 by *Weeks v. United States* (232 U.S. 383 [1914]), which set the stage for the later all-inclusive decision in Mapp by holding that evidence acquired by officials of the federal government in violation of the Fourth Amendment must be excluded in a federal prosecution.

Mapp is clear in its application of the exclusionary rule to state and federal prosecutions. The question is, does the exclusionary rule apply to private parties? The determining case in this area is *Burdeau v. McDowell.*

Unlike illegal searches conducted by public law enforcement officers, evidence secured by a private security officer conducting an illegal search is still admissible in either criminal or civil proceedings. In *Burdeau v. McDowell* (256 U.S. 465 [1921]), the United States Supreme Court said, "It is manifest that there was no invasion of the security afforded by the Fourth Amendment against unreasonable searches and seizures, as whatever wrong was done was the act of individuals in taking property of another." If such evidence is admissible, why should private sector employees concern themselves with the legality of searches? Even though the evidence is admissible, security officers who conduct illegal searches may be subject to liability for other actions, including battery and invasion of privacy.

There is considerable controversy over Burdeau because some people fear that constitutional guarantees are threatened by the acceptance of evidence illegally obtained by private security personnel. It is clear that any involvement by government officials constitutes "state action" or an action "under color of law" and is limited by the constitutional restrictions that apply to public police actions. In *State v. Scrotsky* (39 NJ 410, 416, 189 A.2d 23 [1963]) the court excluded evidence obtained when a detective accompanied a theft victim to the defendant's apartment to identify and recover stolen goods. The court held: "The search and seizure by one served the purpose of both and must be deemed to have been participated in by both." The exclusionary rule is applied in this case, as in many others, to discourage government officials from conducting improper searches and from using private individuals to conduct them.

In cases where private parties act independent of government involvement, the courts have not been so clear. In a significant case, *People v. Randazzo* (220 Cal. App. 2d 268, 34 Cal. Rptr. 65 [1963]), the court admitted evidence obtained by a merchant in a shoplifting case. The court did not deal with any questions of Fourth Amendment violation since there was no state action involved. The court held that redress for the victim of an unreasonable search conducted by a private individual not under cover of law is a tort action, and thus the exclusionary rule does not apply. In *Thacker v. Commonwealth* (310 Ky. 701, 221 SW 2d 682 [1949]), the court held that a private party acts for the state when making an arrest in accordance with the state's arrest statute, and thus would be subject to the exclusionary rule. On the other hand, following the Burdeau precedent, a federal district court found no state action in a case where the plaintiff alleged she was wrongfully detained, slapped, beaten, harassed, and searched by the manager and an employee of the store. (*Weyandt v. Mason's Stores, Inc.*, 279 F. Supp. 283, 287 [W.D. Pa. 1968]). The plaintiff sued, alleging among other things that the employee, a security officer, was acting "under cover of law" because he was licensed under the Pennsylvania Private Detective Act. The court rejected this argument and found that the Pennsylvania law "invests the licensee with no authority of state law."

In summary, although public police are clearly limited by constitutional restrictions, generally, private police are not. Provided that they act as private parties, and are in no way involved with public officials, they are limited by criminal and civil sanctions, but are not, at this time, bound by most constitutional restrictions.

Use of Force

On occasion security personnel must use force to protect someone or to accomplish legitimate purposes. In general, force may be used to protect oneself or others, to defend property, and to prevent the commission of a criminal act. The extent to which force may be used is restricted; no more force may be used than is reasonable under the circumstances. This means that deadly force or force likely to create great bodily harm would not be allowable unless the force being used by the assailant was also deadly force or force likely to create great bodily harm. If the force exceeds what is deemed reasonable, then the officer and his employer are liable for the use of excessive force, which can range from assault and battery to homicide.

Self Defense

In general people may use reasonable force to protect themselves. The amount of force may be equal to but not greater than the force being used against them. In most states a person can protect him- or herself, except when the person was the initial aggressor.

Defense of Others

Security officers may protect others just as they protect themselves. However, two different approaches to defense of others are evident. In the first the officer must try to identify with the attacked person. In this position, the officer is entitled to use whatever force would be appropriate if he or she were the person being attacked. If the officer happens to protect the wrong person—that is, the aggressor—the officer is liable regardless of his or her good intentions. In the second approach, the defender may use force when able to reasonably believe that such force is necessary. In this case, the defender is protected from liability as long as he or she acted in a reasonable manner.

Defense of Property

In defense of property, force may be applied, but it must be short of deadly force. Deadly force is generally allowable only in cases involving felonious attacks on property in which loss of life is likely. As noted by Schnabolk, "one may use deadly force to protect a home against an arsonist but the use of deadly force against a mere trespasser would not be permitted."[4] Security officers acting in the place of their employers are empowered to use the same force that their employers are entitled to use.

Force Used during Arrest or Detention

Like the police, the private citizen security officer has the right to use reasonable force in detaining or arresting someone. Many states still follow the common law principles which allow deadly force in the case of fleeing felons, but many others have restricted the use of deadly force. This restriction allows the use of deadly force only in cases where the felony is violent and the felon is immediately fleeing.

Prevention of Crimes

To determine the amount of force which a security officer may use to prevent crimes, the courts have considered the following: the circumstances, the seriousness of the crime prevented, and the possibility of preventing the crime by other means. Under common law a person can use force to prevent a crime. The courts have ruled, however, that the use of force is limited to situations involving felonies or a breach of the peace, and that nonviolent misdemeanors do not warrant the use of force. Deadly force is justifiable only if it is necessary to protect a person from harm. Some states have broadened this concept to permit the use of deadly force to prevent any felony.

Use of Firearms

Most states regulate the carrying of firearms by private citizens. Almost all states prohibit the carrying of concealed weapons, while only half prohibit carrying an exposed handgun. Although all states excuse police officers from these restrictions, some states also exempt private security officers. Even in states which prohibit carrying concealed or exposed handguns, there are provisions for procuring a license to carry weapons in this manner.

Civil Liability

In the past few years the number of suits filed against security officers and companies has increased dramatically. The increase may be partly attributed to the growth of the security industry, and to the public's demand for accountability and professionalism in the security area. Most of the cases filed against private security officers and operations belong in the tort category, which was mentioned earlier in this chapter. The individual who commits a tort is called the *tortfeasor*, while the accuser is called the *plaintiff*. The plaintiff may be either a person, a corporation, or association. Torts are classified as either intentional, negligent, or strict liability. An intentional tort is a wrong perpetrated by someone who intends to break the law. In contrast, a negligent tort is a wrong perpetrated by someone who fails to exercise sufficient care in doing what is otherwise permissible. A strict tort or willful tort combines intentional and negligent torts; it involves elements of intent and malice or ill will. Malice or ill will may be attributed to someone who is aware of danger to others, but who is indifferent to their safety or fails to use ordinary care to avoid injury.

In most cases of negligence the jury considers awarding damages to compensate the plaintiff. The awards generally take into account the physical, mental, and emotional suffering of the plaintiff, and future medical payments may be allowed for. Punitive damages are also possible, but are more likely to be awarded in cases of intentional liability. Punitive damages are designed to punish the tortfeasor and to deter future inappropriate behavior. Punitive damages are also possible in negligence cases where the actions of the tortfeasor were in total disregard for the safety of others.

The area of civil liability is of great importance to the security industry, since the courts have been more willing to hold the industry legally responsible for protection. This trend is particularly noticeable in the hotel and motel industry, where owners are liable for failure to adequately protect guests from foreseeable criminal activity. In some circumstances a landlord or hotel or motel owner might be held accountable for failure to provide adequate protection from criminal actions. In *Klein v. 1500 Massachusetts Avenue Apartment Corporation* (439 F 2d 477 DC Cir [1970]), a tenant who was criminally assaulted sued the corporation. The decision centered on the issue that the landlord had prior notice of criminal activity (including burglary and assault) against his tenants and property. In addition, the landlord was aware of conditions which made it likely that criminal activities would continue. The court ruled that the landlord had failed in his obligation to provide adequate security, and was thus liable. A similar case was made against Howard Johnson's by Connie Frances. Frances alleged that the hotel had failed to provide adequate locks on the doors. The jury awarded her over a million dollars. (*Garzilli v. Howard Johnson's Motor Lodge, Inc.* 419 F Supp 1210 U.S. DCT EDNY [1976]).

Another legal trend is to prevent corporations from divesting themselves of liability by assigning protection services to a contractor. Under the principle of agency law, such an assignment transferred the liability for the service from the contractee to the contractor. However, the courts have held that some obligations cannot be entirely transferred. This principle is called *nondelegable duty*. Based on this principle, contractual provisions that shift liability to the contractor have not been recognized by the courts. These contractual provisions are commonly called *hold harmless* clauses.

Vicarious Liability

The concept of vicarious or imputed liability (also called *respondeat superior*) arises from agency law, in which one party has the power to control the actions of another party involved in the contract. The master or principal is thus responsible for the servant or agent. In short, the employers are liable for the actions of their employees while they are employed in the firms' business. Employers are liable for the actions of their agents even if the employers do nothing to directly cause the actions. The master is held liable for any intentional tort committed by the servant where its purpose, however misguided, is wholly or in part to further the master's business. Employers may even be liable for some of the actions of their employees when the employees are not at work or engaged in company business. For example, consider the position of an employer who issues a firearm to an employee. While at home and off duty, the employee plays with the firearm, which discharges and injures a neighbor. The neighbor may sue the employer for negligently entrusting a dangerous instrument to an employee or for the negligence in selecting a careless employee.

Vicarious liability requires a direct employer-employee relationship; it does not apply to cases in which an independent contractor is working for a firm. This is because the employer has no way of controlling the way that an independent contractor performs his or her work. However, there are many exceptions to this rule. For example, the employer may be liable for the negligent selection of the contractor; or the employer may have exercised some day-to-day control over the employee.

Criminal Liability

Criminal liability is most frequently used against private security personnel in the cases of assault, battery, manslaughter, and murder. Other common charges include burglary, trespass, criminal defamation, false arrest, unlawful use of weapons, disorderly conduct, extortion, eavesdropping, theft, perjury, and kidnapping. Security officers charged with criminal liability have several options in defending their actions. First, they might try to show that they were entitled to use force in self defense, or that they made a reasonable mistake, which would negate criminal intent. Other defenses include entrapment, intoxication, insanity, consent (the parties involved concurred with the actions), and compulsion (the officer was forced or compelled to commit the act). As noted in previous discussions a corporation or an association could be charged with criminal liability as well as an individual officer.

The reporting of crime is an area in which security officers are liable for criminal prosecution. In general, private citizens are no longer obliged to report crime or to prevent it. But some jurisdictions still recognize the concept of misprision of felony—that is, concealing knowledge of a felony. Such legislation makes it a crime to not report a felony. To be guilty of misprision of felony, the prosecution must prove beyond a reasonable doubt that: 1) the principal had committed and completed the alleged felony, 2) that the defendant had full knowledge of that fact, 3) that the defendant failed to notify the authorities, and 4) that the defendant took affirmative steps to conceal the crime of the principal.

Security officers may also be liable for failure to perform jobs that they have been contracted or employed to perform. If guards fail to act in a situation in which they have the ability and obligation to act, the courts suggest that they could be criminally liable for failure to perform a duty.

Another issue in security work involves undercover operations. Many times security operatives are accused of soliciting an illegal act. Where the security officer clearly intended for crimes to be committed, he or she may be charged with solicitation of the illegal act or conspiracy in the illegal act. This is in contrast to the public sector, where most police officers are protected by statute from crimes they commit in the performance of their duty. Thus, only the private citizen may be charged with such an offense, and the only issue that can be contested is the defendant's intent.

Entrapment, which is solicitation by police officers, is another charge that may be levelled against security officers. While entrapment does not generally apply to private citizens (the case of *State v Farns* (542 P.2d 725 Kan. [1975]) is frequently cited to prove that entrapment does not apply to private citizens), several states have passed legislation which extends entrapment statutes to cover private persons as well as police officers. Until the issue is resolved in the courts in the next few years, security officers involved in undercover operations should be careful to avoid actions that might lead to entrapment charges.

Contract Security and Liability

The principle of *respondeat superior* ("let the master respond") is well established in common law. It is not, in itself, the subject of any substantial dispute, and at those times when it becomes an issue in a dispute, the area of contention is factual rather than the doctrine itself. As noted earlier, in the doctrine of *respondeat superior*:

> A servant is a person employed by a master to perform service in his affairs, whose physical conduct in the performance of the service is controlled or is subject to the right of control by the master. This court has stated that the right of control and not necessarily the exercise of that right is the test of the relation of master and servant. Basically, it is distinction between a person who is subject to orders as to how he does his work and one who agrees only to do the work in his own way. (*Graalum v. Radisson Ramp*, 245 Minn. 54, 71 NW 2d 904, 908 [1955].)

There is no question that an employer (master) is liable for injuries caused by an employee (servant) who is acting within the scope of his or her employment. This is not to say that the employee is relieved of all liability. He or she is in fact the principal in any action, but since the employee rarely has the financial resources to satisfy a third-party suit, an injured person will look beyond the employee to the employer for compensation for damages.

Clearly, the relationship between master and servant under *respondeat superior* needs definition. Under the terms of the *Graalum v. Radisson Ramp* ruling, in-house security officers are servants, whereas contract security personnel may not be. In the latter case, as discussed previously, contract personnel are employees of the supplying agency and, in most cases, the hiring company will not be held liable for their acts. The relationship is a complex one, however.

If the security officer is acting within the scope of his or her employment and commits a wrongful act, then the employer is liable for the actions. The matter then turns on the scope of the officer's employment and the employer-employee relationship.

One court described the scope of employment as depending on:

1. The act being of the kind the offender is employed to perform
2. It occurring substantially within the authorized time and space limits of the employment
3. The offender being motivated, at least in part, by a purpose to serve the master. (*Fournier v. Churchill Downs—Latonia* 292 Ky. 215, 166 SW 2d 38 [1942])

This is further refined in another case in which the court stated that, with respect to an officer:

> If he acts maliciously or in pursuit of some purpose of his own, the defendant is not bound by his conduct, but if, while acting within the general scope of his employment, he simply disregards his master's orders or exceeds his powers, the master will be reasonable for his conduct. (*Hayes v. Sears, Roebuck Co.*, 209 P. 2d 468, 478 [1949].)

Liability, then, is a function of the control exercised or permitted, in the relationship between the security officer and the hiring company. If the hiring company maintains a totally hands-off posture with respect to personnel supplied by the agency, it may well avoid liability for wrongful acts performed by such personnel. On the other hand, there is some precedent for considering the hiring company as sharing some liability, simply by virtue of its underlying rights of control over its own premises, no matter how it wishes to exercise that control. Many hiring companies are, however, motivated to contractually reject any control of security personnel on their premises in order to avoid liabilities. This, as pointed out in *Private Police*, works to discourage hiring companies from regulating the activity of security employees and "the company that exercises controls, e.g., carefully examines the credentials of the guard, carefully determines the procedures the guard will follow, and pays close attention to all his activities, may still be substantially increasing its risk of liability to any third persons who are, in fact, injured by an act of the guard."[4]

It is further suggested in this excellent study that there may be an expansion of certain nondelegable duty rules into consideration of the responsibilities for the actions of security personnel. As discussed previously, the concept of the nondelegable duty provides that there are certain duties and responsibilities which are imposed on an individual and for which he or she remains responsible, even though an independent contractor is hired to implement them. Such duties currently encompass keeping a safe place to work and keeping the premises reasonably safe for business visitors. It is also possible that the courts may find negligence in cases where hiring companies, in an effort to avoid liability, have neglected to exercise any control over the selection and training of personnel and that such negligence on the part of the hiring company has led to injury to third-party victims.

References

1. Wayne Siatt and Sally Matteson, "Special Report: Trends in Security," *Security World* (Jan. 1982), p. 31–33.
2. *Florida Statutes Annotated*, Section 35-3-2-1.
3. *Wisconsin Statutes Annotated*, Section 943.50.
4. Charles Schnabolk, *Physical Security: Practices and Technology* (Stoneham, MA: Butterworth Publishers, 1983), p. 74.

Chapter 19
Public Relations and the Media

KEVIN P. FENNELLY

Media Relations

One of two ways to antagonize the media begins with the mail. Difficult as it is, maintaining up-to-date mailing lists is a vital task. Editors and program directors have long moaned about receiving news of an organization addressed to their predecessors twice removed, and in some instances retired or dead for as long as five years. Not meaning to be humorous, the fact remains that the average practitioner in a small organization cannot afford to purchase any of the several good media directories on the market and thus tends to lean heavily on the media list available.

Media position changes, promotions, replacements, and resignations are frequent in the communication field. Moreover, newspapers do go out of business now and then. A reliable mailing list is brought up to date at least every six months.

Should you get to know someone at any medium whom you can refer to as a personal contact, it would be safe to use that person's name on your addressing, but if your relationship is anything less than personal, it is safer to address your release to the editor. Another reason for this precaution is to ensure that the release does not sit idle, unopened by the person, when addressed to an individual by name who may be out sick, assigned to coverage out of town, or on vacation. Many a deadline has been missed for these reasons. A good story missed is not disappointing only for the practitioner, but for the medium as well.

The second aggravation is the attitude of the practitioner. There is little more annoying than dealing with a publicity hound who is untrained in the ways of media persons who write the rules, call the plays, and sit in supreme judgment of your efforts.

The following are a few *Don'ts* of the newspapers editors:

DON'T . . . plead for print of your release by sending a cover letter to exploit the importance of the news to you or your organization. The release must stand on its own merit.

DON'T . . . send a release without checking the facts, names with correct spelling, or statistics which are contrary or unbelievable.

DON'T . . . push a piece of news which is more than a day old; newspapers are not interested in history.

DON'T . . . send a photograph without protective cardboard in the envelope; a cracked or bent photo cannot be reproduced.

DON'T . . . write on the back of a photograph with anything but a grease pencil (china marking lead) since the writing (especially with a ballpoint pen) can be seen on the print side and renders the photo worthless.

DON'T . . . send an editor an instant or color photograph (unless it's the only photo available) of earth-shattering value.

DON'T . . . ever offer a newsperson a gift of any kind to use bona fide news.

Having covered some of the ground rules on how not to get along with media personnel, let's explore the good habits of a practitioner. Since the goal is to see the results of efforts used, what does it take to build good media relations?

To begin with, consider honesty as your best asset. When media representatives inquire for information, the truth is your best protection from embarrassment; don't ever lie to the media. If you

Table 19–1. Basic B's for Publicity

1. BE THE ONLY PERSON from your department to contact news media. Two individuals calling the same newspaper editor or program director are bound to bring conflict or confusion.
2. BE QUICK to establish personal contact with the right persons at each newspaper, radio, and television station in your area.
3. BE SURE to write everything down. Train your memory, but don't trust it.
4. BE PROMPT in meeting every deadline.
5. BE LEGIBLE. Type news releases. Erase and correct errors. Don't use carbons, except for your own file copy.
6. BE ACCURATE. Doublecheck dates, names, places before you submit your copy.
7. BE HONEST AND IMPARTIAL. Give credit where due.
8. BE BRIEF. Newspaper space and air time are costly.
9. BE BRAVE. Don't be afraid to suggest something new if you honestly believe you have a workable idea. Media people welcome original ideas when they're practical and organized logically.
10. BE BUSINESSLIKE. Never try to obtain publicity by pressure of friendship or business connections. Never ask when a story will appear. Never ask for clippings.
11. BE APPRECIATIVE of all space and time given your department's publicity. The media giving it also have space and time for sale.
12. BE PROFESSIONAL. Members of the press are always invited guests. Arrange a special *Press Table* for large banquets.

cannot give them the facts, tell them so; they are understanding to some extent. Putting them off is a challenge to their tenacity. If you promise to call back with information, then call back *within 15 minutes*, even if you have not as yet all the information requested. Call anyway and let them know you're still hanging in there trying. Make yourself available to the media—not only from 9-5, but around the clock. File your home phone number with the editors and news directors. If an executive in your organization is involved in an auto accident or plane crash or dies suddenly of a heart attack at 3 A.M., the media will want a biographical sketch, even a photograph. Just make sure the next of kin have been notified first; not by calling a spouse—it is not your job to relay the unfortunate news—but by checking with police, carrier officials, or a hospital from whom the media learned of the incident. Keeping media persons happy means keeping current biographies and photos on file of all your organization VIPs.

Building good media relationships takes time. You have to take the time to get to know whom you are dealing with on a one-to-one basis, whether it is the editor or news director of medical, educational, theatre, sports, art, or whatever field you represent. When in the area of the various media, drop in for a few minutes—make certain it's only a few minutes—to make yourself known, or to further establish relationships by just saying "hello." If the only time you visit a media contact is when you have a release in your hand, the acquaintanceship can wear thin. You will be accused of using people. Table 19–1 lists additional suggestions on how an effective media relationship can be established and maintained.

One the other hand, when dealing with the broadcast media, many news directors vow, "If you're going to use the mail for radio, it can't be very important." Hand carry or telephone news intended for radio; if you can't provide the video portion of an event on the same day; don't bother wasting postage and time. You may not only lose a personal contact, but even worse, your integrity as a practitioner.

Exclusives

When does an exclusive news item lose its exclusiveness? Everyone knows a mail release is hardly exclusive news meant for one and only one medium. However, when a mail release is sent to one media representative, you are guilty of playing favorites—unless that representative happens to be the news bureau of a wire service—the only one in town.

Your information is exclusive when it has been asked for and given to one reporter. When a second reporter calls, mention of the first inquirer is made and the news is equally released. If a third reporter calls, notify the first two callers that a general release is being made on the subject of inquiry. This is not only fair standard practice, but it is a traditional method of operation.

Incidentally, the wire services have no use for a release of localized news which has been mailed on a saturation pattern. If you have interesting copy which would also be acceptable out of town and within the region, it is recommended you hand carry the release to the wire service bureau news editor and ask to have your release placed on the regional *B* wire.

Learning to work with the press is like planning a cocktail party: Have plenty to offer to suit the tastes of all who attend, and be sure not to forget anyone who should be included. Once a year, if the budget allows, or when planning a special function of entertainment, make it a point to send a note to those who have favored your organization, enclosing a pair of tickets suggesting they leave their pencil and notebook at the office and just enjoy with their guest. Then follow up with a telephone call a week later to show your sincerity and explain that a special table will be set aside for press colleagues to be together.

Your role is as more than a resource person, and the press is more than an outlet for your releases; together you play the parts of a team serving the public's interest. Ideally, your attitude toward one another should be one of understanding with cooperation on both sides. While the press may be less cooperative over a period of time, or seem so, this is when you must be forgiving and understanding, because while you are one person with one purpose (to see your news in print or aired on the electronic broadcast systems), press persons have multiple reasons for rejecting your efforts. Don't kill a good relationship by showing anger with these gods in any manner, just try harder.[1]

News Conferences

News conferences should only be called if the issue or personality warrants. It is embarrassing to call a conference and have nobody from the media show.

The schedule of newspaper, radio, and television assignments is such that the best time is at 10 A.M. Location is also important. Newspaper, radio, and television reporters and camerapeople go into the field from 8:30 to 9:30 A.M. and must report back between 11 and 12 noon. They go out again about 1 and must report back by 4 P.M.

If your news conference is not centrally located, editors will not be inclined to send out anybody. They try to cover two events at the same time.

Before calling a conference, telephone one or two city editors and one or two television assignment editors. You may decide that it is better to take your guest directly to one newspaper or one radio or television station for a longer exclusive interview or *walk on*.

One public relations official wanted to set up a conference to publicize a panel discussion on an important community problem. The only time the group could gather was at 2 P.M. He didn't want to embarrass the panelists or the chairperson. He suggested that the chairperson call the group together to discuss their program for the following evening. If anybody showed, okay. If not, they still had met for a purpose.

One weekly newspaper sent a reporter and one radio station sent a reporter who taped some of the conversation. But a television station telephoned, and invited the panel to appear on a talk show the following afternoon. The station's moderator stepped aside and allowed the organization's chairperson to handle the discussion for a full forty minutes.[2]

Handouts and Press Kits

The public relations official must always have material available which will give background information to reporters and editors. This can include copies of previously issued releases as well as specially-prepared articles, such as:

- History of organization, local and national
- Current officers, past presidents
- Major projects and accomplishments
- Calendar of events for current year
- Editorials, in draft form or reports
- Features, releases, or reprints
- Reprints of short clippings, articles, or notes by columnists
- Photographs of current president, guest speaker
- Photographs of organization activities
- Photographs of recipients of service
- Biographies of president and/or guest speaker

These materials can be stacked on a table at a convention or community conference. They can be used to give background information to reporters or to an interviewer on a radio talk show or television walk on. Some of these items can be given to the introducer at a guest appearance of your president.

When put together in a binder or folder, these materials make up a press kit for a news conference.[3]

What Is News?

When preparing a news release on your organization's activities, ask yourself: "Would this interest me even if I didn't belong to the organization?" If your answer is "yes," you're on the right track. Another good guide is the newspaper itself. Check the paper to see how news of other similar organizations is presented. You will see the kinds of stories they frequently publish, and you'll learn the form these stories take.

Writing Your News Release

Having gathered the facts for your news release, the next job is to put them in order. Read them over until you are quite familiar with them, and then start organizing the release story. Make sure your release contains the five famous Ws of journalism: who, what, where, when, and why—and often, how. Include all the essential details. If in doubt, give more information, not less. They can always leave out any unnecessary details you have provided, but they cannot include essential details they do not have.

Send your release early. News about an upcoming event should arrive at the newspaper at least a week before the day of the event. Don't worry about getting a release in too early. They won't lose it or throw it away.

Preferred Format

Make sure your name and phone number are at the top of your release, in case the media want to expand it or clarify a point. If you have an office phone number, include that, too. Give them some idea when they can call—not after 11 P.M., until 12:30 A.M., etc. Type doublespaced on one side of the page with good margins. End each page with a complete paragraph.

When you make a mistake, cross out the word and retype it correctly. A typeover leaves them in doubt. If you have an unusual spelling, or anything which looks peculiar but is correct, write "CQ" above it, which means "correct."

How to Draft a Press Release

The old idea that a news story should quickly get across to the reader who, what, when, where, why, and how still holds true. Best of all, try to get all these points in the first sentence or two. A good, factual, interesting lead captures the attention of the reader and creates interest in your program.

Use short words, write short sentences and short paragraphs. Two sentences make a good-sized news release paragraph. Never use ten words when five would do.

Be brief. Most news releases can be written on one or two double-spaced typewritten pages. Two pages, doublespaced, will fill about 12 inches, one newspaper column. Try to follow regular journalistic style in your releases. Don't use an adjective unless it is absolutely necessary to describe something.

When you have finished the first draft of your news release, go over it carefully and critically and cross out all the adjectives and unnecessary words.

Make sure your sentences are short and punchy and your paragraphs are not too long.

News releases should always be typewritten. Use 8½ by 11″ plain white paper. Do not send out carbon copies. Leave margins of at least an inch and a half on either side of the paper and leave some room at the top and at the bottom of the page. Write out clearly from whom the release is coming and write the name and telephone of the person to be contacted for further information. Stipulate *for immediate release* if at all possible. Always start your copy about one-third of the way down the first page. The editor needs this space left open to write a head for your story and give other instructions to subsequent handlers of the copy. If your story runs more than one page write *more* at the bottom of each page except the last, where you will type -30-.

Know Your Paper

Newspapers, as well as other forms of mass media, are excellent sources to communicate with the public to help combat crime. A newspaper informs, interprets, entertains, and serves its communities. In preparing material for publication, be specific and keep it simple. If it appears newsworthy, you have a chance that the story will be printed. List the facts —who, what, when, where, why, and how. Writing styles vary from one newspaper to another. Compare what you've written with other stories that have appeared in that newspaper. Publishing a draft for a press release or a news release article is a skill in written communications, used by many to influence public opinion. Publicize good newsworthy articles which will help build a favorable image within the community. Prepare and present your news in a clear, interesting and persuasive way. Examples of news are promotions, departmental changes, programs expanded, public interest articles, lectures, conferences, seminars, future plans, public safety, and crime prevention tips.

The first step in establishing a sound publicity policy is to become acquainted with your local newspaper personnel and style. Public relations involves interaction with people. People who can help your organization are those who plan and write the news stories, editorials, local columns, feature articles and news broadcasts; managing editors, city editors, editorial writers, business columnists, feature editors, reporters, and photographers. Have only one person designated to handle press contacts. This eliminates duplication and establishes a workable understanding on newspapers' policies and procedures.

Public Speaking

Proficiency in public speaking cannot be gained by mere positive thinking or self-deception; nor does one become the *compleat orator* by mouthing someone else's phrases. The message could be given in either a formal or informal manner. Each of us would do well to learn how to structure a message, how to verbalize our ideas with interest and clarity, and how to polish our presence so that we will communicate with the audience.

Think Before You Do Anything

You've been invited to speak to the local community. Now is the time to stop and plan. Do you know their needs and wants? Do you know what they think of you and your department? Ask, explore, probe, struggle with ideas until you find an idea which satisfies the interests of your audience, and is within the range of your own knowledge and capability to handle.

Limit Your Objectives

Pick a topic that allows you to adequately cover the ideas you have in the time available. The broader the topic, the more difficult it is to handle. It is better to speak clearly about "Here is an idea for our community" than obliquely on "Some of the things our city needs are . . ." Don't try to solve all the world's problems in thirty minutes.

Covering Your Ideas

Aim to leave your audience convinced and ready to act. Visual aids and back-up material help to support some basic ideas. "This is a good idea" is meaningless. "This is a good idea because" makes sense. Before you attempt to outline or write a speech, consider whether you have adequate support for your ideas.

Lecturing and Making Community Presentations

The most effective way to communicate with a group is to talk with them face-to-face. Public speakers are often sought after from business and community groups alike. If your department is able to supply speakers for such occasions, it would enable you to tailor your talk, and deliver the message in an informal and personalized manner. You will find a climate of acceptance at each spoken word, because people are more relaxed in their own surroundings.

A good speaker knows that in order to be able to convey a message, the audience must be listening. If we show an honest and sincere interest in our audience, they will be interested in what we have to say. In order to have an understanding of your audience, you should determine the age and sex of the group, and their knowledge of the subject matter you plan to discuss. Remember, you must talk on their level, and try to keep things simple. You do not want to talk at the audience, but rather with them. Be enthusiastic. If you are excited about your subject matter, they will be, too. Enthusiasm must be honest, genuine, and heartfelt in order to be accepted.

Prepare what you are going to say. Don't use too many facts and figures verbally. Use good visual aids to help reinforce your message on these matters. Deliver your message in such a way as to motivate the group. This can be accomplished by following a working outline and highlighting the key points of interest.

Appearance and Behavior

What do appearance and behavior have to do with giving a community presentation? You have prepared your presentation well and practiced until your voice carries to the back of the hall. If your manner and mannerisms make a good impression, you will engage the interest of your listeners. Audiences react to your mood as well as to your message. The words and subject of your talk are only part of the contact you make with them. Their impression of you as a person is important, and much of this is based on your appearance and behavior. Good behavior contributes a great deal toward the impression made by you the speaker, as does your appearance. The audience will interpret both as manifestations of your personality. A well-groomed and neatly dressed person is looked upon as being an orderly, businesslike person. If you appear untidy and unkempt, it will be assumed that you are lazy and slovenly.

Bodily action should not be overused. Most of us habitually utilize certain actions for emphasis, but if not careful, we overwork those actions to the point that the listener begins to notice the actions rather than the ideas they are supposed to emphasize. Keep in mind your vocal behavior plays a big part in a good speaker's qualities of expression. The speaker who fails to vary pitch level, volume, or tone quality may have difficulty sustaining audience attention for any appreciable length of time, unless the verbal message is especially compelling. The delivery technique should not distract from the message.

Good vs. Bad Delivery

Characteristics of good delivery:

1. Promptness—set up all props and/or visual aids. Test equipment for accuracy and tone.
2. As the guest speaker, you are introduced with stature, a professional, giving good credentials—a person of authority. This creates interest and makes the audience want to listen.
3. Your personal appearance and behavior add to the program. Do not detract from professionalism.
4. You show the ability to communicate clearly, thus delivering an informative message.
5. Demonstrate by the use of props and visual aids.
6. Encourage audience participation. You show the ability of being a good listener, as well as being a good speaker.
7. Give handouts related to subject matter to back up presentation.

Characteristics of bad delivery:

1. Being late and not starting on time
2. Poor introduction
3. Poor appearance
4. A general vagueness, indefiniteness, and lack of clarity
5. Leaving your audience hanging because they do not know or understand the significance of your props
6. Not making your point
7. Leaving without gaining their support or raising their attention

Dale Carnegie wrote,

> Talk about something that you have earned the right to talk about through long study or experience. Talk about something that you know and know that you know. Don't spend ten minutes or ten hours preparing a talk—spend ten weeks or ten months. Better still, spend ten years. Talk about something that has aroused your interest. Talk about something that you have a deep desire to communicate to your listeners. In addition to earning the right to speak, we must have a deep and abiding desire to communicate our convictions and transfer our feelings to our listeners.

Here are eight rules that will help immensely in preparing your talk:

Rule 1. Make brief notes of the interesting things you want to mention.
Rule 2. Don't write out your talk—use easy, conversational language instead.
Rule 3. Never, never, never memorize a talk word for word.
Rule 4. Fill your talk with illustrations and examples.
Rule 5. Know forty times as much about your subject as you can use.
Rule 6. Rehearse your talk by conversing with your friends.
Rule 7. Instead of worrying about your delivery, get busy with the causes that produce it.
Rule 8. Don't imitate others—be yourself.[4]

Visual Aids

The sense of sight is far more efficient than the sense of hearing. Good communications require the combined efforts of visual communication as well as group communication. I encourage those of you who expect to be speaking or lecturing before an audience to develop workable skills in audiovisual communications. In selecting the right visual aids, you must have an understanding of the function, advantages, and limitations of each. Once selected, learn how to use them to your best advantage. You must determine which type of illustration will best communicate the idea to your listeners, which selection will best tell the story.

Some of the most commonly used visual aids are: slide projectors, overhead or opaque projectors, chalkboards, easel pads, charts, flash cards, photographs, magnetic boards, cloth-covered boards, display tables, models and mock-ups, film strips, motion pictures (16mm or 35mm) and flip charts.

Some of the most commonly used audio aids are: record players, tape recorders, cassette players, and telephone hookups.

Some of the most commonly used audiovisual aids are: motion pictures with sound, closed-circuit television, and slide presentation with a cassette player carrying a voice over.

The use of visual aids incorporated into your program in a professional manner enhances the presentation. The presentation will be more effective by following these key points:

1. Thoroughly rehearse your presentation.
2. Be sure to set up ahead of time.

3. Test all electrical equipment.
4. Check out your visibility, and be sure your charts or aids are in full view.
5. Be sure your body does not block your visuals.
6. Keep good eye contact with your audience.
7. Talk with people—not *at* them.
8. Time the presentation.
9. Use good showmanship when presenting the visuals.
10. Have a sense of enthusiasm in your program.

More Audiovisuals

The use of audiovisuals in selling security is not limited to new employee orientation presentations. One large hotel and restaurant chain uses the media described and, in addition, short motion pictures produced in-house to dramatize security and safety problems and procedures, ranging from the handling of bomb threats to fire prevention.

One retail store's organization has made effective use of an audio tape of an interview between the security director and a professional shoplifter, who consented to the interview in return for dismissal of a case pending against him in the local courts. The original tape was a high quality reel-to-reel recording, later reproduced many times on cassette tapes for wide distribution throughout the company.

The shoplifter responded frankly to questions about his trade and skills as they applied to the company. He was unquestionably a ham, but his precise answers, his obvious knowledge of the company's merchandising techniques, methods of presenting goods, use of fixtures, floor layouts of individual stores, exact location of stores, one store's laxity in following a given policy compared to another store, what he liked about stealing from this organization, and what he feared—all have a hypnotic interest for employees viewing the program.

Capturing this thief on tape has made the threat of shoplifting truly credible to the people who can do the most to thwart such activities. He has made literally thousands upon thousands of employees conscious of their role in preventing shoplifting. He has helped to sell the necessity and importance of security.

There are also a wide range of commercially produced 16mm motion pictures and videocassettes aimed at industrial and business consumers. Even films that do not specifically apply to the work scene, and security's role there, can help sell security—for example, a film on rape prevention presented by the security department for the education of female employees.[5]

Proper Use of Flyers and Handouts

Handouts can be one of the smartest tools in any community activity. A handout can reinforce your message. If your audience has some poor listeners, the material can be used to clarify your points. In addition to serving as an advertisement to promote your theme, it can be reread or passed along. Distribution of handouts should be done after your meeting; therefore your audience is not reading during presentation.

Flyers are basically an advertising circular for mass distributions. They should be simple and to the point, conveying your message as well as your department's logo and phone number. Flyers, as well as handouts, should reflect the professionalism of your department. Reproduction stock should be multicolored, different shapes and sizes.

How the Community Rates and Evaluates Your Presentations

Your purpose is to create a long-range favorable public attitude. We must recognize the communities in which we work as our public and center our goals and purposes around their needs, wants, and safety. Feedback can provide us with this information. We should make it part of our program to solicit constructive feedback. This could be accomplished during a question and answer period. Ask what they felt were the three key points of the talk, or areas where they would like a better clarification on material presented. A good community program must show true concern and long-range commitment. End each talk so that you draw out questions and motivate listeners to take action. Use of various arts and skills of communications can build better public support, as well as persuading them to act in a positive manner. If we conduct an attitudinal survey before our formal program and follow-up with another survey later, this would give us comparison on the effects of such a community program. We are evaluated by our appearance, facial signals, voice, clarity, and our knowledge of the subject matter. Develop the ability to hold interests, and clarify all major points. Remember: be prepared in a proper way; speak about something which you have earned the right to talk about; be excited and then speak with enthusiasm; share your personal experiences and talk to each listener.

Selling Security within the Organization

Good sense dictates that there is an ongoing need to *sell* the necessity and importance of the security function to the company as a whole. Employees at all levels of the organization must first be made aware of, then understand, and then come to appreciate the fact that the security function is a viable and integral part of the business, whatever that business or industry may be, and as such contributes to its overall success.

Why is there an ongoing need to sell security? Turnover of employees, including those in the managerial ranks, is one reason. A second is a result of the selling effect itself: that is, as security is understood and accepted, its role expands or takes on new internal dimensions which require new selling.

A final reason is the ever-changing external factors which necessitate change in the security function. For example, race riots appeared on the American scene in the 1960s, followed in rapid succession by civil rights demonstrations, antiwar and general antiestablishment demonstrations, airplane hijackings, executive kidnapping, and hostage ransoming, all having a dramatic impact on the private as well as the public sector. The impact in the private sector, of course, fell directly on the security forces. Shifts in security procedures and new security requirements to meet challenges require selling.

The changes listed above are only some of the most obvious. There have been myriad lesser changes in business and industry, causing a security reaction. In retailing, for example, there have been significant changes in criminal attacks. Not long ago retailers were plagued with hide-in burglars. Today, that problem has abated, but retailers now face a marked increase in *grab and runs*—thefts in which a culprit enters the store, grabs an armload of hanging merchandise from a stand and runs out of the store to a waiting vehicle with driver, and speeds away. Grab and runs were very rare until about 1974. To combat this new theft technique effectively, retail security must sell employees and management on the scope of the problem, what must be done to combat it, and everyone's role (not only security's) in that combat strategy.

Selling security, then, is indeed an important security management responsibility.

Security First

The security executive cannot sell the necessity and importance of the security function to others if your own people do not understand it. The average security department employee has a rather limited view of the security function, seeing it only as it relates to their particular assignment. They do not see the bigger picture. This tunnel vision has a predictable influence on one's attitude, and one's attitude affects one's job performance and relationship with others, in and out of the department.

The single most important aspect of retail security is shrinkage or inventory shortage. Inventory shrinkage, the difference between the inventory of merchandise on the books and the actual physical presence of goods confirmed by a physical count (inventory count), is the one very visible and tangible measurement of a security department's effectiveness in protecting assets.

In one retail organization, for example, the shrinkage percentage figure, causes of shrinkage, and goals are discussed on posters, in handouts, and in the security department's own publication. Yet, at a recent training meeting in the main office and warehouse facility for security officers assigned to that location, not one officer, including those with years of service, could explain the process whereby the company identified the shrinkage percentage. And not one officer knew what the shrinkage percentage meant in terms of dollars. They were staggered when told that the company, like all major retailers, suffers an annual loss of millions of dollars. When they were told how important they were in the overall efforts to protect merchandise, the light of comprehension seemed to come on. The company's error was in assuming the employees understood shrinkage and assuming they knew how important their respective jobs were. Today, these security officers are thoroughly convinced of the need and importance of the department, as well as of their respective jobs.

New Employee Inductions

There is certainly no better opportunity to sell security than that afforded at new employee induction sessions. Not only is there a captive audience, but this is an audience eagerly receptive to information about their new work environment.

Some believe that the presentation on the security organization during the induction program should be made by a member of line management. Even with a prepared script, however, managers tend to deviate from the material, emphasizing those things they think are important (which may not be) and omitting information which they feel is better left unsaid because it is distasteful, such as the consequences of internal dishonesty.

Consequently, to ensure that new employees are exposed to the information deemed necessary and appropriate, it must be presented by either a security employee or some form of audiovisual media.

The personal presentation is by far the better technique, if—and that is an important if—the security employee is a personable, interesting and effective speaker. The higher the rank of the employee making the presentation, the better. Ideally, such presentations should be made by the security director. The further down the chain of command this task is delegated, the lower priority it will be given by the inductees. Thus, the very objective of the exposure—to stress the necessity and importance of security within the organization—is defeated.

In a very large organization, spread over a wide geographical area, the director's personal appearances may necessarily be limited to special events such as the opening of a new facility. Under such circumstances, use of audiovisuals is a good alternative.

Executive Orientations

It is as important, if not more so, to deliver the security message to the management team as it is to the line employees. To ensure this, one organization requires all new incoming middle-management hires to come through the security department for a two-hour orientation (which contrasts with the average one-hour appointment in other departments). Their visit with security, usually within their first month on the job, is part of an overall company orientation. The new controller or unit manager thus becomes acquainted with department heads and their philosophies. This is certainly not an innovative practice, yet security is not always included in this type of executive orientation, and it should be.

Consider the impression made on the new executive, who meets the security director in the latter's office and is given an organizational overview of security. The executive is provided with an organizational chart on which he can fill in the names of key supervisors and their phone extensions for future reference. Using the security functions of the manager's previous employer as a comparison, the security director emphasizes the differences, pointing out the merits and virtues of the new company's program over what the new executive is accustomed to. Following that, an assistant spends time discussing operational practices and problems. Then the executive is introduced to the balance of the department's staff personnel and is given a tour of the security offices.

These new managerial personnel are partially convinced of the importance of security when they arrive, due to the importance attached to the orientation schedule and the two hours devoted to security. There is no question in their minds when they leave the security building that the security function is in the mainstream of the business and has its place in the sun.

Security Tours

Tours of the security facility are a dramatic way to sell security at all levels in the organization. The behind-the-scenes look is intriguing to most people, comparable to the fascination capitalized upon by the television and movie industry in cops-and-robbers entertainment.

To take a class of line supervisors out of their supervisory training school and give them a tour of the security department usually proves to be a highlight of their entire program. Seeing the proprietary alarm room, the communication center, the armory, the fraud investigators at their desks, the banks of files and indices referred to in background investigations—all this makes a lasting impression on employees.

Bulletins

An important aspect of selling is advertising. The power of a strong ad campaign is well known. Advertising copy has to be directed toward its market, must be interesting, and must have some regularity or consistency in terms of exposure. Given these criteria, the security newsletter for management constitutes part of the security department's ad program.

This monthly publication not only keeps company management informed of what contributions the security organization makes, it is also used as a source document for meetings and loss prevention discussions.

This type of bulletin is a natural selling and communication tool. People are curious about crime and the unusual (look at your newspapers), and when such events occur in their neighborhood or workplace, their interest is intensifield. Unless the dissemination of security events compromises security, why not share interesting aspects with other employees? Doing so highlights the necessity and importance of the security function.[6]

References

1. Stephen J. Allen A.P.R., *The Image Engineer.*
2. *Ibid.*
3. *Ibid.*
4. *A Quick And Easy Way To Learn To Speak In Public,* Dale Carnegie & Associates, Inc., (Garden City, New York).
5. Charles A. Sennewald, "*Selling Security With The Organization,*" *Effective Security Management.* (Security World Books) p. 230, 1985.
6. *Ibid.* pp. 231, 232, 233.

Index

Access control
    cargo theft, 70, 72
    computers, 209, 185, 186–187, 190, 192–194
    high-rise office buildings, 174, 176
    hotels, 115–116
    vital records, 2541
Accounts payable and employee theft, 247–248, 253
Activity profiles and computer security, 194–195
ADABAS database system, 199
Administrative offices and campus security, 129
Advertising Council, 8
Affirmative action programs, 237
AFL-CIO and polygraphs, 236
After-hours depositories, 94–95, 98–99
Agency, principle of, 300, 301–302
Air conditioners
    bomb search, 227
    computer security, 209, 214, 215, 218
    fire protection, 209
Air carriers, 68
    cargo theft, 77, 78
Aircraft
    bomb threat, 232
    kidnapping security, 270, 276
Airport terminal
    bomb threat, 232
    guards, 274, 281
Alarms
    banks, 96, 99, 100
    campus security, 130–131
    cargo theft, 81–84
    CCTV, 175
    computer facilities, 215
    false, 99, 129–130, 295
    high-rise office buildings, 173–174
    home, 16
    hotels, 121, 122
    museums, libraries, and archives, 168–170
    vaults, 99, 103
Alcohol server liability, 111, 114
American Civil Liberties Union (ACLU), 236
American District Telegraph (ADT), 11
American Medical Association and hospital security, 151–152
American Polygraph Association (APA), 235, 236
American Society for Industrial Security (ASIS), 8 11
    certification program, 11
Amherst College campus security manual, 136–148
Amphitheaters, bomb threats, 232
Antidiscrimination laws and employee screening, 35, 240–241, 243, 244, 168–170
Applicant screening. *See* Employee screening
Archives. *See* Museums, libraries, and archives
Area protection
    cargo theft, 69–70
    lighting, 69
Arlington, Co., Va. parking ordinance, 164
Armed robbery
    banks, 94, 99
    hospitals, 156
    hotels and motels, 116–119, 121
Armored cars, 11
Arrests. *See also* False arrest citizen's power, 293–295
    force, use of, 299
    guards, 294–295
    hotels, 107
    probable cause, 294–295
    shoplifting, 38, 39, 48, 54, 57–61
Arrest records and applicant screening, 241, 242
Arson, 159

Article surveillance system. *See* Electronic article surveillance (EAS)
Assaults
    hospitals, 152
    liability , 300
    tort law, 292
Athletic facilities and campus security, 142–143
Audiovisual aids and public education, 307, 308–309
Auditoriums, bomb threats, 232
Audits
    computers, 194, 195–196
    employee theft, 248–249, 252, 253
    payrolls, 252
Authorization table and computer security, 192, 193
Automatic teller machines (ATM), 94–95, 98

Back-end machines, 191
Background investigations, 32, 98, 167–168, 241–243
    See also Employee screening
Baker Industries, 11
Bank Administration Institute, 97
Bank deposits, procedural controls, 249, 262
Bank Protection Act of 1968, 93, 97, 102, 104–105
Bankruptcy Act of 1979, 239
Banks, 93
    after-hour deposits, 94–95, 98–99
    alarms, 94, 98, 99
    automatic teller machines (ATM), 94–95, 98
    bullet-resistant barriers, 94, 98
    burglary, 94–95, 98
    check fraud, 94, 95, 104
    camera surveillance, 98, 99, 100
    CCTV, 100
    computers, 100, 103

313

critical problems, 102, 103
embezzlement, 93, 95–97, 103
financial records, privacy laws and, 238
kidnap-extortion, 94
regulatory agencies failures, 93, 101, 102–103
robbery, 94, 99
safe deposit boxes, 94, 99–100
security devices, 98–101
security officer, 97–98
security training, 101–104
tear gas/dye packs, 94
time delay locks, 99
vaults, 94, 98, 99, 100–101, 103
Barriers
cargo theft, 81–82
Bars and restaurants, 114
Bicycle
campus security, 128–129, 143–144
Block watch, 10
Bomb threats, 221–233
buildings, problems with, 231–233
CCTV, 230
evacuation, 222–224
hospitals, 158
hotels, 119
letter and package bombs, 229–230
panic, controlling, 221–222
purpose of calls, 221–222
preparation for, 221–224
search, 223–224, 226–230
vital records, 223
windows, 226–227
written, 226
Booster. See Shoplifter, professional
Buddy alarm system, 15
Building site security. See Environmental design
Bullet-resistant materials
banks, 94, 98
Bulletins, 311
Burdeau v. McDowell, 298
Burglar alarms, 11, 12
Burglary
banks, 94–95, 98
hide-in, 310
hospitals, 156–157
retail, 310

California, 10
merchant privilege, 296
Supreme Court, 295
Cameras, 14
banks, 98, 99, 100
Campus security, 126–148
administrative offices, 129
alarms, 129–130
bicycle theft, 128–129, 143–144
directed patrol, 131–135

dormitories, 138–141
flyers and posters, 131–132
freshman programs, 127
library theft, 128
Lock Your Door program, 130
neighborhood watch, 128
Operation Identification, 129, 131, 133
rape awareness program, 127–128, 146
Safety in Numbers program, 130
security manual, 136–146
summer school program, 127
upper classmen and graduate students, 127
whistle program, 127
Capacitance/proximity detector application, 169
Carbon dioxide fire extinguisher, 209
Cards and identification systems
high-rise office buildings, 176
Cargo theft, 34–35, 63–65
access control, 70, 72
alarms, 81, 82
barriers, 85, 86
checklist, 81–88
communications, 80, 83–84
computer security, 202–203
container control, 76–77
controlled areas, 69–70
disposal of stolen goods, 66
documentation, 73–77
employee screening, 89–90
fences, 71, 79, 82
gatehouses, 79
guards, 69, 80, 87–88
high-value, low-value cargo, 69, 72, 81
inspection report forms, 92
in-transit storage, 69–70
key control, 79, 81
liability for losses, 76–78
lighting, 79, 80, 82–83
locks, 79, 86–87
organized crime, 67–68
package control, 70, 76–77, 85
pallet patterns, 75, 76
personal identification and control, 84–85
personnel security checklist, 72–73, 88–90
physical security checklist, 91–92
pilferage, 64–65
prevention plan, 64
railroad carriers, 66–88, 68
seals, 68–69, 74, 76
security standards, 78–81
storage areas, 71–72, 78–79
theft, 65–67
trailer security, 76

transfer points, 69
trash, 67, 75–76
trucking operations, 68
vehicle control, 70–71, 80–81
warehousing and storage, 71
Cash
employee theft, 35–37, 247, 249–250, 260–262
in mail, 249, 261–262
Cash offices, 36–37
Cash registers
audits, 36
employee theft, 248, 262
Cashiers
employee theft, 35–36, 123
guidelines for, 42–43
hotel, 123
CCTV, 1
banks, 100, 103
bomb threat, 230
coaxial cable, 179
computer facilities, 214, 216
high-rise building, 174–175
hotels, 121, 122–123
monitor, 175
motion detector, 169
pan and tilt system, 175
parking lots, 174, 175
shoplifting, 39, 44, 61
switchers, 175
zoom lenses, 175
Ceilings
computer facilities, 209, 210, 213–214
Cervantez v. J. C. Penney Co., 295
Chauffeurs and executive protection, 269
Checks
bank fraud, 94, 95, 102
employee theft, 35–36, 248, 251, 252, 253, 260
forgeries, 43
retail theft, 36, 42–47, 45
verification procedures, 42
Children and kidnapping threat, 270
Citizen arrest, 293–295
Civil Rights Act of 1964, 31, 237, 240
Cleaning personnel and theft, 34
Coaxial cable, 175
Coffee General Hospital (Douglas, Ga.), 157
Collectibles, transportation of, 171–172
Collyer v. S. H. Kress & Co., 296
Combinations
changing, 37
Commission on Accreditation for Law Enforcement Agencies, 7
Communications
cargo theft security, 80 83–84
Community General Hospital, 158
Community programs and crime prevention, 6–8

Index **315**

Computers, 182–203
 access control, 185, 186–187, 190, 192–194, 209
 alarms, 215
 backup files, 161–162
 banks, 100, 103
 bomb threat, 223
 CCTV, 214, 216
 central processor, vulnerability, 183
 checklist, 202–203
 data communications security, 187–188
 data security, 197–198
 detection and surveillance, 190, 194–196
 emergency power, 187, 209
 encrypting, 189, 191
 facilities in emergency, 208–212, 213–214
 fire protection, 187, 209, 214–215
 guards, 219
 integrity, 196–197
 isolation, 190–192
 logs, 200
 media storage, 186
 personal identification, 184, 186, 189–190
 personnel screening, 182
 physical and environmental security, 185–187
 remote terminals, 183–187
 rooms, 185–187
 security approaches, 183
 security management, 185
 security mechanisms, 198–200
 security organization, 187–188
 storage devices, 183
 vulnerability of components, 182–183
Confidential Information Sources: *Public and Private* (Carroll), 243
Connie Francis case, 110, 300
Consultative selling, 20–22
Controlled cargo, 69
Conventions, security and, 117
Convention hall, bomb threat, 232
Conversion, 293
Conviction records and applicant screening, 31, 241, 242
Cook County Hospital, 153
CPR, 115
Credit cards
 employee theft, 36
 fraud checklist, 45–46
 hotels, 114, 123
 security procedures, 43
Credit information and privacy laws, 32, 238–239, 242
Crime analysis, 14
 risk assessment, 14
Crime prevention and crime analysis
 defined, 3–4
 English heritage, 5–6
 future of, 15–16
 glossary, 18–19
 history, 4–5
 national education campaign, 8–10, 16
 organization and administration, 8
 private security and, 10–12
 risk assessment, 13–14
 selling, 310–311
 specialists, 133
 standards, 7
 status of, 4
Crime scene analysis, 10

Damages, 299–300
Data Encryption Standard (DES), 189
Data Systems Survey, 154–155
Database machines, 191
Database management system (DEMS), 194, 197–198
Davidson v. Madison Corp., 110
Debt collection, 238, 239
Dedication and computer isolation, 190
Defamation, 293
Demerol addiction, 160
Demographics and crime, 16
Denver General Hospital, 153
Department stores
 cash thefts, 36
 inventory control, 253
 shoplifting, 50, 52, 59–60
DES Dorland Worldwide, 8
Detention, 53–54
 and force, use of, 299
 and shoplifting, 53–55, 56, 57, 59, 244–245, 292–293, 295–296
Detroit Police Department, 15
Directed patrol, 131–139
 criminal investigation, 135
 police officers, 132–133
 supervisors, 133–134
 training and materials, 134–135
Disasters
 hospitals, 157–158
 hotels, 117
 museums, libraries, and archives, 170–171
Documentation
 cargo theft, 73–75
 museum, library, and archives, 166–167
Doors
 bomb threat, 228–229
 computer facilities, 209, 213, 216, 217
 glass, 121
Dormitories and campus security, 138–141
Double buys, 36
Drugs
 employee testing, 114
 hospitals, 159–160
 hotels, 113, 114–115, 120
Electronic article surveillance (EAS), 52, 59, 61
Electronic security, 14–15
Elevator security, 14–15
 campus security, 144
Embezzlement, 250–251
 banks, 93, 95–97, 103
 hospitals, 162
 losses, 93, 97, 103
Emergencies, 204, 220
 checklist, 213–220
 computer center, 208–212, 213–214
 equipment location, 212
 evacuation plan, 205–206
 first aid, 217
 manual, 204
 public information, 212, 219
 team, 204–205, 206, 218
 training programs, 213, 215–217
 vital records, protection of, 207–213
Emergency exits
 marking, 206
Emergency medical assistance, 217
 hotels, 105, 115
Emergency power
 computer facility, 187, 209
 high-rise office building, 177
Employee screening, 4, 15, 234–242
 antidiscrimination laws, 31, 237–238, 240, 241
 arrest and conviction records, 31, 241, 242
 background investigation, 32, 241–243
 cargo theft, 89–90
 computer security, 182, 183
 employee theft, 31–32
 hospitals, 150
 information sources, 31–32, 242–243
 laws affecting, 236–239
 museums, libraries, and archives, 167–168
 polygraphs, 32, 234–236, 240–241
 privacy laws, 32, 238–239, 242, 311
 process, 239–242
 psychological testing, 32, 234–236
 theft prevention, 246, 262–263
Employees, selling security to, 310–311
Employee theft, 31–38, 245–265
 audits, 35–36, 248–249, 252, 253
 banks, 93, 95–97, 103
 cash, 37, 247, 249–250, 260–262
 cash register, 35, 248, 262
 checks, 248, 251, 252
 checklist, 258–265
 detaining employees, 37–38
 discharging, 256–257

dishonesty signals, 31–32, 246–248, 250–251, 258–259
employee screening, 31–32
guards, 33
hospitals, 153–154
hotels, 111–113, 123–124
interrogation, 255–256
lockers, 34
losses, 31, 245
methods, 244–245
motives, 243–244, 257–258
museums, libraries, and archives, 166, 167
payrolls, 244–245, 252, 262–263
procedural controls, 31–38, 248–254
promotion and rotation, 250–251
prosecution, 37–38, 256
undercover investigation, 37, 254–255
Encryption and computer security, 189, 191
England
   crime prevention, 5–6
Entrapment, 300, 301
Environmental design, 5, 16
Equal Credit Opportunity Act of, 238
Equal Employment Opportunity Commission (EEOC), 237, 240
Equipment Theft Information Program, Inc. 243
Escobedo v. State of Illinois, 296
Evacuation
   bomb threat, 222–223
   emergency, 205–206
   hotel, 103–107, 117–119
Even money buy, 36
Exchange buys, 36
Exclusionary rule, 297–298
Executive protection, 266–272, 310
   family awareness, 270
   home security, 270
   law enforcement, 271–272
   plan formulation, 267
   in transit, 268–270
Executives, selling security to, 310
Explosives
   gas tanks, 269

Factual Services Bureau, Inc., 161–162
Fair Credit Reporting Act of 1971, 238, 239, 242
Fair Debt Collection Practices Act of, 238
False arrest, 293
   liability for, 38, 41–42, 48, 54, 57–61, 295
False imprisonment, 293, 295
Farraday Cage, 209
Federal Bank Robbery and Incidental Crimes Statute, 95
Federal Bureau of Investigation, 8–9

National Crime Information Center, 243
Felonies
   force, use of, 298
Fences
   cargo storage areas, 71, 79, 82
Fidelman-Danziger, Inc. v. Statler Management, 109
Financial records, privacy law protection of, 238
Fire detectors
   computer facilities, 214–215
Fire department
   bomb threat, 223
   hotel inspection, 124–125
Fire exits
   bomb threat, 229
Fire extinguisher
   bomb threat, 229
   computer facilities, 209, 215
Fire protection
   computer facilities, 187, 209, 214–215
   high-rise office buildings, 176–177
   hospitals, 158–159
   hotels, 105–106, 107, 117–119
   museums, archives, and libraries, 170
Firearms, *See also* Weapons
   liability, 300
   negligence, 292
   state regulation, 299
Fireflies, 269
Fitting rooms and theft, 38–39
Fixed temperature, rate-of-rise detector, 176–177
Florida detention statute, 295–296
Flyers, 309
   campus security campaign, 131–132
Food and beverage control, 123, 124
Force, use of guards, 292, 298–299
   arrests, 209
Forms and employee theft, 251
Fournier v. Churchill Downs-Latonia, 301
Fourth Amendment rights, 297–298
Fraud
   banks, 94, 95, 104
   hospitals, 162
   hotels, 123–124
Freedom of Information Act of 1966, 238, 242
Fronczak baby kidnapping, 157
Front-end machines, 191

Gambling and hospitals, 163
Garzilli v. Howard Johnson's Motor Lodges, 110, 300
Gas Cap security, 269
Gatehouses
   cargo security, 79
   lighting, 79

Gorden State Community Hospital, 159
Graalum v. Radisson Ramp, 301
Grab and run, 310
Grady Memorial Hospital, 157
Griggs v. Duke Power Co., 237, 238
Guards
   arrest powers, 243–247, 294–295
   cargo theft security, 73, 91–92
   computer facilities, 219
   force, use of, 292, 298–300
   high-rise office buildings, 175–176
   liability, 292, 295, 299–302
   Miranda rights and, 296
   museums, archives, and libraries, 171, 173
   searches, 298
   shoplifting security, 38–39, 40, 52
   training, 80

Hahnemann Hospital (Philadelphia), 157
Hallcrest Report, 245
Halon 1301 fire extinguisher, 209, 215
Hammurabi's Code, 5
Handouts and employee theft, 35
Hartford hospital fire, 159
Hayes v. Sears, Roebuck Co., 301
High-integrity position, testing for, 241
High-rise office buildings, 173–177
   access control, 174, 176
   alarms, 173–174
   CCTV, 174–175
   emergency power, 177
   fire, 176–177
   guard patrols, 175–176
Hijacking, 35, 69
Hobbs Act crimes, 94, 158
Home Office Crime Prevention Training Center (HOCPTC), 6
Home security, 270
   alarms, 16
Hospitals, 147–166
   armed robbery, 156
   assaults, 152
   bomb threat, 158
   burglary, 156–157
   disasters, 157–158
   drug abuse, 159–160
   embezzlement and fraud, 162
   employee screening, 148
   employee theft, 153–154
   fires, 158–159
   gambling, 163
   homicides, 1656
   imposters, 163
   information loss, 161–162
   internal disturbance, 157
   kidnapping, 157
   liability, 149–150, 155–156, 157
   parking, 162–164
   patient property losses, 155–156

rationale for security, 149–151
safety, 162–163
security expansion, 151–152
strikes, 158
theft, 152–156
unique aspects of hospitals, 160–161
vandalism, 160
Hostage behavior, 239
Hotels, 105–125
access control, 115–116
alarms, 121, 122
alcohol server liability, 111, 114
arrest/prosecution of criminals, 107
bars and restaurants, 114
bomb threats, 119
CCTV, 121, 122–123
conventions, 117
credit cards, 114, 123
disaster, 117
doors, 121
drugs, 113, 114–115, 120
emergency medical assistance, 105, 115
employee cooperation and security, 113–114
employee theft, 111–113, 123–124
fire protection and evacuation, 105–106, 107, 117–119
food and beverage controls, 124
frauds, 123–124
holdups, 116–117, 121
injuries to guests, 110, 300
key control, 106, 113, 121–122
liability, 108–111, 112, 125, 300
lighting, 121
locks, 112–113, 120–121
loss prevention, 106, 112
physical security, 105
property theft, 109–110, 111–113, 119–123
prostitution, 113–114
public figures, protection of, 115, 117
reputation of hotel, 108–109
security department organization, 107–115
security training, 106
valuables, protection of, 123–124
windows, 121
Human parts, illegal sale of, 153
Hyatt Regency Hotel tragedy, 158
Hydrosphygmograph, 234

Imposters in hospitals, 163
IMS database management, 200
Indiana detention statutes, 296
Information sources and employee screening, 31–32, 242–243,
Infrared detectors
high-rise office buildings, 173–174
In re Deborah C., 297

Insurance companies
medical record thefts, 161–162
safety engineer inspection, 124
Integrity shopping, 36
Interference and computer communications, 189
Internal Revenue Service, 238
Internal theft. *See* Employee theft
International Association of Police Chiefs, 7
International Society of Crime Prevention Protection, 7
International Society of Stress Analysts, 235
International Union of Police Associations, 8
Interrogation, 296–297
employee theft, 255–256
Inventories and employee theft, 253, 259
Ionization smoke detector, 176
Isolation and computer security, 190
J. C. Penney Co., 295
John E. Reid school, 235
Joint Commission on Accreditation of Hospitals (JCAH), 150, 159

Key control, 229
cargo security, 79, 81
hotels, 106, 113, 121–122
museums, libraries, and archives, 169
security survey, 268
Keys
employee theft, 34
master, 79, 106, 121–122
Kidnapping, 266–272, 310
bank security, 94
hospitals, 157
law enforcement officers, 271–272
Klein v. 1500 Massachusetts Avenue Apartments Corporation, 300

Labor disputes
hospitals, 158–160
hotels, 117
Laboratories and campus security, 141–142
Ladies Home Journal, 47–48
LaGuardia Hospital (New York), 157
Law Enforcement Assistance Administration, 6
Law Enforcement Intelligence Unit, 243
Law enforcement officers
executive protection, 271–272
kidnapping, 271–272
private security forces, use of, 12
Lecturing and community relations, 307–309
Lee Hospital (Johnstown, Pa.) 158
Lens (CCTV)
zoom, 170

Letter bomb, 230–231
Lexan, 170
Liability
assaults, 300
cargo loss, 76–78
civil, 299301
contract security, 301–302
criminal, 300–301
employee screening, 293
false arrest, 38, 41–42, 48, 54, 57–61, 295
false imprisonment, 295
guards, 243–244, 247, 292, 295, 299–302
hospitals, 149–150, 155–156, 157
hotels, 108–111, 112, 125
tear gas/dye pack, 94
vicarious, 300
Libel, 293
Library theft, *See also* Museums, archives and libraries
campus security, 128
Lighting
area, 69
cargo security, 79, 80, 82–83
hotel security, 121
storage areas, 69
Lincoln, Nebraska Police, 7
Lockers for employees, 33, 113
searches, 38
Locks
cargo security, 79, 86–87
employee theft, 34
hotels, 112–113, 120–121
safe deposit boxes, 99–100
time, 99
Logs and computer security, 200
Louisiana detention statutes, 296

McGruff, 8
Macomb Community College, 6
Malicious prosecution, 293
Mailroom and employee theft, 247, 254
Maine detention statute, 296
Manholes
bomb search, 231
Mapp v. Ohio, 297–298
Marietta, Ohio nursing home fire, 159
Markdowns, unauthorized, 34
Master keys
cargo theft control, 79
hotels, 106, 121–122
Medical records theft, 161–162
Media. *See* News media
Medicare/Medicaid program and hospital security, 151, 159
Mental distress, infliction of, 293
Merchant's privilege, 53–55, 56, 57, 59, 292–293, 295–296
Metal detectors

bomb threats, 230
Microwave detectors
  application, 174
Military Police, 8, 9
Milton S. Hershey Medical Center, 158
Minneapolis
  hospital pilferage survey, 153–154
Minnesota Multiphasic Personality Inventory (MMPI), 241
Miranda warning and guards, 296
Mirrors, 39, 44
Misdemeanors
  arrest powers, 295
  force, use of, 299
Monitoring
  CCTV, 175
  computers, 189
Montana detention statutes, 296
Monterey Park, California, 129
Morris v. Hotel Riviera, 110
Motion detectors
  CCTV, 169
  high-rise buildings, 173, 174
  museums, libraries, and archives, 169
Motor vehicles and campus security, 142–144
Museum of Fine Arts (Boston), 170
Museums, archives, and libraries security, 165–177
  collectibles, 164
  disaster, 170–171
  documentation, 166–167
  employee screening, 167, 168
  employee theft, 166, 167
  fire, 170
  guards, 171, 173
  personnel, 167–168
  physical security, 168–170
  security responsibility, 168
  transportation, 171–172
National Bureau of Standards encrypting system, 189
National Citizens Crime Prevention Campaign, 8–10
National Crime Information Center, 243
National Crime Prevention Coalition, 8–9
  Computerized Information Center, 8, 10
National Crime Prevention Institute, 6, 12–13
National Fire Protection Association standards, 159
National Neighborhood Watch Program, 8
National Organization of Black Law Enforcement Executives, 8
National Retail Merchants Association, 8
National Sheriff's Association, 8
Negligence, 292, 299–300

firearms, 292, 300
Neighborhood watch groups, 7–8, 15
  campus security, 128
News conferences, 304
News media. *See also* Public relations
  bomb threat, 232–233
  emergency, 212–219
  relations with, 303–304
News releases, 304
Night depositories. *See* After-hours depositories
Nondelegable duty, 300
Nuclear explosion and computer records, 208
Nursing home fires, 159
Occupational Safety and Health Act (OSHA), 237
  hospitals, 150
Offices and campus security, 141–142
Office buildings. *See also* High-rise office buildings
  bomb threat, 232
Office supplies, theft of, 247
Oklahoma, search legislation, 297
Operation Identification
  campus security, 129, 131, 133
Organized crime and cargo theft, 67–68

Package bomb, 230–231, 270
Package control and cargo security, 70, 76–77, 85
Parking security
  cargo theft, 70, 81
  CCTV, 174, 175
  hospitals, 162–163
  personal identification, 184, 186, 199–200
Payrolls
  audits, 252
  employee theft, 247–248, 252, 262–263
Pedestrians
  campus security, 136
  kidnapping security, 269
Pennsylvania substandard nursing homes, 159
People v. Randazzo, 298
Personal identification systems
  campus security, 184, 186, 189–190
  cargo security, 72–73
  checks, 43
  hotels, 113
Personnel manager and employee screening, 239–240
Petty cash security control, 249–250, 260
Photoelectric cells
  smoke detectors, 176
Physical security
  cargo theft, 91–92
  hotels, 105

Pilferable cargo, 69
Pilferage
  cargo, 64–65, 69
  hospitals, 152–154
Police. *See also* Law Enforcement officers
  British, 5–6
Police Executive Research Forum, 8
Polygraphs, 15
  employee screening, 32, 234–236, 240–241
  history of, 234
  opposition to, 236
  training, 235
  validity, 235–236
Postal Inspection, 9
Posters and campus security program, 131–132
Primary prevention, 4
Prison population, 15
Privacy Act of 1974, 238, 239, 242
Privacy, invasion of, 293
Privacy laws
  banks, 238
  employee screening, 32, 238–239, 242, 293
  searches, 297
Privacy locks and campus security, 198–199
Private Police, 302
Private Sector, The (O'Toole), 242–243
Probable cause and arrest and detention, 295, 296
Prosecution decisions
  employee theft, 256
  hotel crime, 107
  malicious, 293
  shoplifting, 53–57
Prostitution and hotels, 113–114
Protected cargo, 69
Psychological stress evaluator (PSE), 32, 234–236
Public education and crime prevention, 6, 8–9
Public figures, protection of, 115, 117
Public relations, 15, 303–311
  bulletins, 311
  exclusives, 304–305
  flyers and handouts, 309
  media relations, 303–304
  news conferences, 304
  news releases, 305–306
  public speaking, 307–309
  selling security, 310–311
Public service campaigns, 8–10
Public speaking, 307–309
Punitive damages, 299–301
Purchasing agent, theft by, 247, 251–252, 265
  competitive bids, 251–252

Railways, 11
  cargo theft, 60–61, 62
Rape
  campus security, 127–128, 146
  hospitals, 152
Receiving departments and theft, 34, 35, 253–254, 263–264
  cargo theft, 72–74, 79
  personnel control, 72–73
References, checking, 242
Refunds
  employee theft, 36, 43
  shoplifting, 43
Regulatory agencies and bank crime, 93, 101, 102–103
Repair personnel and theft, 34
Residential security. *See* Home security
Respondeat superior, 300, 301
Responsibility, separation of, and employee theft prevention, 250
Retail security, 31–45, 310
  also Employee theft; Shoplifting, 35–37, 43
  checks and credit cards, 36, 42–43, 45
  loss prevention techniques, 44–45
  prosecution, 37–38
  refunds, 36, 43
  undercover investigation, 37
Retail/Services Labor Report 237
Rhode Island detention statute, 295
Right to Financial Privacy Act of 1978, 238
Rings of protection, 190–191
Riots, 310
Risk assessment
  executive protection, 267–268
Robbery. *See* Armed robbery
  museums, libraries, and archives, 173
Room search techniques, 226–230
Rubbish. *See* Trash

Sacred Heart General Hospital (Eugene, Ore.) 158
Safe deposit boxes
  banks, 94, 99–100
  hotels, 122
  locks, 99–100
Safety
  hospitals, 162–163
Safes
  alarms, 99
  hotels, 122
St. Louis
  hospital loss study, 153
St. Paul Hospital (Dallas), 162
Sales employees and shoplifting prevention, 50–52, 59–60
Sales invoices and theft control, 264–265
Schools, bomb threat, 231–232
Scrub suits, theft of, 153

Seals and cargo security, 68–69, 74, 76
Searches
  admissible evidence, 298
  battery liability, 292–293
  bombs, 223–224, 226–230
  lockers, 38
  privacy laws, 297
  and seizure, 297
  shoplifting, 292–293, 295, 297
  warrants, 297
Secondary crime prevention, 4
Security approach, 183
Security director
  bombs, 97–98
Security kernel, 191
Security surveys, 7–8
  executive protection, 267
  hotels, 124
  key control, 268
Security table and computer security, 192
Security tours, 311
Self-defense and use of force, 298
Selling security, 20–26, 310–311
  employees, 310–311
  training, 20
Sensitive cargo, 69, 72
Sexual assaults
  campus, 127–128, 146
  hospitals, 152
Shipping and employee theft, 34, 253–254, 263–264. *See also Cargo theft*
Shoplifters, 38–45, 47–61
  apprehension, 38–40
  arrests, 38, 39, 48, 54, 57–61
  cameras, 38
  CCTV, 39, 44, 61
  civil liability for false arrest, 38, 41–42, 54, 57–61
  detention, 53, 55, 56, 57, 59, 292–293, 295–296
  employee vigilance, 38
  guards, 38–39, 40, 52
  loss prevention techniques, 38, 44–46
  losses, 31. 49–50, 61
  methods, 39
  mirrors, 39, 44
  professional, 38, 48, 49
  prosecution decision, 53–57
  refunds, 46
  searches, 292–293, 295, 297
  types of, 47–48
  weapons, 38, 40
Shopping malls, shoplifting, 49, 50
Short loading, 66
Skipper, 123
Skylights
  museums, libraries, and archives, 169, 170
Slander, 293
Smoke detectors

  computer facilities, 187, 214–215
  high-rise office buildings, 106
Southwest Texas State University, 6
Specialty stores, shoplifting, 50, 51–52, 59–60
Spoofing, computers, 189
Sprinkler systems
  computer facilities, 187, 209
  museums, libraries, and archives, 170
State v. Farns, 301
State v. Scrotsky, 298
State switching and computer security, 190
States
  crime prevention programs, 7
  detention statutes, 295–296
  firearm laws, 299
  hotel liability laws, 110, 112
  medical record laws, 162
  polygraph and psychological testing laws, 236, 237, 239
  private security systems, upgrading of, 12
  security force training requirements, 12
Stockroom security, 33–34
Stop Crime Coalition (California), 9
Storage areas
  cargo theft, 71–72, 78–79
  lighting, 69, 71
Store detectives, ,38
Suspect rights, 296–297
Sweeping and computer security, 183
Symbolic addressing and computer isolation, 190
SYSTEM 2000 database management, 199, 200

Tear gas/dye pack, 94
Telephone
  campus security, 128
  kidnapping threats, 271
Terrorism, 21. *See also* Bomb threats
  executive protection, 266–272
  rise in United States, 266
Tertiary crime prevention, 4
Testing
  employee screening, 32, 234–236, 240–241
  drugs, 114
  honesty, 241
Texas Crime Prevention Institute, 6
Thacker v. Commonwealth, 298
Theft. *See also* Cargo theft/Employee theft; Retail theft
  employee screening, 246, 262–263
  hospitals, 152–156
  hotels, 109–110, 111–113, 119–123
Theft triangle, 246
Thermal detectors, 176–177

Threat monitoring and computers, 194
Three Mile Island disaster, 157–158
Time locks, 99
Tort law, 292–293
Tortfeasor, 299, 300
Touro Infirmary fire, 159
Trade associations and security, 11
Transportation-distribution theft. *See* Cargo theft
Trash
  bomb search, 231
  cargo theft, 75–76
  employee theft, 34, 248, 254
Travelers checks, 123
Trespass, 293
Trucks and employee theft, 34–35

Ultrasonic motion detector, 174
Undercover investigation
  employee theft, 41, 254–255
  liability, 301
Underwriters Laboratories
  bank security devices, 98, 100, 101
  fire protection devices, 176, 209
  locks, 100
  safes, 101
  vaults, 101
Uneven buy, 36
Uniform Guidelines on Employee Selection Procedures, 237
U. S. Department of Justice
  Bureau of Justice Assistance, 9
University of Minnesota dishonesty study, 32–33
University of Louisville National Crime Prevention Institute, 6, 132–135
User directives, 192
User violation profiles (computers), 195

Vacations and employee theft, 250–251
Vandalism
  hospitals, 160
Vaults
  alarms, 103, 107
  banks, 94, 98, 99, 100–101, 103
  hotels, 122
Vehicles. *See also* Parking security
  cargo theft control, 70–71, 80–81, 85–86
  executive protection, 269
  computer facility security, 209
Virginia Commonwealth University study of bank barriers, 94, 98
Visitors
  registration, 229
Vital records, protection of
  access control, 251
  bomb threat, 223
  emergency, 207–213
  testing of program, 211–212

Walkouts and hotel security, 114
Warehouse security, 71. *See also* Cargo theft
  employee theft, 35, 253–254
  shipping and receiving, separation of, 15
Water carriers and cargo theft, 78
Weapons, 295. *See also* Firearms
Weeks v. United States 297–298
Weyandot v. Mason's Stores, Inc. 298
Wesley Memorial Hospital (Chicago), 158
West Virginia detention statutes, 296
Where's What, 242
Whistle program and campus security, 127
William J. Burns International Detective Agency, 11
Windows
  alarms, 169
  bomb threat, 228–229
  hotels, 121
Wisconsin detention statute, 297

X-ray film, theft of, 162, 164